Vectors, Tensors, and the Basic Equations of Fluid Mechanics

Vectors, Tensors, and the Basic Equations of Fluid Mechanics

Rutherford Aris

Regents' Professor,
University of Minnesota

DOVER PUBLICATIONS, INC.
New York

Published in Canada by General Publishing Company, Ltd., 30 Lesmill Road, Don Mills, Toronto, Ontario.
Published in the United Kingdom by Constable and Company, Ltd.

This Dover edition, first published in 1989, is an unabridged and corrected republication of the work first published by Prentice-Hall, Inc., Englewood Cliffs, New Jersey, in 1962.

Manufactured in the United States of America
Dover Publications, Inc., 31 East 2nd Street, Mineola, N.Y. 11501

Library of Congress Cataloging-in-Publication Data

Aris, Rutherford.
Vectors, tensors, and the basic equations of fluid mechanics / Rutherford Aris.
p. cm.
ISBN 0-486-66110-5
1. Fluid dynamics. 2. Calculus of tensors. 3. Vector analysis. I. Title.
QA911.A69 1989
532—dc20 89-23501
CIP

To Pat

whose good humor is as perfect
as her typing

Preface

"What! another book on vectors and tensors?" The cry goes up alike from the student searching the bookstore for a beginning text as from the savant who learned his stuff years ago from some tome on *der Ricci Calcul*. "What conceivable reason can this fellow have for inflicting another book on us?"

First let it be said that this is a text intended for the engineering scientist, for the physicist or applied mathematician perhaps, but not for the differential geometer or pure mathematician. Second, it is an introductory text, intended for a reader with some acquaintance with the calculus of partial differentiation and multiple integration but nothing more. Therefore, being for a reader with interest in the physical world, it sets out to show that the calculus of tensors is the language most appropriate to the rational examination of physical field theories. Of these theories I have selected the theory of fluid mechanics as being of central importance and wide appeal. The idea behind this treatment is to use the physical theory to motivate the thorough study of the mathematical subject and, conversely, to show how the mathematical theory can give a truer insight into the physical situation.

The day is quickly passing when it is necessary to make any apology for giving engineers the "straight stuff" in mathematics. It is realised increasingly that his knowledge of mathematics must go beyond a nodding acquaintance with its notions and notation. Even if mathematics is to remain merely a tool for him, he will never be its master until he has understood why it is so formed and is practiced in its manipulations. I have attempted therefore to keep the presentation both elementary and physically motivated but at the same time not to shun the more difficult ideas or applications. Some of the topics introduced are close to the present frontiers of research and it is hoped that the development of fluid mechanics that has been followed is thoroughly in accord with the best current understanding.

In the last decade there has been a renascence of interest in rational mechanics in the mathematical world. It has been fairly widespread and attracted the attention of many mathematicians whose abilities are of the first order. If one name is to be singled out, it is probably not unfair to the others to select that of Truesdell, whose deep scholarship and extensive writing have been of great influence. The work that has been done, and is still proceeding, on the foundations of continuum mechanics will be the basis for future advances in the engineering of continuous media and the sooner the engineer becomes acquainted with it the better. He should not be put off by a certain astringency of aside or *hauteur de mathématicien* which sometimes marks the style of this work. It is not easy reading, nor would one wish it to be, but if this introduction makes the literature appear less formidable, one of its chief purposes will have been fulfilled. If it whets the reader's appetite for more substantial fare, I shall be more than content.

Many applied mathematical texts treat only of Cartesian tensors since these suffice for the principal applications. More purely mathematical texts will properly regard Cartesian tensors as a special case. For certain applications of current importance, Cartesian tensors are not sufficient, and since the understanding of tensors is intended to serve the student in fields other than fluid mechanics, I have not hesitated to treat them generally. However, the ideas are first introduced in the Cartesian framework and then redeveloped more generally. Though this may appear to involve duplication, it is a sound pedagogical principle* to introduce the basic ideas in their most elementary form and to go over the ground again building on the understanding that has been gained. This has the effect of dividing the book into two parts and the first six chapters form a complete course in themselves, which may be suitable at an undergraduate level. The whole book is founded on a course of lectures given to graduate students, and, in as much as their knowledge of matrix algebra is sometimes in need of refreshment and to make the book more self-contained for the independent reader, a short appendix gives the necessary background. The last chapter may be read after Chapter 6, since it in no way requires the intervening chapters. Indeed it is less permeated with the ideas of tensor analysis and is a topic whose foundations are still being strengthened. The treatment I have attempted is therefore not a very deep one, but in a book emanating from a chemical engineering department its entire omission would have

* It is in some sort a converse of "Ockham's razor" and might be called "Ior's ears." Like its more famous predecessor it can now be stated in impeccable Latin. " 'Inmitte istud in alteram auriculam,' dixit. 'Age!' " (From the book *Winnie Ille Pu:* A Latin Version of A. A. Milne's *Winnie the Pooh*, translated by Alexander Lenard. Reprinted by permission of E. P. Dutton & Co., Inc.)

been unpardonable. To have gone further and treated of relativity and magnetohydrodynamics would have been to enlarge the book beyond the bounds of an introduction. The exercises, few of which are at all difficult, are an integral part of the text. They provide practice in manipulation and extensions of the preceding sections. Results which are to be used later are frequently given as exercises and they are not to be regarded as less important than the equations of the text. They are always obvious, in the technical sense of lying right in the way, and are the type of minor hurdle that has to be cleared in reading the literature.

I am indebted to many colleagues for helpful suggestions, but particularly to L. E. Scriven for his careful criticism and timely insistence that a vector is a vector is a vector. I am particularly grateful to J. Serrin, whose lectures at this university first really showed me the structure of fluid mechanics. It is good that the substance of his lectures is available to a wider public in the *Handbuch der Physik* article referred to frequently below. As usual I have received most valuable encouragement from N. R. Amundson. I need hardly add that the book's faults are entirely my own.

Only those who have typed such a manuscript as this will properly appreciate the care and patience of my sister-in-law, Mrs. A. Blair, who penetrated my scribbling and scratchings to produce a first rate typescript. Perhaps only the wives of authors will understand that growing irascibility which mine has had to tolerate and her relief at finally seeing the thing in print.

The compositors who work on a text so burdened with affixes deserve the gratitude of an author even if they are unknown to him: they certainly have mine. I would also like to thank Mr. D. Yesberg for his valuable help in proofreading and in compiling the index.

RUTHERFORD ARIS

Contents

Vectors, Tensors, and the Basic Equations of Fluid Mechanics

Introduction 1

1.1. Mathematical theories and engineering science

At the turn of the century Bertrand Russell described the mathematician as one who neither knows what he is talking about nor cares whether what he says is true. The engineer sometimes prides himself on being the man who can do for a reasonable cost what another would expend a fortune on, if indeed he could do it at all. Between such extremes of abstraction and practicality it would hardly seem possible that there should be much commerce. The philosopher and artisan must tread diverging paths. Yet the trend has been quite otherwise and today the engineer is increasingly aware of his need for mathematical insight and the mathematician proves more and more the stimulation of physical problems. Russell was referring to the logical foundations of pure mathematics, to which he had made his own contributions, and constructing a paradox which would throw into relief the debate that was then at its height. There are, of course, regions of pure mathematics which have developed into such abstraction as to have no apparent contact with the commonplace world. Equally, there are engineering skills that have been developed for particular purposes with no apparent application to other situations. The progress of the moment, at any rate in the science of engineering, lies in the region where the two disciplines have common interests: engineering education, worthy of the name, has always lain there.

If the mathematician has little care whether what he says is true, it is only in the sense that his primary concern is with the inner consistency and deductive consequences of an axiomatic theory. He is content with certain undefined quantities and his satisfaction lies in the structure into which they can be built. Even if the engineer regards himself as dedicated to doing a job economically he cannot rest content with its particular details and still retain a reputation for economy. It is his understanding of the common features of diverse problems that allows him to be economical and hence he must be concerned with abstraction and generalization. It is the business of mathematical theory to provide just such an abstraction and generalization, but it will do it in its own fashion and use the axiomatic method. From what at first seem rather farfetched abstractions and assumptions, it will produce a coherent body of consequences. In so far as these consequences correspond to the observable behavior of the materials the engineer handles, he will have confidence in the mathematical theory and its foundations. The theory itself will have been used to design the critical experiments and to interpret their results. If there is complete discordance between the valid expectations of the theory and the results of critically performed experiments the theory may be rejected. Some measure of disagreement may suggest modification of the theory. Agreement within the limits of experimental error gives confidence in the mathematical model and opens the way for further progress. Continuum mechanics in general, and fluid mechanics in particular, provide mathematical models of the real world in which the engineer can have a high degree of confidence.

The idea of a continuum is an abstraction. Modern physics leads us to believe that matter is composed of elementary particles. For many purposes we need not look within the molecule, but this is to be regarded as an entity of small but finite dimensions which interacts with its fellows according to certain laws. Matter is thus not continuous but discrete and its gross properties are averages over a large number of molecules. The equations of fluid motion have been obtained from this viewpoint, but, though at first sight it seems a much more fundamental one, it stands on the same footing as continuum mechanics—a mathematical model worthy of a certain degree of confidence. For many purposes it is not necessary to know much of the molecular structure, and the continuum hypothesis is an equally satisfactory basis for a mathematical model. In this model the material is not regarded as aggregated at certain points within the medium, so that at most one can speak of the probability of a molecule being at a particular point at a particular time. Rather, we think of the material as continuously filling the region it occupies or, more precisely, that the transformation between two regions it may occupy at different times is a continuous transformation. With this abstraction we can speak of the velocity at a point in a way that is inherently more satisfying than with the molecular model. For with the latter it is necessary to take the average velocity of molecules in the neighborhood of the point. But how large should this neighborhood be? If it is too large its relevance to the point is in question; if it is too small the validity of the average is destroyed. We might hope that there is some range of intermediate sizes for which the average is virtually constant, but this is an unsatisfactory compromise and, in fact, much more sophisticated averages must be invoked to link the molecular and continuum models. In the continuum model velocity is a certain time derivative of the transformation.

The reader may wonder why we start with such a discussion in a book primarily devoted to vectors and tensors. It is because tensor calculus is the natural language of continuum or field theories and we wish to motivate the study of it by considering the basic equations of fluid mechanics. As any language is more than its grammar, so the language of tensor analysis is more than a mere notation. It embodies an outlook or cast of thought just as surely as the speech of a people is redolent with their habit of mind. In this case it is the idea that the "physical" entity is the same though its mathematical description may vary. It follows that there must be a relation between any two mathematical descriptions if they refer to the same entity, and it is this relation that gives the language its character.

1.2. Scalars, vectors, and tensors

There are many physical quantities with which only a single magnitude can be associated. For example, when suitable units of mass and length have been adopted the density of a fluid may be measured. This density, or mass per unit volume, perhaps varies throughout the bulk of a fluid, but in the neighborhood of a given point is found to be sensibly constant. We may associate this density with the point but that is all; there is no sense of direction associated with the density. Such quantities are called *scalars* and in any system of units they are specified by a single real number. If the units in which a scalar is expressed are changed, the real number will change but the physical entity remains the same. Thus the density of water at 4°C is 1 g/cm^3 or 62.427 lb/ft^3; the two different numbers 1 and 62.427 express the same density.

There are other quantities associated with a point that have not only a magnitude but also a direction. If a force of 1 lb weight is said to act at a certain point, we can still ask in what direction the force acts and it is not fully specified until this direction is given. Such a physical quantity is a *vector*. A change of units will change the numerical value of the magnitude in precisely the same way as the real number associated with a scalar is changed, but there is also another change that may be made. Direction has to be specified in relation to a given frame of reference and this frame of reference is just as arbitrary as the system of units in which the magnitude is expressed. For example, a system of three mutually perpendicular axes might be constructed at a point O as follows. Take $O1$ to be the direction of the magnetic north in a horizontal plane, $O2$ to be the direction due west in this plane, and $O3$ to be the direction vertically upwards. Then a direction can be fixed by giving the cosine of the angle between it and each of the three axes in turn. If l_1, l_2, and l_3 are these direction cosines they are not independent but $l_1^2 + l_2^2 + l_3^2 = 1$. If F is the magnitude of the force, the three numbers $F_i = l_i F$ allow us to reconstruct the force, for its magnitude $F = (F_1^2 + F_2^2 + F_3^2)^{1/2}$ and the direction cosines are given by $l_i = F_i/F$, $i = 1, 2, 3$. Thus the three numbers F_1, F_2, F_3 completely specify the force and are called its components in the system of axes we have set up. However, this system was quite arbitrary and another system with $O1$ due south, $O2$ due east, and $O3$ vertically upwards would have been just as valid. In this new system the same direction would be given by direction cosines equal respectively to $-l_1$, $-l_2$, and l_3 and so the components of the force would be $-F_1$, $-F_2$, and F_3. Thus the components of the physical entity, force, change with changing description of direction though the entity itself remains the same just as the real number representing the scalar, density, changed with changing units though the density remained the same. We distinguish therefore between the vector as an entity and its components which allow us to reconstruct it *in a particular system of reference.* The set of components is meaningless unless the system of reference is also prescribed just as the magnitude 62.427 is meaningless as a density until the units are also prescribed.

If then the components of the same entity change with changing frame of reference we need to find out how they will change so as to be sure that the same entity is retained. In three-dimensional space a reference frame consists of three different directions which do not all lie in the same plane. We should also specify the units in which measurements are made in these directions for these need not be the same. These base vectors need not be the same at different points in space and any transformation of base vectors is valid provided the transformed vectors do not lie in a plane. A plane is a space with only two dimensions so that three vectors lying in a plane cannot get a grip on three-dimensional space. Without trying to define things precisely at this point, let us denote the three base vectors by **a**, **b**, **c** then the components of a vector **v** with respect to this frame of reference are the three numbers α, β, and γ such that

$$\mathbf{v} = \alpha\mathbf{a} + \beta\mathbf{b} + \gamma\mathbf{c}.$$

If the base vectors are transformed to **x**, **y**, **z** the new components ξ, η, ζ must satisfy

$$\mathbf{v} = \zeta\mathbf{x} + \eta\mathbf{y} + \zeta\mathbf{z}.$$

If then we know how the base vectors of the new system can be expressed in terms of the old, we shall be able to see how the components should transform. In ordinary three-dimensional space the system defined by three mutually orthogonal directions with equal units of measurement is called

Cartesian. The base vectors may be thought of as lines of unit length lying along the three axes. The cardinal virtue of this system is that these base vectors can be the same everywhere. In the first few chapters we will consider Cartesian vectors, that is, vectors whose components are expressed with a Cartesian frame of reference and the only transformation we shall consider is from one Cartesian system to another. Later we consider more general systems of reference in which new features arise because of the variability of the base vectors. If the space is not Euclidean, as for example the surface of a sphere, the variability of base vectors is inevitable.

We shall construct an algebra and calculus of vectors showing how a sum, product, or derivative may be defined. In fact, two distinct products of two vectors can be defined both of which have great significance. We cannot, however, define the reciprocal of a vector in a unique way, as can be done with a scalar. A scalar can be thought of as a vector in one dimension and its one component gives it a grip on its one-dimensional space and the reciprocal of a scalar a is simply $1/a$. A single vector does not have sufficient grip on the three-dimensional space to allow its reciprocal to be defined, but we can construct an analog of the reciprocal for a triad of vectors not all in one plane. In particular, it will be found that the reciprocal of the triad of base vectors has great importance.

Although the quotient of two vectors cannot be defined satisfactorily, tensors arise physically in situations that make them look rather like this. For example, a stress is a force per unit area. We have seen that force is a vector and so is an element of area if we remember that we have to specify both its size and orientation, that is, the direction of its normal. If $\mathbf{f}$ denotes the vector of force and $\mathbf{A}$ the vector of magnitude equal to the area in the direction of its normal, the stress $\mathbf{T}$ might be thought of as $\mathbf{f}/\mathbf{A}$. However, because division by a vector is undefined, it does not arise quite in this way. Rather we find that the stress system is such that given $\mathbf{A}$ we can find $\mathbf{f}$ by multiplying $\mathbf{A}$ by a new entity $\mathbf{T}$ which is like $\mathbf{f}/\mathbf{A}$ only in the sense that $\mathbf{f} = \mathbf{A}\mathbf{T}$. This new mathematical entity corresponds to a physical entity, namely, the stress system at a point. It is a quantity with which two directions seem to be associated and not just one, as in the case of the vector, or none at all, as with the scalar. In fact it needs nine numbers to specify it in any reference system corresponding to the nine possible combinations of two base vectors. Again we want to be sure that the same physical entity is described when we change the system of reference and hence must require that the components should transform appropriately. What we do is to lay down the transformation rules of the components and when confronted by a set of nine components satisfying these requirements we know they are components of a single mathematical entity. This entity is called a *tensor* (or more properly a second order tensor) and is thus a suitable representation of the kind of physical entity with which two directions can be associated.

As with a vector, we distinguish carefully between the tensor as an entity and its components which must be with reference to a specified system of reference. Some writers speak of the entity as a dyadic whose components form a tensor, a usage soundly grounded in the history of vector analysis. However, we have felt it sufficient to maintain the distinction where necessary by speaking of the tensor and its components. The word tensor is quite general and where necessary its order must be specifically mentioned, for it will appear that a scalar is a tensor of order zero and a vector a tensor of order one. Physical quantities are rarely associated with tensors of higher order than the second but tensors up to the fourth order will arise. Progress in a language is marked by those small liberties taken with strict construction, that impart a certain style without detracting from the meaning. It is hoped that no real ambiguity has been concealed by the free use of the word tensor, but to emphasize this point as much as is possible at this stage we repeat: the distinction must always be borne in mind between the tensor, as a mathematical entity representing a physical entity, and the components of the tensor which are only meaningful when the system of reference has been specified.

1.3. Preview

In the next two chapters we shall develop first the algebra and then the calculus of Cartesian vectors and tensors. The general theorems and ideas of this are applicable in the whole of continuum mechanics as well as in electricity and other fields. Our application of them to fluid mechanics begins with a discussion of kinematics in Chapter 4. Chapter 5 considers the relations between stress and strain in a fluid and allows the basic equations of fluid motion to be discussed in the following chapter. This part of the book is essentially complete in itself. The last chapter on reaction and flow may be read at this point since it does not depend on more general tensor methods.

The groundwork laid in Cartesian tensors allows us to take up the more general calculus of tensors fairly concisely in Chapter 7 and apply these notions to fluid flow in space in Chapter 8. The discussion of flow in a surface given in Chapter 10 requires some understanding of the geometry of surfaces which is given in the preceding chapter.

BIBLIOGRAPHY

1.1. A relatively elementary exposition of the molecular approach to the theory of fluids is well given in

Patterson, G. N., Molecular flow of gases. New York: John Wiley, 1957.

The standard reference is

Hirschfelder, J. O., C. F. Curtiss and R. B. Bird, Molecular theory of gases and liquids. New York: John Wiley, 1954.

A further discussion of the averaging involved in passing from the molecular to the continuum models may be found in

Morse, P. M., and H. Feshbach, Methods of theoretical physics, Vol. 1, pp. 1-3. New York: McGraw-Hill, 1953.

The viewpoint we have briefly outlined is morc fully and forcefully presented in the opening sections of

Truesdell, C., and R. Toupin, The classical field theories, in Handbuch der Physik III/1, ed. S. Flugge. Berlin: Springer-Verlag, 1960.

Cartesian Vectors and Tensors: Their Algebra 2

Until Chapter 8 the unqualified words vector and tensor will refer to the Cartesian vector and tensor about to be defined and if it is necessary to refer to the more general concept of the tensor this will be specifically stated. In this chapter we shall develop the algebra of Cartesian vectors and tensors.

2.11. Definition of a vector

We shall define a vector by first giving an example of one and isolating the particular feature of its behavior that characterizes it as a vector. We may then say that a vector is anything which has this characteristic behavior. To establish that any quantity is a vector we shall have to show that it behaves in this way.

In the ordinary three-dimensional space of everyday life, known technically as Euclidean [3]-space, the position of a point may be specified by three Cartesian coordinates. To determine these we must first establish a frame of reference by taking any point O as the origin and drawing through it three mutually perpendicular straight lines $O1$, $O2$, $O3$. These will have positive senses disposed according to Fig. 2.1 for the right-handed Cartesian coordinates which we adopt as standard. The name right-handed is used since the disposition is the same as the thumb, first, and second fingers of the right hand, or, alternatively because a right-handed screw turned from $O1$ to $O2$ would travel in the positive $O3$ direction. The coordinates of a point P are the lengths of the projections of OP on to the three axes $O1$, $O2$, and $O3$. Let these three lengths be x_1, x_2, and x_3 respectively. We call them the Cartesian

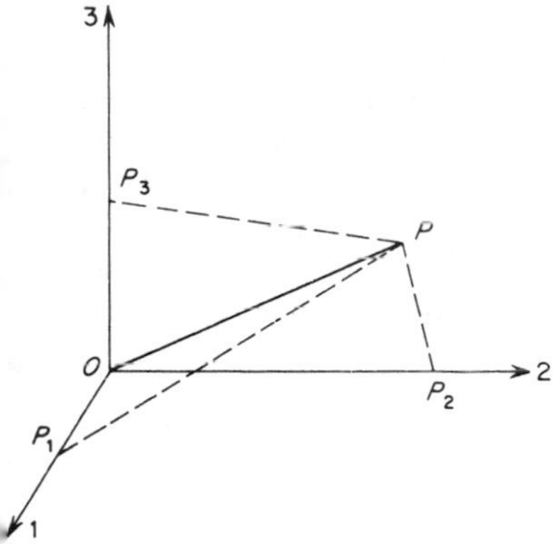

Fig. 2.1

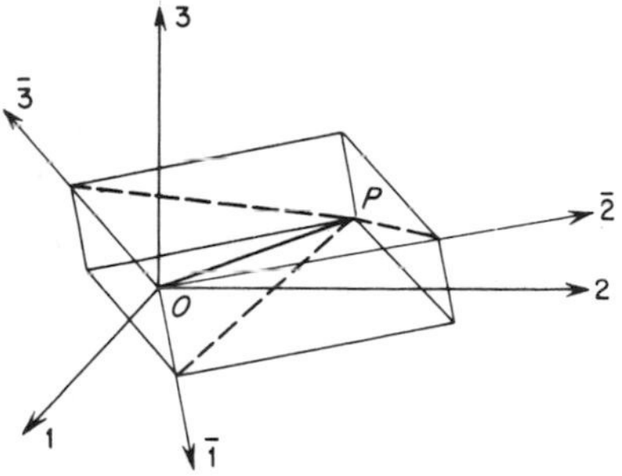

Fig. 2.2

coordinates of P or the components of the position vector of P with respect to the Cartesian frame of reference $O123$.

Now suppose that the coordinate system is rigidly rotated to a new position $O\bar{1}\bar{2}\bar{3}$ as shown in Fig. 2.2 and the new coordinates of P are $\bar{x}_1$, $\bar{x}_2$, $\bar{x}_3$. The rotation can be specified by giving the angles between the old and new axes. Let l_{ij} be the cosine of the angle between the old O axis Oi and the new one $O\bar{j}$, then the new coordinates are related to the old by the formulae

$$\bar{x}_j = l_{1j}x_1 + l_{2j}x_2 + l_{3j}x_3, \qquad j = 1, 2, 3, \tag{2.11.1}$$

and conversely

$$x_i = l_{i1}\bar{x}_1 + l_{i2}\bar{x}_2 + l_{i3}\bar{x}_3, \qquad i = 1, 2, 3. \tag{2.11.2}$$

The reader unfamiliar with this transformation should consult the appendix. Another way of describing this transformation is to say that if $\bar{x}_1$, $\bar{x}_2$, $\bar{x}_3$ are to be the coordinates (or components of the position vector) in the new coordinate system of the *same point* P as x_1, x_2, x_3 are in the old then they must be related by Eqs. (2.11.1) and (2.11.2).

We now introduce a very valuable abbreviation, the *Cartesian summation convention.*

In any product of terms a repeated suffix is held to be summed over its three values, 1, 2, and 3. A suffix not repeated in any product can take any of the values 1, 2, or 3. Thus the equations above can be written

$$\bar{x}_j = l_{ij}x_i, \tag{2.11.1}$$

$$x_i = l_{ij}\bar{x}_j, \tag{2.11.2}$$

i being the repeated suffix in the first case and j in the second. The other suffix (j in the first case and i in the second) is called the free suffix and may take any of the values 1, 2, or 3 so that it is unnecessary to write i or $j = 1, 2, 3$ after the formula. The repeated or dummy suffix as it is sometimes called may be assigned any letter; thus $l_{ij}x_j$ and $l_{ip}x_p$ mean the same thing and it is sometimes convenient in manipulating these formulae to make such changes.

The position vector is our standard Cartesian vector and its components are the coordinates of P. The directed line segment OP gives a convenient geometrical representation of this vector. Since the position vector is to represent the same physical point P its components x_i in the frame of reference $O123$ must transform into $\bar{x}_j = l_{ij}x_i$ in the rotated frame of reference $O\bar{1}\bar{2}\bar{3}$. Accordingly, we make the following definition.

Definition. A Cartesian vector, **a**, in three dimensions is a quantity with three components a_1, a_2, a_3 in the frame of reference $O123$, which, under rotation of the coordinate frame to $O\bar{1}\bar{2}\bar{3}$, become components $\bar{a}_1$, $\bar{a}_2$, $\bar{a}_3$, where

$$\bar{a}_j = l_{ij}a_i. \tag{2.11.3}$$

The vector **a** is to be regarded as an entity, just as the physical quantity it represents is an entity. It is sometimes convenient to use the bold face **a** to show this. In any particular coordinate system it has components a_1, a_2, a_3 and it is at other times convenient to use the typical component a_i.

It is convenient here to introduce the Kronecker delta, denoted by δ_{ij}. It is such that

$$\delta_{ij} = \begin{cases} 1, & i = j, \\ 0, & i \neq j, \end{cases} \tag{2.11.4}$$

and represents the identical transformation. If δ_{ij} occurs in any formula with a repeated suffix all it does is to replace the dummy suffix by the other suffix of the Kronecker delta. For example,

$$\delta_{ij}a_j = \delta_{i1}a_1 + \delta_{i2}a_2 + \delta_{i3}a_3 = a_i$$

since only the term in which the second suffix is i is not zero.

2.12. Example of vectors

The position vector is the prototype of Cartesian vectors and much of our terminology is drawn from it. Thus we may speak of the *length* or *magnitude* of a vector **a**

$$|\mathbf{a}| = (a_i a_i)^{1/2}. \tag{2.12.1}$$

If $|\mathbf{a}| = 1$ it is called a *unit vector* and its components may be thought of as direction cosines. Thus for any vector $\mathbf{a}$ with components a_i the vector with components $a_i/|\mathbf{a}|$ is a unit vector and so represents the *direction* of the vector. Since there are only two arbitrary elements to a unit vector (the third being fixed by the requirement of unit length), the specification of the magnitude and direction of a vector involves three quantities and is equivalent to specifying the three components.

If the position of a point P is a function of time, we may write $x_i = x_i(t)$ and $\bar{x}_j = \bar{x}_j(t)$ where $\bar{x}_j(t) = l_{ij}x_i(t)$. The l_{ij} connecting two coordinate frames are of course independent of time so we may differentiate these formulae with respect to t as many times as we wish

$$\frac{d^n\bar{x}_j}{dt^n} = l_{ij}\frac{d^n x_i}{dt^n} \tag{2.12.2}$$

This equation shows that all the derivatives of position are vectors and in particular the velocity ($n = 1$) and the acceleration ($n = 2$).

A force is specified by its magnitude F and the direction (n_1, n_2, n_3) in which it acts. Let $f_i = Fn_i$ then f_1, f_2, f_3 are the components of a vector. To see this we observe that since direction cosines are simply coordinates of a point on the unit sphere, they transform as coordinates. The direction cosines of this line in the frame of reference $O\bar{1}\bar{2}\bar{3}$ will therefore be $\bar{n}_1$, $\bar{n}_2$, $\bar{n}_3$ where

$$\bar{n}_j = l_{ij}n_i$$

Hence

$$\bar{f}_j = F\bar{n}_j = Fl_{ij}n_i = l_{ij}(Fn_i) = l_{ij}f_i$$

so that force is indeed a vector.

Exercise 2.12.1. If ρ is any scalar property per unit volume of a fluid in motion, show how to define a flux vector $\mathbf{f}$ such that f_i is the rate of flow of ρ per unit area across a small element perpendicular to the axis $0i$.

2.13. Scalar multiplication

If α is any scalar number the product of this scalar and the vector $\mathbf{a}$ is a vector with components αa_i. We see that the length or magnitude of $\alpha\mathbf{a}$ is simply α times the length of $\mathbf{a}$ and the direction of $\alpha\mathbf{a}$ is the same as that of $\mathbf{a}$. Thus scalar multiplication really amounts to a change of length scale as its name suggests. By multiplying both sides of (2.11.3) by the scalar α it is evident that the vector character of $\mathbf{a}$ is unchanged by scalar multiplication.

2.21. Addition of vectors—Coplanar vectors

If $\mathbf{a}$ and $\mathbf{b}$ are two vectors with components a_i and b_i their sum is the vector with components $a_i + b_i$. By phrasing the definition in this way we have really begged the question of whether the sum of two vectors is a vector. However, this is very easily answered; for if we say that the sum of **a** and **b** is an entity with components $a_i + b_i$, then, if the definition of the sum is to be independent of rotation, it must have components $\bar{a}_j + \bar{b}_j$ in the frame of reference $O\bar{1}\bar{2}\bar{3}$. However,

$$\bar{a}_j + \bar{b}_j = l_{ij}a_i + l_{ij}b_i = l_{ij}(a_i + b_i), \tag{2.21.1}$$

which shows that the sum is indeed a vector.

Geometrically we see in Fig. 2.3 that if **a** and **b** are represented by two directed line segments OP and OQ, then their sum is represented by OR

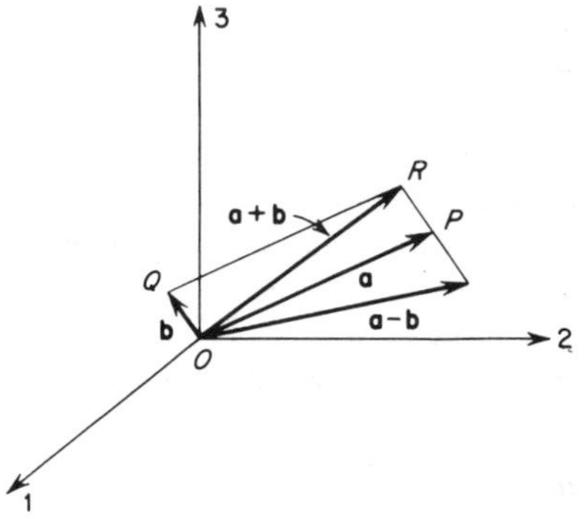

Fig. 2.3

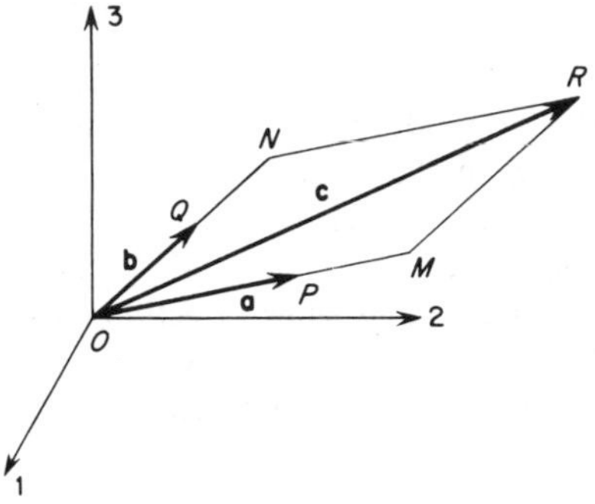

Fig. 2.4

where PR is OQ translated parallel to itself. This is sometimes known as the parallelogram rule of addition. The geometrical representation makes it clear that the order of addition is immaterial, so that

$$\mathbf{a} + \mathbf{b} = \mathbf{b} + \mathbf{a} \tag{2.21.2}$$

This procedure can be continued for the addition of more than two vectors, the sum of **a**, **b**. . ., **k** being the vector with components $a_i + b_i + \ldots + k_i$. The order and association of vectors in addition is immaterial, for example

$$(\mathbf{a} + \mathbf{b}) + \mathbf{c} = \mathbf{a} + (\mathbf{b} + \mathbf{c}) \tag{2.21.3}$$

Subtraction may be defined by combining addition with scalar multiplication by -1. Thus,

$$\mathbf{a} - \mathbf{b} = \mathbf{a} + (-1)\mathbf{b} \tag{2.21.4}$$

is evidently the vector with components $a_i - b_i$. We also see that any vector **c** which is in the same plane as **a** and **b** can be represented in the form

$$\mathbf{c} = \alpha\mathbf{a} + \beta\mathbf{b}. \tag{2.21.5}$$

For in Fig. 2.4 let OR represent the vector **c** in the plane of OP and OQ. Draw RM parallel to OQ and RN parallel to OP and let $\alpha = OM/OP$,

$\beta = ON/OQ$. Then OM and ON represent the vectors $\alpha\mathbf{a}$ and $\beta\mathbf{b}$ and $\mathbf{c}$ is their sum. Geometrically it is clear that if OR points out of the plane of OP and OQ then its component out of the plane cannot be represented as a combination of $\mathbf{a}$ and $\mathbf{b}$. Thus the condition for $\mathbf{c}$ to lie in the plane of $\mathbf{a}$ and $\mathbf{b}$ is that it can be expressed in the form (2.21.5). (An analytical demonstration of this is given as an exercise later.) These equations written in component form are

$$\alpha a_i + \beta b_i = c_i$$

and if α and β are eliminated between the three equations we have the relation

$$\begin{vmatrix} a_1 & b_1 & c_1 \\ a_2 & b_2 & c_2 \\ a_3 & b_3 & c_3 \end{vmatrix} = 0 \tag{2.21.6}$$

between the components of coplanar vectors.

2.22. Unit vectors

The three unit vectors that have only one nonvanishing component are of special importance. They are

$$\begin{aligned} \mathbf{e}_{(1)} &= (1, 0, 0) \\ \mathbf{e}_{(2)} &= (0, 1, 0) \\ \mathbf{e}_{(3)} &= (0, 0, 1). \end{aligned} \tag{2.22.1}$$

The suffixes to the $\mathbf{e}$ are enclosed in parentheses to show that they do not denote components. The jth component of $\mathbf{e}_{(i)}$ is denoted by $e_{(i)j}$ and

$$e_{(i)j} = \delta_{ij} \tag{2.22.2}$$

The components a_1, a_2, and a_3 of $\mathbf{a}$ are themselves scalars and the sum of the scalar products $a_i\mathbf{e}_{(i)}$ is a vector. But by comparing components we see that

$$\begin{aligned} \mathbf{a} &= a_i\mathbf{e}_{(1)} + a_2\mathbf{e}_{(2)} + a_3\mathbf{e}_{(3)} \\ &= a_i\mathbf{e}_{(i)}. \end{aligned} \tag{2.22.3}$$

In the last expression we allow the summation convention to apply also to the parenthetical index.

2.23. A basis of non-coplanar vectors

The three unit vectors are said to form a basis for the representation of any vector. They are not the only basis, though they are the natural one. We shall show that any three vectors $\mathbf{a}$, $\mathbf{b}$, and $\mathbf{c}$ can be used as a basis provided they do not lie in the same plane. For let **d** be any other vector and suppose that it can be expressed as

$$\alpha\mathbf{a} + \beta\mathbf{b} + \gamma\mathbf{c} = \mathbf{d}. \tag{2.23.1}$$

Then to determine the three scalars α, β, and γ we have three equations

$$\begin{aligned} \alpha a_1 + \beta b_1 + \gamma c_1 &= d_1 \\ \alpha a_2 + \beta b_2 + \gamma c_2 &= d_2 \\ \alpha a_3 + \beta b_3 + \gamma c_3 &= d_3 \end{aligned} \tag{2.23.2}$$

and these can certainly be solved provided that the determinant of the coefficients does not vanish. But this determinant is just that of (2.21.6) and that it should not vanish is equivalent to asserting that **a**, **b**, and **c** should not be coplanar.

If M is any nonsingular matrix and we write the components of **a**, **b**, and **c** as column vectors a, b, and c. Then $\bar{a} = Ma$, $\bar{b} = Mb$, and $\bar{c} = Mc$, are three non-coplanar vectors, for

$$\begin{vmatrix} \bar{a}_1 & \bar{b}_1 & \bar{c}_1 \\ \bar{a}_2 & \bar{b}_2 & \bar{c}_2 \\ \bar{a}_3 & \bar{b}_3 & \bar{c}_3 \end{vmatrix} = |M| \begin{vmatrix} a_1 & b_1 & c_1 \\ a_2 & b_2 & c_2 \\ a_3 & b_3 & c_3 \end{vmatrix}$$

does not vanish if $|M| \neq 0$. The vector **d** could be expressed in the form

$$\bar{\alpha}\bar{\mathbf{a}} + \bar{\beta}\bar{\mathbf{b}} + \bar{\gamma}\bar{\mathbf{c}} = \bar{\mathbf{d}}$$

so that $\bar{\mathbf{a}}$, $\bar{\mathbf{b}}$, $\bar{\mathbf{c}}$ form a new basis for the vector space. So general a transformation as this takes us outside the scope of Cartesian vectors, which are concerned with a basis of mutually orthogonal vectors and transformations not by any nonsingular matrix M but by an orthogonal matrix L. (See Appendix, paragraph A6.)

If $\mathbf{e}_{(1)}$, $\mathbf{e}_{(2)}$, $\mathbf{e}_{(3)}$ the three unit vectors are subjected to a rotation given by an orthogonal matrix L whose typical element is l_{ij}, they become three new unit vectors $\bar{\mathbf{e}}_{(j)}$ with components l_{1j}, l_{2j}, l_{3j} in the old reference frame. Thus

$$\bar{\mathbf{e}}_{(j)} = l_{ij}\mathbf{e}_{(i)}, \tag{2.23.3}$$

where summation over bracketed index is assumed. Then

$$\mathbf{a} = a_i\mathbf{e}_{(i)} = \bar{a}_j\bar{\mathbf{e}}_{(j)},$$

and substituting for the new base vectors

$$\mathbf{a} = \bar{a}_j l_{ij}\mathbf{e}_{(i)}.$$

Comparing this with the preceding equation we see that

$$a_i = l_{ij}\bar{a}_j \tag{2.23.4}$$

This is precisely the law of transformation of vector components.

Exercise 2.23.1. Show how to find the vector which lies in the intersection of the plane of **a** and **b** with the plane of **c** and **d**.

Exercise 2.23.2. Let the vectors $\mathbf{b}_{(i)}$ be a basis and form a new basis $\bar{\mathbf{b}}_{(j)} = m_{jk}\mathbf{b}_{(k)}$. Show that if

$$\mathbf{d} = \alpha_i \mathbf{b}_{(i)} = \bar{\alpha}_j \bar{\mathbf{b}}_{(j)}$$

then

$$(\bar{\alpha}_1, \bar{\alpha}_2, \bar{\alpha}_3) = (\alpha_1, \alpha_2, \alpha_3)M^{-1}$$

where M is the matrix whose ij^{th} element is m_{ij}.

2.31. Scalar product—Orthogonality

The scalar product of two vectors **a** and **b** is defined as

$$\mathbf{a} \cdot \mathbf{b} = a_i b_i \tag{2.13.1}$$

and read as "*a* dot *b*." It is invariant under rotation of axes and so it is a scalar. For let $\bar{a}_j = l_{ij}a_i$ and $\bar{b}_q = l_{pq}b_p$ be the components in a new frame of reference. Then

$$\begin{aligned}\bar{\mathbf{a}} \cdot \bar{\mathbf{b}} &= \bar{a}_j \bar{b}_j = l_{ij}a_i l_{pj} b_p = l_{ij} l_{pj} a_i b_p \\ &= \delta_{ip} a_i b_p = a_i b_i = \mathbf{a} \cdot \mathbf{b},\end{aligned} \tag{2.31.2}$$

since $l_{ij}l_{pj} = \delta_{ip}$ by the orthogonality of the rotated axes [see Appendix, Eqs. (A6.6) and (A6.7)].

If **m** and **n** are unit vectors in the directions of **a** and **b** respectively

$$\mathbf{m} \cdot \mathbf{n} = \cos\theta,$$

where θ is the angle between the two directions [see Appendix, Eq. (A2.5)]. Then since $\mathbf{a} = |\mathbf{a}|\,\mathbf{m}$ and $\mathbf{b} = |\mathbf{b}|\,\mathbf{n}$ we have

$$\mathbf{a} \cdot \mathbf{b} = |\mathbf{a}|\,|\mathbf{b}| \cos\theta. \tag{2.31.3}$$

The scalar product

$$\mathbf{a} \cdot \mathbf{n} = |\mathbf{a}| \cos\theta$$

is the projection of the vector **a** on the direction of the vector **b**. If the angle between the two vectors is a right angle, $\theta = \pi/2$ and $\cos\theta = 0$. Such vectors are said to be *orthogonal* and the condition for orthogonality is

$$\mathbf{a} \cdot \mathbf{b} = 0. \tag{2.31.4}$$

The unit vectors of 2.22 are mutually orthogonal.

We may remark that the scalar product can be written as $\delta_{ij}a_i b_j$ and that it is commutative, that is, $\mathbf{a} \cdot \mathbf{b} = \mathbf{b} \cdot \mathbf{a}$.

Exercise 2.31.1. Show that if $\mathbf{c} = \alpha\mathbf{a} + \beta\mathbf{b}$ it is coplanar with **a** and **b**.

Exercise 2.31.2. If **f** is the flux vector of some scalar property of a fluid in motion and **n** the unit normal to an element of area dS, show that $\mathbf{f} \cdot \mathbf{n}\, dS$ is the flux of that property through the element of area.

2.32. Vector product

Of the nine possible products of the components the scalar product is a linear combination of three of them to form a scalar. The other six can be combined in pairs to form the components of a vector. We shall define it as a vector and show how its components can be calculated. The vector or cross product $\mathbf{a} \wedge \mathbf{b}$ (read "**a** cross **b**") is the vector normal to the plane of **a** and **b** of magnitude $|\mathbf{a}|\,|\mathbf{b}| \sin \theta$. To fix the sense of the vector product we require that **a**, **b** and $\mathbf{a} \wedge \mathbf{b}$ should form a right-handed system. The notation $\mathbf{a} \times \mathbf{b}$ is also commonly used for the vector product. We notice that $|\mathbf{a}|\,|\mathbf{b}| \sin \theta$ is the area of the parallelogram two of whose sides are the vectors **a** and **b**. Also it is clear that if we reverse the order of the factors the vector product must point in the opposite direction. Therefore the vector is not commutative, but

$$\mathbf{b} \wedge \mathbf{a} = -\mathbf{a} \wedge \mathbf{b}. \tag{2.32.1}$$

Consider now the vector products of the unit vectors. They are all of unit length and mutually orthogonal so their vector products will be unit vectors. Remembering the right-handed rule we therefore have

$$\begin{aligned} \mathbf{e}_{(2)} \wedge \mathbf{e}_{(3)} &= -\mathbf{e}_{(3)} \wedge \mathbf{e}_{(2)} = \mathbf{e}_{(1)} \\ \mathbf{e}_{(3)} \wedge \mathbf{e}_{(1)} &= -\mathbf{e}_{(1)} \wedge \mathbf{e}_{(3)} = \mathbf{e}_{(2)} \\ \mathbf{e}_{(1)} \wedge \mathbf{e}_{(2)} &= -\mathbf{e}_{(2)} \wedge \mathbf{e}_{(1)} = \mathbf{e}_{(3)} \end{aligned} \tag{2.32.2}$$

Now let us write $\mathbf{a} \wedge \mathbf{b}$ in the form

$$(a_1\mathbf{e}_{(1)} + a_2\mathbf{e}_{(2)} + a_3\mathbf{e}_{(3)}) \wedge (b_1\mathbf{e}_{(1)} + b_2\mathbf{e}_{(2)} + b_3\mathbf{e}_{(3)})$$

and using the relations (2.32.2) to collect together the nine products we have

$$\mathbf{a} \wedge \mathbf{b} = (a_2b_3 - a_3b_2)\mathbf{e}_{(1)} + (a_3b_1 - a_1b_3)\mathbf{e}_{(2)} + (a_1b_2 - a_2b_1)\mathbf{e}_{(3)}. \tag{2.32.3}$$

This shows how the components of the vector product are obtained from the products of the components of the two factors. The symbolic determinant

$$\begin{vmatrix} \mathbf{e}_{(1)} & \mathbf{e}_{(2)} & \mathbf{e}_{(3)} \\ a_1 & a_2 & a_3 \\ b_1 & b_2 & b_3 \end{vmatrix}$$

is sometimes used to represent this, for expanding on the elements of the first row we have (2.32.3).

A very valuable notation can be introduced with the permutation symbol ϵ_{ijk}. This is defined by

$$\epsilon_{ijk} = \begin{cases} 0, \text{ if any two of } i, j, k \text{ are the same} \\ 1, \text{ if } ijk \text{ is an even permutation of } 1, 2, 3 \\ -1, \text{ if } ijk \text{ is an odd permutation of } 123 \end{cases} \tag{2.32.4}$$

Then

$$\mathbf{a} \wedge \mathbf{b} = \epsilon_{ijk} a_i b_j \mathbf{e}_{(k)} \tag{2.32.5}$$

If $\mathbf{a} \wedge \mathbf{b} = 0$ the two vectors are parallel. The cross product of a vector with itself always vanishes.

Exercise 2.32.1. Show by enumerating typical cases that

$$\epsilon_{ijk}\epsilon_{klm} = \delta_{il}\delta_{jm} - \delta_{im}\delta_{jl}$$

Exercise 2.32.2. Show that the condition for the vectors **a**, **b**, and **c** to be coplanar can be written

$$\epsilon_{ijk} a_i b_j c_k = 0$$

Exercise 2.32.3. Show that if $\mathbf{d} = \alpha\mathbf{a} + \beta\mathbf{b} + \gamma\mathbf{c}$, where **a**, **b**, and **c** are not coplanar then $\alpha = \epsilon_{ijk} d_i b_j c_k / \epsilon_{ijk} a_i b_j c_k$, and find similar expressions for β and γ.

Exercise 2.32.4. If **a** and **b** are any vectors, show that

$$(\mathbf{a} \wedge \mathbf{b}) \cdot (\mathbf{a} \wedge \mathbf{b}) + (\mathbf{a} \cdot \mathbf{b})^2 = |\mathbf{a}|^2\,|\mathbf{b}|^2$$

2.33. Velocity due to rigid body rotation

A rigid body is one in which the mutual distance of any two points does not change. Suppose such a body rotates about an axis through the origin of coordinates with direction given by a unit vector **n**. If ω is the angular velocity we can represent this rotation by a vector

$$\boldsymbol{\omega} = \omega\mathbf{n}. \tag{2.33.1}$$

Let P be any point in the body at position **x** (see Fig. 2.5). Then $\mathbf{n} \wedge \mathbf{x}$ is a vector in the direction of PR of magnitude $|\mathbf{x}| \sin\theta$. However, $|\mathbf{x}| \sin\theta = PQ$ is the perpendicular distance from P to the axis of rotation. In a very short interval of time δt the radius PQ moves through an angle $\omega\ \delta t$ and hence P through a distance $(PQ)\omega\ \delta t$. It follows that the very short distance PR is a vector $\delta\mathbf{x}$ perpendicular to the plane of OP and the axis of rotation and hence

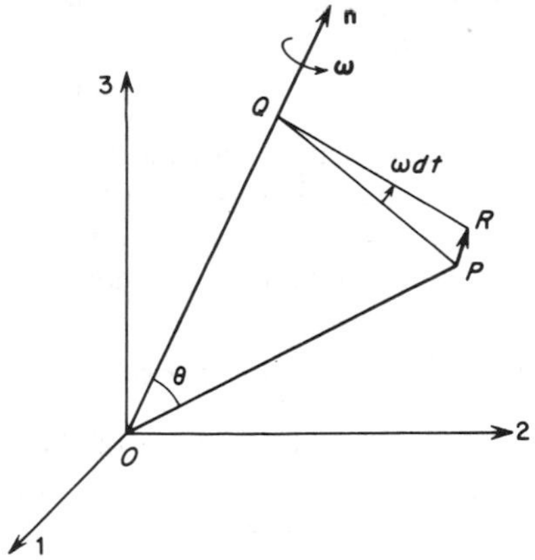

Fig. 2.5

$$\delta\mathbf{x} = (\omega\mathbf{n} \wedge \mathbf{x})\,\delta t = (\boldsymbol{\omega} \wedge \mathbf{x})\,\delta t.$$

However, the limit as $\delta t \to 0$ of $\delta\mathbf{x}/\delta t$ is the velocity **v** of the point P. Thus the linear velocity **v** of the point **x** due to a rotation $\boldsymbol{\omega}$ is

$$\mathbf{v} = \boldsymbol{\omega} \wedge \mathbf{x}. \tag{2.33.2}$$

Whenever the velocity of the point at position **x** can be represented as a vector product of the position vector with a constant vector then the motion is due to a solid body rotation. If **x** and **y** are two points with velocities **v** and **w** due to a rigid body rotation, then also $\mathbf{w} = \boldsymbol{\omega} \wedge \mathbf{y}$ and by subtraction

$$(\mathbf{v} - \mathbf{w}) = \boldsymbol{\omega} \wedge (\mathbf{x} - \mathbf{y}) \tag{2.33.3}$$

This shows that if the relative velocity of two points is related to their relative position in this way then their motion is due to a rigid body rotation.

Exercise 2.33.1. If a force **f** acts at a point **x** show that its moments about the three coordinate axes are the components of a vector, $\mathbf{x} \wedge \mathbf{f}$.

Exercise 2.33.2. If **a**, **b**, and **c** are three non-coplanar vectors forming three edges of a tetrahedron, show that the vectors normal to each face of magnitude equal to the area of the face are

$$\mathbf{n}_1 = \tfrac{1}{2}(\mathbf{b} \wedge \mathbf{c}), \quad \mathbf{n}_2 = \tfrac{1}{2}(\mathbf{c} \wedge \mathbf{a}), \quad \mathbf{n}_3 = \tfrac{1}{2}(\mathbf{a} \wedge \mathbf{b}), \quad \mathbf{n}_4 = \tfrac{1}{2}(\mathbf{a} - \mathbf{b}) \wedge (\mathbf{c} - \mathbf{b}).$$

Exercise 2.33.3. If **n** is the unit outward normal at any point of the surface of a tetrahedron and dS the element of surface area, show that $\iint \mathbf{n}\, dS$ taken over the whole surface is zero. Extend this result to a polyhedron and interpret it geometrically.

Exercise 2.33.4. Show that Snell's law of the refraction of light can be written

$$\mu_1 \mathbf{m}_1 \wedge \mathbf{n} = \mu_2 \mathbf{m}_2 \wedge \mathbf{n}$$

where μ_1 and μ_2 are the refractive indices of the two media either side of the interface to which the normal at the point of incidence is **n**. $\mathbf{m}_1$ and $\mathbf{m}_2$ are unit vectors in directions of the incident and refracted beams. Find a vectorial expression for the law of reflection of light.

2.34. Triple scalar product

Of the possible products of three vectors **a**, **b**, and **c** the simplest is the scalar product of one with the vector product of the other two. This is known as the triple scalar product

$$\mathbf{a} \cdot (\mathbf{b} \wedge \mathbf{c}) = \epsilon_{ijk} a_i b_j c_k. \tag{2.34.1}$$

We observe that the vanishing of this is just the condition for coplanarity of **a**, **b**, and **c**, for this says that **a** is orthogonal to the normal to the plane of **b** and **c** and so in it. (Cf. Ex. 2.32.2). Physically it may be interpreted as the volume of the parallelepiped with sides **a**, **b**, and **c** for $\mathbf{b} \wedge \mathbf{c}$ has magnitude equal to the area of one face and direction **n** normal to it and $\mathbf{a} \cdot \mathbf{n}$ is the height. Notice that an even permutation may be applied to **a**, **b**, and **c** without changing the triple scalar product but that an odd permutation will change its sign. The notation [**a**, **b**, **c**] or (**abc**) is sometimes used for the triple scalar product.

2.35. Triple vector product

Another product may be formed from three vectors $\mathbf{a} \wedge (\mathbf{b} \wedge \mathbf{c})$. This is known as the triple vector product. Since $\mathbf{b} \wedge \mathbf{c}$ is a vector normal to the plane of **b** and **c** and $\mathbf{a} \wedge (\mathbf{b} \wedge \mathbf{c})$ is a vector normal to $\mathbf{b} \wedge \mathbf{c}$, the triple vector product must be in the plane of **b** and **c** and so can be expressed in the form

$$\mathbf{a} \wedge (\mathbf{b} \wedge \mathbf{c}) = \beta\mathbf{b} + \gamma\mathbf{c}.$$

In component notation this is

$$\epsilon_{ijk} a_j \epsilon_{klm} b_l c_m = \beta b_i + \gamma c_i$$

However, by Ex. 2.32.1,

$$\begin{aligned} \epsilon_{ijk}\epsilon_{klm} a_j b_l c_m &= (\delta_{il}\,\delta_{jm} - \delta_{im}\,\delta_{jl}) a_j b_l c_m \\ &= b_i(a_j c_j) - c_i(a_j b_j) \end{aligned}$$

and hence

$$\mathbf{a} \wedge (\mathbf{b} \wedge \mathbf{c}) = (\mathbf{a} \cdot \mathbf{c})\mathbf{b} - (\mathbf{a} \cdot \mathbf{b})\mathbf{c}. \tag{2.35.1}$$

Permuting the letters in this equation we have the identity

$$\mathbf{a} \wedge (\mathbf{b} \wedge \mathbf{c}) + \mathbf{b} \wedge (\mathbf{c} \wedge \mathbf{a}) + \mathbf{c} \wedge (\mathbf{a} \wedge \mathbf{b}) = 0. \tag{2.35.2}$$

There are a number of other identities and extensions to products of a larger number of factors. Some of these are given as exercises and are valuable practice in these elementary manipulations.

Exercise 2.35.1. Show $\mathbf{a} \cdot (\mathbf{b} \wedge \mathbf{c})$ vanishes identically if two of the three vectors are proportional to one another.

Exercise 2.35.2. If **e** is any unit vector and **a** an arbitrary vector show that

$$\mathbf{a} = (\mathbf{a} \cdot \mathbf{e})\mathbf{e} + \mathbf{e} \wedge (\mathbf{a} \wedge \mathbf{e}).$$

This shows that **a** can be resolved into a component parallel to and one perpendicular to an arbitrary direction **e**.

Exercise 2.35.3. Prove that

(i) $(\mathbf{a} \wedge \mathbf{b}) \cdot (\mathbf{c} \wedge \mathbf{d}) = \begin{vmatrix} \mathbf{a} \cdot \mathbf{c} & \mathbf{b} \cdot \mathbf{c} \\ \mathbf{a} \cdot \mathbf{d} & \mathbf{b} \cdot \mathbf{d} \end{vmatrix}$

(ii) $\begin{aligned}[t] (\mathbf{a} \wedge \mathbf{b}) \wedge (\mathbf{c} \wedge \mathbf{d}) &= [\mathbf{c} \cdot (\mathbf{d} \wedge \mathbf{a})]\mathbf{b} - [\mathbf{c} \cdot (\mathbf{d} \wedge \mathbf{b})]\mathbf{a} \\ &= [\mathbf{a} \cdot (\mathbf{b} \wedge \mathbf{d})]\mathbf{c} - [\mathbf{a} \cdot (\mathbf{b} \wedge \mathbf{c})]\mathbf{d} \end{aligned}$

(iii) $[\mathbf{a} \cdot (\mathbf{b} \wedge \mathbf{c})]\mathbf{d} = [\mathbf{d} \cdot (\mathbf{b} \wedge \mathbf{c})]\mathbf{a} + [\mathbf{a} \cdot (\mathbf{d} \wedge \mathbf{c})]\mathbf{b} + [\mathbf{a} \cdot (\mathbf{b} \wedge \mathbf{d})]\mathbf{c}$

Exercise 2.35.4. Show that the two lines

$$\mathbf{x} = \mathbf{a} + \mathbf{l}s, \qquad \mathbf{x} = \mathbf{b} + \mathbf{m}s,$$

where s is a parameter and $\mathbf{l}$ and $\mathbf{m}$ are two unit vectors, will intersect if $\mathbf{a} \cdot (\mathbf{l} \wedge \mathbf{m}) = \mathbf{b} \cdot (\mathbf{l} \wedge \mathbf{m})$ and find their point of intersection.

2.36. Reciprocal base systems

We have seen (Section 2.23) that any triad of vectors can serve as a basis provided that they are not coplanar. Suppose $\mathbf{b}_{(1)}$, $\mathbf{b}_{(2)}$, and $\mathbf{b}_{(3)}$ are three vectors (the parenthetical index applies to the vector and not to the component but the summation convection will still apply) and

$$B = \mathbf{b}_{(1)} \cdot (\mathbf{b}_{(2)} \wedge \mathbf{b}_{(3)}) \neq 0. \qquad (2.36.1)$$

They provide a basis and any vector a can be expressed in the form

$$\mathbf{a} = a_i \mathbf{b}_{(i)}, \qquad (2.36.2)$$

where a_i are the components of the vector with respect to this basis.

From the triad of base vectors, three new base vectors can be constructed by putting

$$\mathbf{b}^{(1)} = \frac{\mathbf{b}_{(2)} \wedge \mathbf{b}_{(3)}}{B}, \qquad \mathbf{b}^{(2)} = \frac{\mathbf{b}_{(3)} \wedge \mathbf{b}_{(1)}}{B}, \qquad \mathbf{b}^{(3)} = \frac{\mathbf{b}_{(1)} \wedge \mathbf{b}_{(2)}}{B}. \qquad (2.36.3)$$

This new triad has the property that the scalar product of $\mathbf{b}^{(i)}$ with $\mathbf{b}_{(i)}$ is unity but its scalar product with $\mathbf{b}_{(j)}$, $i \neq j$, is zero. We may write

$$\mathbf{b}^{(i)} \cdot \mathbf{b}_{(j)} = \delta_{ij}. \qquad (2.36.4)$$

This proves the statement made earlier that no consistent definition of a single vector can be constructed but that a reciprocal triad to a triad of non-coplanar vectors does exist.

If the components of a with respect to the reciprocal base system are denoted by a^j then

$$\mathbf{a} = a^j \mathbf{b}^{(j)}. \qquad (2.36.4)$$

However, comparing with Eq. 2.36.1 we have

$$a_i \mathbf{b}_{(i)} = a^j \mathbf{b}^{(j)}$$

and scalar multiplying each side by $\mathbf{b}_{(j)}$ gives

$$a^j = a_i \mathbf{b}_{(i)} \cdot \mathbf{b}_{(j)}. \qquad (2.36.5)$$

The components with respect to the reciprocal basis are thus related rather simply to the original components.

In particular, if the base vectors are a right-handed set of orthogonal unit vectors $B = 1$ and $\mathbf{b}^{(i)} = \mathbf{b}_{(i)}$. Thus for a Cartesian basis the reciprocal set is identical with the original. Therefore, for Cartesian vectors we have no need to make the distinction between reciprocal bases that will prove fruitful in a more general context.

Exercise 2.36.1. Show that $\mathbf{b}^{(1)} \cdot (\mathbf{b}^{(2)} \wedge \mathbf{b}^{(3)}) = B^{-1}$ and that

$$\mathbf{b}_{(1)} = B(\mathbf{b}^{(2)} \wedge \mathbf{b}^{(3)}) \text{ etc.}$$

Notice that the formula

$$\mathbf{b}_{(i)} = \tfrac{1}{2} B \epsilon_{ijk} (\mathbf{b}^{(j)} \wedge \mathbf{b}^{(k)})$$

expresses all three of these relations.

Exercise 2.36.1. If the basis is a right-handed triad of orthogonal vectors not of equal lengths, show that the reciprocal basis vectors have the same three directions but lengths reciprocal to the original vectors.

2.41. Second order tensors

The vector or first order tensor was defined as an entity with three components which transformed in a certain fashion under rotation of the coordinate frame. We define a *second order Cartesian tensor* similarly as an entity having nine components A_{ij}, $i, j = 1, 2, 3$, in the Cartesian frame of reference $O123$ which on rotation of the frame of reference to $O\bar{1}\bar{2}\bar{3}$ become

$$\bar{A}_{pq} = l_{ip} l_{jq} A_{ij}. \tag{2.41.1}$$

By the orthogonality properties of the direction cosines l_{rs} we have the inverse transformation

$$A_{ij} = l_{ip} l_{jq} \bar{A}_{pq}. \tag{2.41.2}$$

To establish that a given entity is a second order tensor we have to demonstrate that its components transform according to Eq. (2.41.1). A valuable means of establishing tensor character is the quotient rule which will be discussed later in Section 2.6.

A second order tensor may be written down as a 3×3 matrix

$$A = \begin{bmatrix} A_{11} & A_{12} & A_{13} \\ A_{21} & A_{22} & A_{23} \\ A_{31} & A_{32} & A_{33} \end{bmatrix}$$

and it is occasionally convenient to treat it as such. In the notation of matrices (see Appendix, paragraphs A5 and A6) the transformations above would be written $L'AL = \bar{A}$ or $A = L\bar{A}L'$. We shall use a boldface $\mathbf{A}$ to denote the tensor as such but more frequently use the typical component A_{ij}.

If $A_{ij} = A_{ji}$ the tensor is said to be *symmetric* and a symmetric tensor has only six distinct components. If $A_{ij} = -A_{ji}$ the tensor is called *antisymmetric* and such a tensor is characterized by only three scalar quantities for the diagonal terms A_{ii} are zero. The tensor whose ij^{th} element is A_{ji} is called the *transpose* $\mathbf{A}'$ of $\mathbf{A}$.

The analogy with a matrix allows us to define a *conjugate* second order tensor. The determinant of a tensor A is the determinant of the matrix A, namely,

$$\det A = \epsilon_{ijk} A_{1i} A_{2j} A_{3k}. \tag{2.41.3}$$

If this is not zero we can find the inverse matrix by dividing the cofactor of each element by the determinant and transposing. This is called the conjugate tensor and is as close as we come to division in tensor analysis. It will have been evident from the very variety of the definitions of multiplication that no definition of the quotient of vectors is possible. If we denote the elements of the conjugate tensor by A^{ij}, then from matrix theory we see that $A^{ij}A_{jk} = \delta_{ik}$. We shall not pursue this topic here but will prove later, in a more general context, that the conjugate is a tensor with these properties (see Section 7.24).

2.42. Examples of second order tensors

A second order tensor we have already encountered is the Kronecker delta δ_{ij}. Of its nine components six vanish and the remaining three are all equal to unity. However, it transforms as a tensor for its components in the frame $O\bar{1}\bar{2}\bar{3}$ are

$$\bar{\delta}_{pq} = l_{ip} l_{jq} \delta_{ij} = l_{ip} l_{iq} = \delta_{pq} \tag{2.42.1}$$

by the orthogonality relations between the direction cosines l_{ij}. In fact, the components of δ_{ij} in all coordinate systems are the same, namely, l if $i = j$ but zero otherwise. δ_{ij} is called an *isotropic* tensor for this reason.

If $\mathbf{a}$ and $\mathbf{b}$ are two vectors the set of nine products $a_i b_j = A_{ij}$ is a second order tensor, for

$$\begin{aligned} \bar{A}_{pq} &= \bar{a}_p \bar{b}_q = l_{ip} a_i l_{jq} b_j = l_{ip} l_{jq} (a_i b_j) \\ &= l_{ip} l_{jq} A_{ij}. \end{aligned} \tag{2.42.2}$$

An important example of this is the momentum flux tensor for a fluid. If ρ is the density and $\mathbf{v}$ the velocity, ρv_i is the ith component in the direction Oi. The rate at which this momentum crosses a unit area normal to Oj is $\rho v_i v_j$.

We will reserve a discussion of two important second order tensors until the next chapter. These are the rate of strain and stress tensors.

Exercise 2.42.1. Prove that for any vector $\mathbf{a}$, $\epsilon_{ijk} a_k$ are the components of a second order tensor.

Exercise 2.42.2. Show that the flux of any vector property of a flowing fluid can be represented as a second order tensor.

Exercise 2.42.3. If $r^2 = x_k x_k$ and $f(r)$ is any twice differentiable function, show that the nine derivatives

$$\left\{f''(r) - \frac{f'(r)}{r}\right\}\frac{x_i x_j}{r^2} + \frac{f'(r)}{r}\delta_{ij}$$

are the components of a tensor.

2.43. Scalar multiplication and addition

If α is a scalar and $\mathbf{A}$ a second order tensor, the scalar product of α and $\mathbf{A}$ is a tensor $\alpha\mathbf{A}$ each of whose components is α times the corresponding component of $\mathbf{A}$.

The sum of two second order tensors is a second order tensor each of whose components is the sum of the corresponding components of the two tensors. Thus the ij^{th} component of $\mathbf{A} + \mathbf{B}$ is $A_{ij} + B_{ij}$. Notice that tensors must be of the same order to be added; a vector cannot be added to a second order tensor. A linear combination of tensors results from using both scalar multiplication and addition. $\alpha\mathbf{A} + \beta\mathbf{B}$ is the tensor whose ij^{th} component is $\alpha A_{ij} + \beta B_{ij}$. Subtraction may therefore be defined by putting $\alpha = 1$, $\beta = -1$.

Any tensor may be represented as the sum of a symmetric part and an antisymmetric part. For

$$A_{ij} = \tfrac{1}{2}(A_{ij} + A_{ji}) + \tfrac{1}{2}(A_{ij} - A_{ji}) \tag{2.43.1}$$

and interchanging i and j in the first factor leaves it unchanged but changes the sign of the second. Thus,

$$\mathbf{A} = \tfrac{1}{2}(\mathbf{A} + \mathbf{A}') + \tfrac{1}{2}(\mathbf{A} - \mathbf{A}') \tag{2.43.2}$$

represents $\mathbf{A}$ as the sum of a symmetric tensor and antisymmetric tensor.

Exercise 2.43.1. Prove that $\alpha A_{ij} + \beta B_{ij}$ are the components of a second order tensor, if A_{ij} and B_{ij} are.

2.44. Contraction and multiplication

The operation of identifying two indices of a tensor and so summing on them is known as contraction. A_{ii} is the only contraction of A_{ij},

$$A_{ii} = A_{11} + A_{22} + A_{33} \tag{2.44.1}$$

and this is no longer a tensor of the second order but a scalar, or tensor of order zero. To show that it is a zero order tensor we must show that it is invariant under rotation of axes. Now in the frame of reference $O\bar{1}\bar{2}\bar{3}$ the contracted tensor is obtained by identifying the suffixes of $\bar{A}_{pq}$. Thus,

$$\bar{A}_{pp} = l_{ip}l_{jp}A_{ij} = \delta_{ij}A_{ij} = A_{ii} \tag{2.44.1}$$

since $l_{ip}l_{jp} = \delta_{ij}$ by the orthogonality of the l_{ij}. The scalar A_{ii} is known as the trace of the second order tensor. The notation tr **A** is sometimes used.

If **A** and **B** are two second order tensors we can form 81 numbers from the products of the 9 components of each, $A_{ij}B_{km}$, $i, j, k, m = 1, 2, 3$. The full set of these products are components of a fourth order tensor, which we have yet to define. In the barred coordinate system the corresponding set of products is $\bar{A}_{pq}\bar{B}_{rs}$; but

$$(\bar{A}_{pq}\bar{B}_{rs}) = l_{ip}l_{jq}A_{ij}l_{kr}l_{ms}B_{km} = l_{ip}l_{jq}l_{kr}l_{ms}(A_{ij}B_{km}). \tag{2.44.2}$$

This is clearly an analogue of Eq. (2.42.1) but now each side has four free indices and is transformed by a product of four direction cosines. This is the definition of a fourth order tensor which will be given formally in Section 2.6. However the contractions of this general product are second order tensors. These are $A_{ij}B_{ki}$, $A_{ij}B_{ik}$, $A_{ij}B_{kj}$, $A_{ij}B_{jk}$. (The contractions $A_{ii}B_{km}$ and $A_{ij}B_{kk}$ are of course scalar products of scalars A_{ii} and B_{kk} with the tensors **A** and **B**.) The suffix notation makes quite clear just which contraction is involved. However, the notation of a scalar product is sometimes useful. In this notation the tensor with components $A_{ij}B_{jk}$ is written $\mathbf{A} \cdot \mathbf{B}$ the summation being over adjacent suffixes. The four forms listed above are thus $\mathbf{B} \cdot \mathbf{A}$, $\mathbf{A}' \cdot \mathbf{B}$, $\mathbf{A} \cdot \mathbf{B}'$, and $\mathbf{A} \cdot \mathbf{B}$ respectively.

The product $A_{ij}a_j$ of a vector **a** and tensor **A** is a vector whose i^{th} component is $A_{ij}a_j$. Another possible product of these two is $A_{ij}a_i$. These may be written $\mathbf{A} \cdot \mathbf{a}$ and $\mathbf{a} \cdot \mathbf{A}$ respectively.

The doubly contracted product $A_{ij}B_{ji}$ is a scalar, and this may be written $\mathbf{A} : \mathbf{B}$.

Exercise 2.44.1. Prove directly that $A_{ij}B_{jk}$ is a second order tensor.

Exercise 2.44.2. Show that the trace of the tensor of Ex. 2.42.3 is

$$\frac{1}{r^2}\frac{d}{dr}\left(r^2\frac{df}{dr}\right).$$

2.45. The vector of an antisymmetric tensor

There is a very important relation between a vector in three dimensions and the antisymmetric second order tensor. Each has three independent components, and the two may be written as follows:

$$\boldsymbol{\omega} = \begin{bmatrix} \omega_1 \\ \omega_2 \\ \omega_3 \end{bmatrix} \qquad \boldsymbol{\Omega} = \begin{bmatrix} 0 & \omega_3 & -\omega_2 \\ -\omega_3 & 0 & \omega_1 \\ \omega_2 & -\omega_1 & 0 \end{bmatrix}. \tag{2.45.1}$$

In the notation we have already developed the components of $\mathbf{\Omega}$ may be written

$$\Omega_{ij} = \epsilon_{ijk}\omega_k. \tag{2.45.2}$$

In deriving the components of $\boldsymbol{\omega}$ from $\mathbf{\Omega}$ we notice that if we form $\epsilon_{ijk}\Omega_{ij}$, then for any fixed k only the terms with $i \neq k$, $j \neq k$ will appear. Thus,

$$\epsilon_{ij3}\Omega_{ij} = \epsilon_{123}\Omega_{12} + \epsilon_{213}\Omega_{21} = \Omega_{12} - \Omega_{21} = 2\omega_3.$$

It follows that

$$\omega_k = \tfrac{1}{2}\epsilon_{ijk}\Omega_{ij}. \tag{2.45.3}$$

By the properties of the permutation symbols the indices on these formulae can be juggled around. For example,

$$\Omega_{ij} = \epsilon_{kij}\omega_k = -\epsilon_{ikj}\omega_j$$

or

$$\omega_k = \tfrac{1}{2}\epsilon_{kij}\Omega_{ij} = -\tfrac{1}{2}\epsilon_{ikj}\Omega_{ij}.$$

The notations $\mathbf{\Omega}_\times$ and $\mathbf{vec}\,\mathbf{\Omega}$ are sometimes used for $2\boldsymbol{\omega}$ and $\boldsymbol{\omega}$ respectively.

The importance of this relation lies in the identity of the cross product of $\boldsymbol{\omega}$ and arbitrary vector $\mathbf{a}$ with the contracted product of the antisymmetric tensor and $\mathbf{a}$. For the ith component of $\mathbf{a} \wedge \boldsymbol{\omega}$ is $\epsilon_{ijk}a_j\omega_k$ whereas the contracted product $\mathbf{\Omega} \cdot \mathbf{a}$ is a vector with i^{th} component $\Omega_{ij}a_j$. Thus,

$$\mathbf{a} \wedge \boldsymbol{\omega} = \mathbf{\Omega} \cdot \mathbf{a} \quad \text{or} \quad \boldsymbol{\omega} \wedge \mathbf{a} = \mathbf{a} \cdot \mathbf{\Omega}. \tag{2.45.4}$$

The results of Section 2.33 may be reinterpreted in terms of an antisymmetric tensor by saying: *If the relative velocity* $\mathbf{v}$ *of any two points is equal to* $\mathbf{\Omega} \cdot \mathbf{x}$ *where* $\mathbf{\Omega}$ *is an antisymmetric second order tensor independent of the relative position vector* $\mathbf{x}$ *of the two points, then the motion is due to a rigid body rotation.* The angular velocity is given by $-\mathbf{vec}\,\mathbf{\Omega}$ in this case. This result is of cardinal importance in interpreting the rate of strain tensor.

Exercise 2.45.1. Interpret the motion when the ith component of the velocity is given by $(\alpha\delta_{ij} + \epsilon_{ijk}x_k)\omega_j$, $\boldsymbol{\omega}$ being a constant vector and $\mathbf{x}$ the position vector.

2.5. Canonical form of a symmetric tensor

The interpretation and manipulation of tensors is often greatly simplified if they can be thrown into a diagonal form. For example, if we are told that after deformation a certain set of points lies on the surface $A_{ij}x_ix_j = 1$, we are told a great deal about the deformation. If $A_{ij} = A_{ji}$, this surface has the equation

$$A_{11}x_1^2 + A_{22}x_2^2 + A_{33}x_3^2 + 2A_{23}x_2x_3 + 2A_{31}x_3x_1 + 2A_{12}x_1x_2 = 1. \tag{2.5.1}$$

This is obviously a quadric surface (ellipsoid, hyperboloid, paraboloid, cone, or pair of planes) with symmetry about the origin, but it is difficult to say more than this without a highly developed geometrical insight. If, however, the tensor were $A_{ii} = a_i^{-2}$, $A_{ij} = 0$ with $i \neq j$, the surface would be

$$\frac{x_1^2}{a_1^2} + \frac{x_2^2}{a_2^2} + \frac{x_3^2}{a_3^2} = 1 \tag{2.5.2}$$

the veriest schoolboy would know that he was dealing with an ellipsoid of semi-axes a_1, a_2, and a_3. Or, supposing he were only acquainted with the equation of the ellipse, he would immediately argue that the surface is symmetrical about all the coordinate planes and that a section by a plane x_3 = constant would be the ellipse with semi-axes

$$a_1\left\{1 - \left(\frac{x_3}{a_3}\right)^2\right\}^{1/2} \quad \text{and} \quad a_2\left\{1 - \left(\frac{x_3}{a_3}\right)^2\right\}^{1/2}.$$

The ratio of these two is constant so the sections are similar ellipses and their magnitude is proportional to $\{1 - (x_3/a_3)^2\}^{1/2}$ which decreases from 1 to 0 as x_3 increases from 0 to a_3. Thus by elementary reasoning a very good impression of the nature of the surface is obtained. We shall show that there always exists a coordinate system $O\bar{1}\bar{2}\bar{3}$ in which the tensor **A** does have diagonal form. The method follows precisely the corresponding steps in the reduction of a matrix to canonical form, treated in Sections 10–12 of the Appendix. For this reason we will give the reduction only in the case of distinct characteristic values and leave the translation into tensor notation of the more general case to the reader.

If **a** is an arbitrary vector, $\mathbf{A} \cdot \mathbf{a}$ is a vector and for certain **a** may have the same direction as **a** itself had. The two vectors $\mathbf{A} \cdot \mathbf{a}$ and **a** would thus differ only in magnitude and we might write $\mathbf{A} \cdot \mathbf{a} = \lambda\mathbf{a}$. Writing this in component form

$$A_{ij}a_j = \lambda a_i = \lambda\delta_{ij}a_j \quad \text{or} \quad (A_{ij} - \lambda\delta_{ij})a_j = 0 \tag{2.5.3}$$

However, this is a set of three homogeneous equations for the unknowns a_j and so has a solution only if the determinant of the coefficients vanishes. Thus the values of λ must be such as to satisfy the cubic equation

$$\det(A_{ij} - \lambda\delta_{ij}) = \Psi - \lambda\Phi + \lambda^2\Theta - \lambda^3 = 0. \tag{2.5.4}$$

In this equation we have, by expanding the determinant,

$$\begin{aligned} \Theta &= A_{11} + A_{22} + A_{33} = \operatorname{tr} A \\ \Phi &= A_{22}A_{33} - A_{23}A_{32} + A_{33}A_{11} - A_{31}A_{13} + A_{11}A_{22} - A_{12}A_{21} \\ \Psi &= \det A_{ij}, \end{aligned} \tag{2.5.5}$$

and these are called the three invariants of the tensor since their values are unchanged by rotation of the coordinate frame. Eq. (2.5.4) is called the characteristic equation of the tensor and the three values of λ its characteristic values. As with matrices the names latent roots or eigenvalues are sometimes used.

If λ satisfies the characteristic equation, the corresponding a_i can be found. However, though there are three equations in (2.5.3), the vanishing of the determinant means that the third is linearly dependent on the first two, for otherwise the only solution would be the trivial one $\mathbf{a} = 0$. The two independent equations can be solved to give the ratios a_1/a_3 and a_2/a_3 and the magnitude of a_1, a_2, a_3 may now be fixed by requiring that

$$\mathbf{a} \cdot \mathbf{a} = a_1^2 + a_2^2 + a_3^2 = 1.$$

If $\lambda_{(k)}$, $k = 1, 2, 3$, are the three characteristic values we may write the corresponding characteristic vectors as $\mathbf{l}_{(k)} = l_{ik}$, where the first of these indices is the index of the component and the second to the characteristic value of which it is the vector. We shall now prove that if $\lambda_{(1)}$, $\lambda_{(2)}$, and $\lambda_{(3)}$ are all distinct then $\mathbf{l}_{(1)}$, $\mathbf{l}_{(2)}$, and $\mathbf{l}_{(3)}$ are mutually orthogonal if A is symmetric.

For consider two latent roots with indices p and q. We have

$$A_{ij}l_{jp} = \lambda_{(p)}l_{ip} \quad \text{and} \quad A_{ij}l_{jq} = \lambda_{(q)}l_{iq} \tag{2.5.6}$$

(N.B. There is no summation on p or q on the right side of these equations: for this reason the suffix q on λ has been put in parenthesis). Multiply the first of these equations by l_{iq} and the second by l_{ip}, then

$$A_{ij}l_{iq}l_{jp} = \lambda_{(p)}l_{ip}l_{iq} \quad \text{and} \quad A_{ij}l_{ip}l_{jq} = \lambda_{(q)}l_{ip}l_{iq}.$$

However, $\mathbf{A}$ is symmetric, so $A_{ij} = A_{ji}$ and we may interchange the dummy suffixes in the first equation without changing the value. Thus

$$A_{ij}l_{ip}l_{jq} = \lambda_{(p)}l_{ip}l_{iq} = \lambda_{(q)}l_{ip}l_{iq} \tag{2.5.7}$$

and $\lambda_{(p)}$ and $\lambda_{(q)}$ have been assumed to be distinct so the equation is only possible if $l_{ip}l_{iq} = 0$. However, this says that $\mathbf{l}_{(p)} \cdot \mathbf{l}_{(q)} = 0$ and so the vectors are orthogonal. Also by their construction $\mathbf{l}_{(p)} \cdot \mathbf{l}_{(p)} = 1$ so that

$$l_{ip}l_{iq} = \delta_{pq}. \tag{2.5.8}$$

However, this is the relation satisfied by the direction cosines l_{ij} that specify a rotation of the coordinate system, and this interpretation may be placed on the characteristic vectors. Thus if the coordinate $\bar{X}_j = l_{ij}x_i$ the components of the tensor $\mathbf{A}$ in the frame of reference $O\bar{1}\bar{2}\bar{3}$ are

$$\bar{A}_{pq} = l_{ip}l_{jq}A_{ij} = \lambda_{(p)}l_{ip}l_{iq} = \lambda_{(p)}\delta_{pq} \tag{2.5.9}$$

by (2.5.7) and (2.5.8). This merely says that $\mathbf{A}$ has diagonal form and its diagonal components are $\lambda_{(1)}$, $\lambda_{(2)}$, and $\lambda_{(3)}$.

The directions given by the normalized characteristic vectors are known as the principal directions or axes of the tensor. It is often possible to prove a tensor relationship very easily when using a coordinate system whose axes are coincident with the principal axes of one of the tensors involved. When a relationship of this kind has been established and expressed in correct tensor form it must hold for all coordinate systems, since the tensors involved transform as tensors and so their relationship is unchanged. This is the great virtue of tensor analysis. A notable example of this will be given in Chapter 4 where Serrin's elegant treatment of stress-rate of strain relations is presented.

Exercise 2.5.1. Show that Θ, Φ, and Ψ are invariants under rotation of axes.

Exercise 2.5.2. Follow the case of equal characteristic values given in the Appendix (Section A.11) translating from matrix notation into tensors.

Exercise 2.5.3. (Cayley-Hamilton Theorem) Prove that

$$A_{ik}A_{km}A_{mj} = \Theta A_{ik}A_{kj} - \Phi A_{ij} + \Psi.$$

Exercise 2.5.4. If A_{ij} and B_{pq} have a principal direction in common, show that this is also a principal direction of the tensor $A_{ik}B_{kj} + A_{jk}B_{ki}$.

2.61. Higher order tensors

We can now state quite tersely the general definition and laws of operation of Cartesian tensors of any order for the reader should be familiar with their behavior through seeing the simpler examples.

Definition. The tensor **A**, of order n, is a quantity defined by 3^n components which may be written $A_{ijk\ldots n}$, provided that under rotation to a new coordinate frame they transform according to the law

$$\bar{A}_{pq\ldots t} = l_{ip}l_{jq}\ldots l_{nt}A_{ij\ldots n}. \tag{2.61.1}$$

Symmetries. If interchange of two of the indices does not change the value of the component the tensor is said to be symmetric with respect to these indices. If the absolute value is unchanged but the sign reversed it is antisymmetric with respect to the indices.

Contraction. The 3^{n-2} quantities formed by identifying two of the indices of an nth order tensor and invoking the summation convention are components of a tensor of order $n - 2$.

Scalar multiplication. If α is any scalar $\alpha\mathbf{A}$ is the tensor with components $\alpha A_{ij\ldots n}$.

Addition. If **A** and **B** are tensors of the same order they may be added to give a tensor with components $A_{ij\ldots n} + B_{ij\ldots n}$.

Multiplication. If $A_{ij...n}$ and $B_{pq...t}$ are tensors of order n and m respectively, the set of products $A_{ij...n}B_{pq...t}$ are the components of a tensor of order $n + m$. This tensor can be contracted in different ways to form tensors of orders $n + m - 2$, $n + m - 4, \ldots 1$ or 0.

The laws of tensor algebra are:

$$\begin{aligned} \alpha\mathbf{A} &= \mathbf{A}\alpha, \\ \alpha(\mathbf{A} + \mathbf{B}) &= \alpha\mathbf{A} + \alpha\mathbf{B}, \\ \alpha(\mathbf{A}) &= \alpha\mathbf{A} = (\alpha\mathbf{A}), \\ \mathbf{A} + \mathbf{B} &= \mathbf{B} + \mathbf{A}, \\ \mathbf{A} + (\mathbf{B} + \mathbf{C}) &= (\mathbf{A} + \mathbf{B}) + \mathbf{C}, \\ \mathbf{A}(\mathbf{B} + \mathbf{C}) &= \mathbf{AB} + \mathbf{AC}. \end{aligned}$$

Since a great variety of products of two tensors are possible, nothing comparable to $\mathbf{AB} = \mathbf{BA}$ can be asserted in general.

Exercise 2.61.1. Prove that contraction preserves the tensor character while lowering the order by two.

Exercise 2.61.2. Prove that ϵ_{ijk} is a third order tensor.

2.62. The quotient rule

We have constantly remarked that to prove that a given set of quantities forms the set of the components of a tensor requires that we show that they transform according to the rule of tensor transformation. A short cut in establishing tensorial character is the so-called quotient rule. The simple case we shall prove is as follows: If A_{ij} $i, j = 1, 2, 3$ are nine quantities and $\mathbf{b}$ and $\mathbf{c}$ are vectors, $\mathbf{b}$ being quite independent of the A_{ij}, and $A_{ij}b_j = c_i$, then the A_{ij} are components of a tensor $\mathbf{A}$. The value of this is that a relation $\mathbf{A} \cdot \mathbf{b} = \mathbf{c}$ may arise in the study of a physical situation in which it is known that $\mathbf{b}$ and $\mathbf{c}$ are vectors. Then the quotient rule establishes that $\mathbf{A}$ is a tensor and we are now assured that the equation holds in all coordinate frames.

To prove the rule we observe that if $\mathbf{A}$ really is a tensor it must satisfy two requirements. First, the equation $\mathbf{A} \cdot \mathbf{b} = \mathbf{c}$ being a tensor equation must transform to $\bar{\mathbf{A}} \cdot \bar{\mathbf{b}} = \bar{\mathbf{c}}$, that is, $\bar{A}_{pq}\bar{b}_q = \bar{c}_p$. Second, the components must transform as tensor components, that is, $\bar{A}_{pq} = l_{ip}l_{jq}A_{ij}$. Let us define $\bar{A}_{pq} = \bar{c}_p/\bar{b}_q$ so that the first relation is satisfied, and see if these quantities have the correct transformation property. Now $\mathbf{b}$ and $\mathbf{c}$ are vectors, so

$$\bar{b}_q = l_{jq}b_j \quad \text{and} \quad \bar{c}_p = l_{ip}c_i. \tag{2.62.1}$$

Hence,

$$\bar{A}_{pq}\bar{b}_q = \bar{c}_p = l_{ip}c_i = l_{ip}A_{ij}b_j = l_{ip}l_{jq}A_{ij}\bar{b}_q$$

or

$$(\bar{A}_{pq} - l_{ip}l_{jq}A_{ij})\bar{b}_q = 0. \tag{2.62.2}$$

However, **b** is independent of **A** so that this relation can only hold if the expression in the brackets vanishes, that is,

$$\bar{A}_{pq} = l_{ip}l_{jq}A_{ij}. \tag{2.62.3}$$

This shows that A_{ij} transforms as a tensor. The proof can be easily adapted to the following more general rule.

If $A_{ij...n}$ is a set of 3^n quantities and $B_{pq...t}$ a tensor of order m independent of **A** and the k times contracted product $A_{ij...n}B_{pq...t}$ is a tensor of order $m + n - 2k$, $1 \leqq k \leqq \frac{1}{2}(m + n)$, then $A_{ij...n}$ is a tensor of order n.

A special case arises if **B** is the product of n vectors and $A_{ij...n}b_i c_j ... e_n$ can be shown to be a scalar. This again establishes that **A** is an nth order tensor.

Exercise 2.62.1. Prove the quotient rule in detail for $A_{ij}B_{jk} = C_{ik}$ and $A_{ij}b_i c_j = \alpha$.

2.7. Isotropic tensors

An isotropic tensor is one whose *components* are unchanged by rotation of the frame of reference. The trivial cases of this are the tensors of all orders whose components are all zero. All tensors of the zeroth order are isotropic and there are no first order isotropic tensors. We have already met the only isotropic second order tensor, namely, δ_{ij}, but it is of interest to prove that it is the only one.

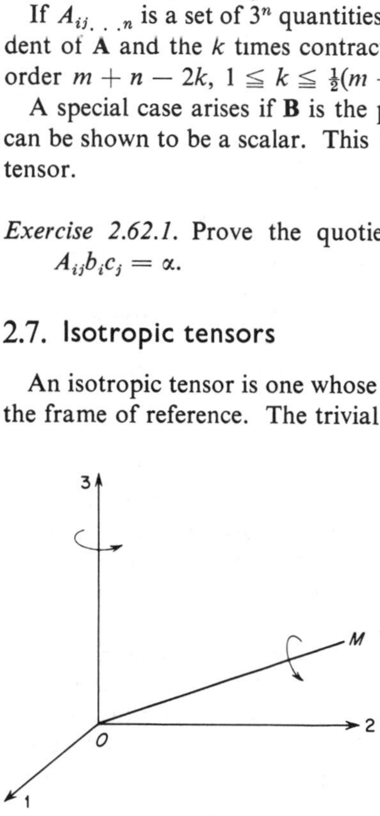

Fig. 2.6

Consider a general second order tensor A_{ij} and apply some particular rotations to it. The first of these is a rotation about a line equally inclined to all three coordinate axes, that is, with direction cosines all equal to $3^{-1/2}$. A rotation of 120° such as is shown in Fig. 2.6 can be made to carry the $O1$ axis into the $O3$ position, the $O2$ into the $O1$, and the $O3$ into the $O2$. Thus,

$$l_{31} = l_{12} = l_{23} = 1$$

and the other l_{ij} are zero. Hence, for example,

$$\bar{A}_{11} = A_{33} \quad \text{and} \quad \bar{A}_{23} = A_{12}.$$

However, if A is isotropic,

$$\bar{A}_{11} = A_{11} \quad \text{and} \quad \bar{A}_{23} = A_{23}$$

and so

$$\bar{A}_{11} = A_{11} = A_{33} \quad \text{and} \quad \bar{A}_{23} = A_{23} = A_{12}.$$

Applying this to each component in turn we see that

$$A_{11} = A_{22} = A_{33} \tag{2.7.1}$$

$$A_{23} = A_{32} = A_{31} = A_{13} = A_{12} = A_{21} \tag{2.7.2}$$

Now apply a rotation through a right angle about $O3$, so that $l_{12} = -l_{21} = l_{33} = 1$ and the other l_{ij} are zero. Then $\bar{A}_{12} = -A_{21}$ by the transformation and $\bar{A}_{12} = A_{12}$ by the requirement of isotropy. Thus by (2.7.2) $A_{12} = A_{21}$ and now $A_{12} = -A_{21}$ and the only way these can be simultaneously true is for them both to be zero. It follows that all the off diagonal components ($i \neq j$) are zero and all the diagonal ones are equal. Clearly a scalar multiple of an isotropic tensor is isotropic and so we may take $A_{ij} = \delta_{ij}$.

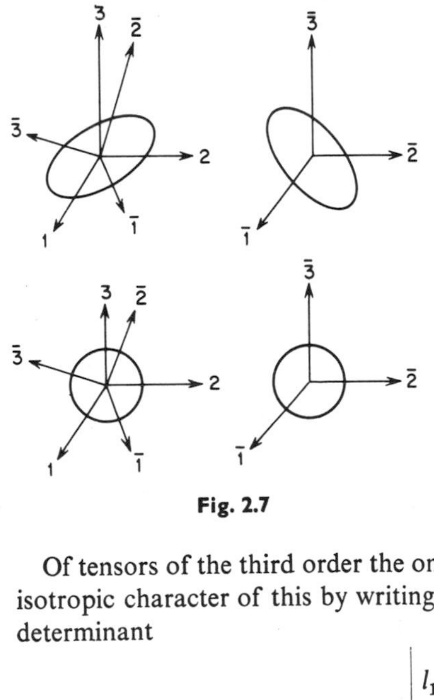

Fig. 2.7

The idea of isotropy for a second order tensor is connected with the geometrical figure of a sphere. We have noticed that $A_{ij}x_ix_j = 1$ is the equation of a quadric surface. The ellipsoid may be regarded as the typical quadric and clearly if its axes are unequal a rotation of the coordinate frame will require a different equation. A sphere, however, is invariant as a whole under rotation of axes and has the equation

$$x_1^2 + x_2^2 + x_3^2 = r^2.$$

This corresponds to the tensor $A_{ij} = \delta_{ij}/r^2$, so that isotropy is to be interpreted as geometrical invariance under rotation. (Cf. Fig. 2.7.)

Of tensors of the third order the only isotropic one is ϵ_{ijk}. We may see the isotropic character of this by writing $\bar{\epsilon}_{pqr} = \epsilon_{ijk}l_{ip}l_{jq}l_{kr}$ which is evidently the determinant

$$\bar{\epsilon}_{pqr} = \begin{vmatrix} l_{1p} & l_{1q} & l_{1r} \\ l_{2p} & l_{2q} & l_{2r} \\ l_{3p} & l_{3q} & l_{3r} \end{vmatrix}. \tag{2.7.3}$$

Now if any of the p, q, r are the same, this is a determinant with two identical columns and so vanishes. If $p = 1$, $q = 2$, $r = 3$, we know that this determinant is $+1$ so $\bar{\epsilon}_{123} = 1$. If p, q, r is an even permutation of 1, 2, 3 the sign of

Class	I	II	III(i)	III(ii)	III(iii)	IV
Operation A	Rotation about OM through 120°. $l_{12} = l_{31} = l_{23} = 1$.					
	$T_{1111} = T_{2222}$	$T_{1112} = T_{2223}$	$T_{1122} = T_{2233}$	$T_{1221} = T_{2332}$	$T_{1212} = T_{2323}$	$T_{1123} = T_{2231}$
Operation B	Rotation about $O3$ through 90°. $l_{12} = -l_{21} = l_{33} = 1$.					
	$T_{1111} = T_{2222}$	$T_{1112} = -T_{2221}$	$T_{1122} = T_{2211}$	$T_{1221} = T_{2112}$	$T_{1212} = T_{2121}$	$T_{1123} = -T_{2213}$
Operation C	Rotation to reverse the direction of OM. $l_{13} = l_{22} = l_{31} = -1$.					
	$T_{1111} = T_{3333}$	$T_{1112} = T_{3332}$ $T_{2223} = T_{2221}$	$T_{1122} = T_{3322}$	$T_{1221} = T_{3223}$	$T_{1212} = T_{3232}$	$T_{1123} = T_{3321}$ $T_{2231} = T_{2213}$
Conclusion Representation	All equal γ_{ijpq}	All zero	All equal $\delta_{ij}\delta_{pq} - \gamma_{ijpq}$	All equal $\delta_{iq}\delta_{jp} - \gamma_{iqjp}$	All equal $\delta_{ip}\delta_{jq} - \gamma_{ipjq}$	All zero

the determinant is unchanged, but if it is an odd permutation, $\bar{\epsilon}_{pqr} = -1$. However, this is just the definition of ϵ_{pqr} and so the components are unchanged by rotation.

The isotropic tensors of the fourth order are of some importance and there are three independent ones. It is clear that a product of isotropic tensors is isotropic so that we can immediately write down two such, namely, $\delta_{ij}\delta_{pq}$ and $\epsilon_{ijk}\epsilon_{kpq}$, but we need to work harder to find all the independent isotropic tensors. To outline the reasoning which is necessary to be sure we have all, it is convenient to divide the 81 components of a fourth order tensor T_{ijpq} into classes as follows.

Class	Character	Typical member
I	All suffixes the same	T_{1111}
II	Three suffixes thc same	T_{1112}
III(i)	Suffixes the same in pairs	T_{1122}
(ii)		T_{1221}
(iii)		T_{1212}
IV	Only two suffixes the same	T_{1123}

(Since the suffixes must be equal to either 1, 2, or 3, it follows that at least two of them are the same.)

We now apply special rotations of the type shown in Fig. 2.6. These are listed in the table opposite by giving the nonzero l_{ij} of the transformation, and their effect on the typical member of each class is shown. We can write the transformation as an equality since the isotropic requirement is for the component to be unchanged. The second line is the effect of the transformation on another member of the class needed for the conclusion. Under class II, for example, we find $T_{1112} = T_{2223} = -T_{2221}$ and $T_{2223} = T_{2221}$ and the conclusion is that $T_{2221} = T_{2223} = 0$ and consequently all members of the class are zero. In the tensor representation of the class we put $\gamma_{ijpq} = 1$ if $i = j = p = q$ but zero otherwise. Then the representation of the subclasses of class III is given by a combination of Kronecker deltas and γ. For example, $\delta_{ij}\delta_{pq}$ will be 1 if $i = j$ and $p = q$, but the definition of the class excludes the possibility $i = j = p = q$ so that $\delta_{ij}\delta_{pq} - \gamma_{ijpq}$ represents it. Suppose then we make a linear combination of these and write

$$T_{ijpq} = a\delta_{ij}\delta_{pq} + b\delta_{ip}\delta_{jq} + c\delta_{iq}\delta_{jp} + (d - a - b - c)\gamma_{ijpq} \quad (2.7.4)$$

This certainly has isotropic properties undcr the rotations A, B, and C of the table but these are rather special since they leave a cube invariant as a whole. If we take a rotation that does not leave the cube invariant, for example, a rotation about $O3$ through an angle θ, then γ_{ijpq} is not invariant. Hence we must put $d = a + b + c$ and the general isotropic tensor can be written as a linear combination of the first three.

We observe that $\delta_{ij}\delta_{pq}$ is the product of two second order isotropic tensors and that the contracted product $\epsilon_{ijk}\epsilon_{kpq}$ is the difference of the other two. It is sometimes convenient to take the general isotropic tensor to be

$$T_{ijpq} = \lambda\delta_{ij}\delta_{pq} + \mu(\delta_{ip}\delta_{jq} + \delta_{iq}\delta_{jp}) + \nu(\delta_{ip}\delta_{jq} - \delta_{iq}\delta_{jp}) \qquad (2.7.5)$$

The second term is symmetric with respect to the first and second or third and fourth indices and the third term is antisymmetric with respect to them.

Exercise 2.7.1. Show that ϵ_{ijk} is the only isotropic tensor of the third order.

Exercise 2.7.2. Show that if T_{ijpq} is an isotropic fourth order tensor then the symmetric tensors A_{ij} and $T_{ijpq}A_{pq}$ have the same principal axes.

Exercise 2.7.3. Show that under a suitable rotation any component of γ_{ijpq} can be made nonzero.

2.81. Dyadics and other notations

There is no need to give a full discussion of the history of vector and tensor notations, but it will be useful to mention some other forms that appear in the literature and to define by comparison the usage of this book.

The notation on which we have largely depended is sometimes known as the kernel-index method.* The tensor is represented by its typical component, A_{ij}, A being the kernel letter and i and j the indices. It is quite useful, but not at all necessary, to use a lower case letter for the tensor of the first order. This notation is entirely adequate but it is often convenient to use a single bold-face symbol without suffixes for the tensor. We have done this more with vectors than with second order tensors, for such expressions as $\mathbf{a} \cdot \mathbf{b}$, $\mathbf{a} \wedge \mathbf{b}$, etc. are unambiguous whereas some care is needed in distinguishing tensor products $\mathbf{A} \cdot \mathbf{B}$, $\mathbf{A}' \cdot \mathbf{B}$, etc. In general it is probably easier in manipulations to have the suffixes written explicitly and so avoid the danger of mistakes that the suppression of suffixes induces. However, the physical meaning of the final formulae is often brought out more clearly by the dyadic notation of bold face symbols.

The name "dyad" is applied to the product of two vectors **ab** and is a tensor with components $a_i b_j$. If

$$\mathbf{C} = \mathbf{a}_{(1)}\mathbf{b}_{(1)} + \mathbf{a}_{(2)}\mathbf{b}_{(2)} + \ldots + \mathbf{a}_{(n)}\mathbf{b}_{(n)} \qquad (2.81.1)$$

is the sum of n dyads, it is called a dyadic. The products $\mathbf{C} \cdot \mathbf{c}$ and $\mathbf{c} \cdot \mathbf{C}$ are defined by

$$\mathbf{C} \cdot \mathbf{c} = \sum_1^n \mathbf{a}_{(r)}(\mathbf{b}_{(r)} \cdot \mathbf{c})$$

* See, for example, J. A. Schouten, Tensor Analysis for Physicists (Oxford University Press, Oxford, 1951), p. 110.

and

$$\mathbf{c} \cdot \mathbf{C} = \sum_{1}^{n} (\mathbf{c} \cdot \mathbf{a}_{(r)})\mathbf{b}_{(r)} \tag{2.81.2}$$

The dyads formed from the three unit vectors $\mathbf{e}_{(i)}$ give a convenient representation of the dyadic or second order tensor

$$\mathbf{A} = \mathbf{e}_{(i)} A_{ij} \mathbf{e}_{(j)} \tag{2.81.3}$$

where A_{ij} is the component of the tensor and the summation convention is invoked even on the bracketed indices. Another standard form of dyadic is composed from three vectors **a**, **b**, and **c** in the form

$$\mathbf{A} = \mathbf{e}_{(1)}\mathbf{a} + \mathbf{e}_{(2)}\mathbf{b} + \mathbf{e}_{(3)}\mathbf{c} \tag{2.81.4}$$

and the transposed tensor is called the conjugate dyadic

$$\mathbf{A}_c = \mathbf{a}\mathbf{e}_{(1)} + \mathbf{b}\mathbf{e}_{(2)} + \mathbf{c}\mathbf{e}_{(3)} \tag{2.81.5}$$

If we replace the dyads in these forms by scalar and vector products, we have the scalar of the dyadic

$$\mathbf{A}_s \quad \text{or} \quad \mathbf{A} = \mathbf{e}_{(1)} \cdot \mathbf{a} + \mathbf{e}_{(2)} \cdot \mathbf{b} + \mathbf{e}_{(3)} \cdot \mathbf{c} \tag{2.81.6}$$

and the vector of the dyadic

$$\mathbf{A}_v \quad \text{or} \quad \mathbf{A}_\times = \mathbf{e}_{(1)} \wedge \mathbf{a} + \mathbf{e}_{(2)} \wedge \mathbf{b} + \mathbf{e}_{(3)} \wedge \mathbf{c}. \tag{2.81.7}$$

The scalar of the dyadic is the same as the trace or spur of the tensor for it is simply the contraction. For an antisymmetric dyadic ($\mathbf{A}_c = -\mathbf{A}$) the vector of the dyadic is twice the vector of the antisymmetric tensor.

In multiplication of dyadics the rule is that summation should always be on adjacent indices. For example, the corresponding dyadic and tensor forms are:

$$\begin{array}{ll} \mathbf{A} \cdot \mathbf{B}, & A_{ij}B_{jk} \\ \mathbf{A}_c \cdot \mathbf{B}, & A_{ij}B_{ik} \\ \mathbf{a} \wedge \mathbf{A}, & \epsilon_{ijk}a_j A_{km} \\ \mathbf{A} \wedge \mathbf{a}, & \epsilon_{ijk}A_{mj}a_k \end{array}$$

The symbol $\times$ is more commonly used with dyadics than is $\wedge$, but we retain the latter to be consistent with the vector notation to which we have given preference.

Exercise 2.81.1. Translate the equations of Section 2.5 into dyadic notation.

Exercise 2.81.2. Show that

$$\mathbf{a} \cdot (\mathbf{b} \wedge \mathbf{A}) = (\mathbf{a} \wedge \mathbf{b}) \cdot \mathbf{A}$$

Exercise 2.81.3. Prove that if $\mathbf{A}_c = -\mathbf{A}$, then $\mathbf{A}_v \wedge \mathbf{a} = 2\mathbf{a} \cdot \mathbf{A}$.

2.82. Axial vectors

The only transformation of coordinates we have considered is the rotation of a right-handed Cartesian system. A slight extension of this would be to allow reflections in the coordinate planes as well as rotations. In this case a right-handed coordinate frame is transformed into a left-handed one. The matrix L of the direction cosines l_{ij} is still an orthogonal one but its determinant is -1 instead of $+1$. Now certain items in our constructions have depended on our maintaining the convention of right-handedness. Thus $\mathbf{a} = \mathbf{b} \wedge \mathbf{c}$, for example, gives a definite sense to the direction of $\mathbf{a}$ by the requirement that $\mathbf{b}$, $\mathbf{c}$, $\mathbf{a}$ should form a right-handed system in that order. If we transform to a left-handed system, the vector product changes sign, for

$$\begin{aligned}(\bar{\mathbf{b}} \wedge \bar{\mathbf{c}})_p &= \epsilon_{pqr} l_{jq} b_j l_{kr} c_k \\ &= -\epsilon_{ijk} l_{kp} b_j c_k = -\bar{a}_p,\end{aligned}$$

since the determinant of the l_{ij} is -1. A vector which does this is called an *axial vector* or *pseudo vector*. It can be thought of as a vector with given direction and magnitude but reversible sense of direction. We do not have much need of the distinction at this stage, but will find later that it is the same as the distinction between an absolute tensor and one of weight 1.

BIBLIOGRAPHY

The wealth of texts on vectors and tensors makes it an invidious task to give a brief list. Those mentioned here are mentioned in connection with the specific point of reference and omission casts no aspersion.

2.1. Many accounts of Cartesian vectors and tensors are included in the various volumes of "Mathematical Methods of . . . " and "Applied Mathematics for . . . ." In particular H. Jeffreys' "Cartesian Tensors" (Cambridge 1931) is largely contained in

Jeffreys, H. and B. S., Methods of mathematical physics, (2nd Ed.). Cambridge: Cambridge University Press, 1950.

The classic text on dyadics is that founded on Gibbs' lectures.

Wilson, E. B., Vector analysis. New Haven: Yale University Press, 1901.

A discussion of dyadics as well as of vector and tensor formalism will be found in

Morse, P. M., and H. Feshbach, Methods of theoretical physics. New York: McGraw-Hill, 1953.

A more elementary account of dyadics is given in

Phillips, H. B., Vector analysis, Chapter 10. New York: John Wiley, 1933.

Often an account of fluid mechanics will open with a brief résumé of the vector notation to be used. See, for example,

Serrin, J., Mathematical principles of classical fluid mechanics, in Handbuch der Physik, Bd. VIII/1, ed. S. Flugge and C. Truesdell. Berlin: Springer-Verlag, 1959.

or

Truesdell, C., The kinematics of vorticity. Bloomington: Indiana University Press, 1954.

Cartesian Vectors and Tensors: Their Calculus 3

3.11. Tensor functions of a time-like variable

In the last chapter we considered only the algebraic manipulations and relationships of tensors. We now want to find out what we can say about the behavior of tensors when they are functions of continuous variables. Certainly this will be necessary in applications, for if we think of an unsteady fluid motion we realise that we shall have to consider its velocity at any point or time. Now velocity is a vector so we shall have a vector $\mathbf{v}$ whose components are functions of the coordinates x_1, x_2, x_3 and time t. Any variable which is independent of the coordinates can be called time-like since in many applications it will be the time. We will consider first tensors whose components are functions of one such variable, t.

Suppose $A_{ij} = A_{ij}(t)$ is a tensor function; then it is clear that all the derivatives of it that exist are themselves tensors, for the l_{ij} of the transformation are independent of t and we may differentiate the relation

$$\bar{A}_{pq}(t) = l_{ip}l_{jq}A_{ij}(t) \tag{3.11.1}$$

as many times as we are able, to give

$$\frac{d^r}{dt^r}[\bar{A}_{pq}(t)] = l_{ip}l_{jq}\frac{d^r}{dt^r}[A_{ij}(t)] \tag{3.11.2}$$

We may integrate the tensor components and preserve their character equally well.

The most important example of this has already been cited. If $\mathbf{x}$, the position vector of a particle, is a function of time, then its first derivative is the velocity

$$\mathbf{v}(t) = \dot{\mathbf{x}}(t), \qquad v_i = dx_i/dt, \tag{3.11.3}$$

and its acceleration is

$$\mathbf{a}(t) = \ddot{\mathbf{x}}(t), \qquad a_i = d^2x_i/dt^2. \tag{3.11.4}$$

The differentiation of products of tensors proceeds according to the usual rules of differentiation of products. In particular,

$$\frac{d}{dt}(\mathbf{a} \cdot \mathbf{b}) = \frac{d\mathbf{a}}{dt} \cdot \mathbf{b} + \mathbf{a} \cdot \frac{d\mathbf{b}}{dt}, \tag{3.11.5}$$

and

$$\frac{d}{dt}(\mathbf{a} \wedge \mathbf{b}) = \frac{d\mathbf{a}}{dt} \wedge \mathbf{b} + \mathbf{a} \wedge \frac{d\mathbf{b}}{dt}. \tag{3.11.6}$$

Exercise 3.11.1. Show that if a particle moves with constant speed its acceleration is normal to its velocity.

Exercise 3.11.2. Show that the acceleration of a particle moving in the surface of a sphere of radius r has a radial component $-v^2/r$, where v is its speed.

Exercise 3.11.3. If $\mathbf{a}(t)$, $\mathbf{b}(t)$, and $\mathbf{c}(t)$ are three mutually orthogonal unit vectors, show that their first derivatives are coplanar.

Exercise 3.11.4. Show that if the position vector of a particle, its velocity, and its acceleration are coplanar, then all the higher derivatives are in the same plane.

Exercise 3.11.5. Show that $\mathbf{x} \wedge (d\mathbf{x}/dt) = 0$ is the condition that $\mathbf{x}(t)$ should remain parallel to itself.

Exercise 3.11.6. If a_i is the component of a vector $\mathbf{a}$ with respect to a system of base vectors $\mathbf{b}_{(1)}$, $\mathbf{b}_{(2)}$, $\mathbf{b}_{(3)}$ show that

$$\mathbf{b}_{(i)} = \partial\mathbf{a}/\partial a_i.$$

3.12. Curves in space

The variable position vector $\mathbf{x}(t)$ describes the motion of a particle. For a finite interval of t, say $a \leqq t \leqq b$, we can plot the position as a curve in space. If the curve does not cross itself (that is, if $\mathbf{x}(t) \neq \mathbf{x}(t')$, $a \leqq t < t' \leqq b$) it is called *simple*; if $\mathbf{x}(a) = \mathbf{x}(b)$ the curve is *closed.* The variable t is now just a parameter along the curve which may be thought of as the time in the motion of the particle only if such picturesqueness* is desired. If t and t' are the parameters of two points, the chord joining them is the vector $\mathbf{x}(t') - \mathbf{x}(t)$. As $t \to t'$ this vector approaches $(t' - t)\,\dot{\mathbf{x}}(t)$ and so in the limit it is proportional to $\dot{\mathbf{x}}(t)$. However, the limit of the chord is the tangent so that $\dot{\mathbf{x}}(t)$ is in the direction of the tangent. If $v^2 = \dot{\mathbf{x}} \cdot \dot{\mathbf{x}}$ we can construct a unit tangent vector $\boldsymbol{\tau} = \dot{\mathbf{x}}/v$.

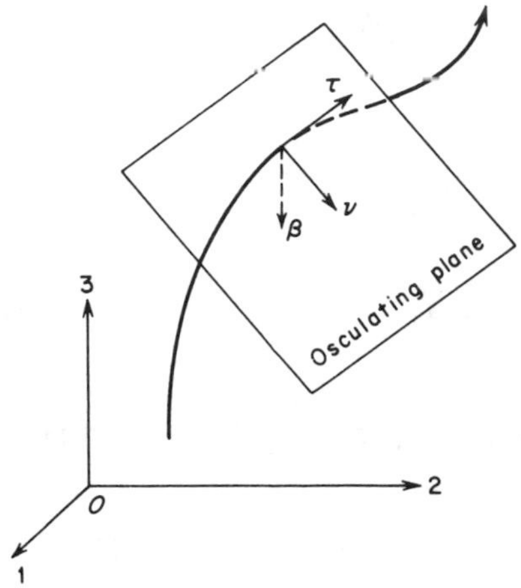

Fig. 3.1

If $\mathbf{x}(t)$ and $\mathbf{x}(t + dt)$ are two very close points,

$$\mathbf{x}(t + dt) = \mathbf{x}(t) + dt\,\dot{\mathbf{x}}(t) + 0(dt)^2$$

and the distance between them is

$$\begin{aligned} ds^2 &= \{\mathbf{x}(t + dt) \\ &\qquad - \mathbf{x}(t)\} \cdot \{\mathbf{x}(t + dt) - \mathbf{x}(t)\} \\ &= \dot{\mathbf{x}}(t) \cdot \dot{\mathbf{x}}(t)\, dt^2 + 0(dt^3). \end{aligned}$$

The arc length from any given point $t = a$ is therefore

$$s(t) = \int_a^t [\dot{\mathbf{x}}(t') \cdot \dot{\mathbf{x}}(t')]^{1/2}\, dt'. \tag{3.12.1}$$

s is the natural parameter to use on the curve, and we observe that

$$\frac{d\mathbf{x}}{ds} = \frac{d\mathbf{x}}{dt}\frac{dt}{ds} = \dot{\mathbf{x}}(t) \Big/ \frac{ds}{dt} = \boldsymbol{\tau} \tag{3.12.2}$$

A curve for which the length can be so calculated is called *rectifiable*. From this point on we will regard s as the parameter, identifying t with s and letting the dot denote differentiation with respect to s. Thus

$$\boldsymbol{\tau}(s) = \dot{\mathbf{x}}(s) \tag{3.12.3}$$

is the unit tangent vector. (Cf. Fig. 3.1.)

Let $\mathbf{x}(s)$, $\mathbf{x}(s + ds)$, and $\mathbf{x}(s - ds)$ be three nearby points on the curve. The plane containing them must also contain the vectors

$$\frac{\mathbf{x}(s + ds) - \mathbf{x}(s)}{ds} = \dot{\mathbf{x}}(s) + 0(ds)$$

and

$$\frac{\mathbf{x}(s + ds) - 2\mathbf{x}(s) + \mathbf{x}(s - ds)}{ds^2} = \ddot{\mathbf{x}}(s) + 0(ds).$$

Thus, in the limit when the points are coincident, the plane reaches a limiting position defined by the first two derivatives $\dot{x}(s)$ and $\ddot{x}(s)$. This limiting plane is called the osculating plane and the curve appears to lie in this plane in the

* For an amusing use of this word by a pure mathematician see the preface to E. C. Titchmarsh's "Theory of Fourier Integrals." Oxford: 1937.

immediate neighborhood of the point. Now $\dot{\mathbf{x}} = \boldsymbol{\tau}$ so $\ddot{\mathbf{x}} = \dot{\boldsymbol{\tau}}$ and since $\boldsymbol{\tau} \cdot \boldsymbol{\tau} = 1$

$$\boldsymbol{\tau} \cdot \dot{\boldsymbol{\tau}} = 0 \tag{3.12.4}$$

so that the vector $\dot{\boldsymbol{\tau}}$ is at right angles to the tangent. Let

$$\frac{1}{\rho^2} = \dot{\boldsymbol{\tau}} \cdot \dot{\boldsymbol{\tau}} \tag{3.12.5}$$

and

$$\boldsymbol{\nu} = \rho \dot{\boldsymbol{\tau}} \tag{3.12.6}$$

Then $\boldsymbol{\nu}$ is a unit normal and defines the direction of the so-called principal normal to the curve.

To interpret ρ, we observe that the small angle $d\theta$ between the tangents at s and $s + ds$ is given by

$$\cos d\theta = \boldsymbol{\tau}(s) \cdot \boldsymbol{\tau}(s + ds)$$

or

$$\begin{aligned} 1 - \tfrac{1}{2} d\theta^2 + \ldots &= \boldsymbol{\tau} \cdot \boldsymbol{\tau} + \boldsymbol{\tau} \cdot \dot{\boldsymbol{\tau}}\, ds + \tfrac{1}{2} \boldsymbol{\tau} \cdot \ddot{\boldsymbol{\tau}}\, ds^2 + \ldots \\ &= 1 - \tfrac{1}{2}(\dot{\boldsymbol{\tau}} \cdot \dot{\boldsymbol{\tau}})\, ds^2 + \ldots \end{aligned}$$

since $\boldsymbol{\tau} \cdot \dot{\boldsymbol{\tau}} = 0$ and so $\boldsymbol{\tau} \cdot \ddot{\boldsymbol{\tau}} + \dot{\boldsymbol{\tau}} \cdot \dot{\boldsymbol{\tau}} = 0$. Thus,

$$\rho = \frac{ds}{d\theta} \tag{3.12.7}$$

is the reciprocal of the rate of change of the angle of the tangent with arc length, that is, the radius of curvature. Its reciprocal $1/\rho$ is the curvature.

A second normal to the curve may be taken to form a right-hand system with $\boldsymbol{\tau}$ and $\boldsymbol{\nu}$. This is called the unit binormal,

$$\boldsymbol{\beta} = \boldsymbol{\tau} \wedge \boldsymbol{\nu}. \tag{3.12.8}$$

Since $\boldsymbol{\beta} \cdot \boldsymbol{\beta} = 1$, $\dot{\boldsymbol{\beta}} \cdot \boldsymbol{\beta} = 0$ and $\dot{\boldsymbol{\beta}}$ is at right angles to $\boldsymbol{\beta}$. However, $\boldsymbol{\beta} \cdot \boldsymbol{\tau} = 0$ so $\dot{\boldsymbol{\beta}} \cdot \boldsymbol{\tau} = -\boldsymbol{\beta} \cdot \dot{\boldsymbol{\tau}} = -\boldsymbol{\beta} \cdot \boldsymbol{\nu}/\rho = 0$ so $\dot{\boldsymbol{\beta}}$ is also at right angles to $\boldsymbol{\tau}$ and so must be in the direction of $\boldsymbol{\nu}$. Let

$$\dot{\boldsymbol{\beta}} = -\boldsymbol{\nu}/\sigma \tag{3.12.9}$$

where $1/\sigma$ is a scalar known as the torsion. Clearly, $1/\sigma$ is the magnitude rate of change of the direction of the binormal, just as $1/\rho$ was the rate of change of the tangent.

Further, since $\boldsymbol{\nu} = \boldsymbol{\beta} \wedge \boldsymbol{\tau}$, we have

$$\begin{aligned} \dot{\boldsymbol{\nu}} = \dot{\boldsymbol{\beta}} \wedge \boldsymbol{\tau} + \boldsymbol{\beta} \wedge \dot{\boldsymbol{\tau}} &= -\frac{1}{\sigma} \boldsymbol{\nu} \wedge \boldsymbol{\tau} + \frac{1}{\rho} \boldsymbol{\beta} \wedge \boldsymbol{\nu} \\ &= \frac{1}{\sigma} \boldsymbol{\beta} - \frac{1}{\rho} \boldsymbol{\tau} \end{aligned} \tag{3.12.10}$$

These three formulae,

$$\dot{\boldsymbol{\tau}} = \frac{1}{\rho}\boldsymbol{\nu}$$

$$\dot{\boldsymbol{\nu}} = -\frac{1}{\rho}\boldsymbol{\tau} + \frac{1}{\sigma}\boldsymbol{\beta} \tag{3.12.11}$$

$$\dot{\boldsymbol{\beta}} = -\frac{1}{\sigma}\boldsymbol{\nu},$$

are known as the Serret-Frenet formulae.

It may be shown that if two curves have the same dependence of curvature and torsion upon arc length, then they are the same curve apart from some translation or rotation in space. These two functions $\rho(s)$ and $\sigma(s)$ are the intrinsic equations of the curve.

Exercise 3.12.1. Interpret the curve given by

$$x_1 = a \cos (s \cos \alpha/a), \qquad x_2 = a \sin (s \cos \alpha/a),$$
$$x_3 = s \sin \alpha$$

and find its curvature and torsion.

Exercise 3.12.2. Show that if the tangent to a curve makes a constant angle with a fixed direction then the ratio of its curvature and torsion is constant. Such a curve is called a helix.

Exercise 3.12.3. Prove that $\rho^2\sigma\dot{\mathbf{x}} \cdot (\ddot{\mathbf{x}} \wedge \dddot{\mathbf{x}}) = 1$, and hence that the curve lies in a plane if $\dot{\mathbf{x}} \cdot (\ddot{\mathbf{x}} \wedge \dddot{\mathbf{x}}) = 0$.

3.13. Line integrals

If $F(x_1, x_2, x_3)$, or more briefly $F(\mathbf{x})$, is a function of position and C is the arc of a simple curve $\mathbf{x} = \mathbf{x}(t)$, $a \leqq t \leqq b$, we can define the integral of F along C as

$$\int_C F(x)\, dt = \int_a^b F[x_1(t), x_2(t), x_3(t)]\, dt \tag{3.13.1}$$

provided this second integral exists. If the curve C is composed of a number of arcs which have to be given by different equations, then an integral like the right-hand side of (3.13.1) must be calculated for each arc. If A and B are the end points of the curve (given by $t = a$ and $t = b$, respectively) the integral above is the integral along C from A to B. The integral in the opposite direction from B to A is obtained by reversing the limits and therefore has the same absolute value but the opposite sign. If $\mathbf{x}(a) = \mathbf{x}(b)$, the curve C is closed and the integral is sometimes written

$$\oint_C F(\mathbf{x})\, dt$$

An important case, and one which we shall frequently use, occurs when the parameter t is the arc length s.

$$\int_C F(\mathbf{x})\, ds = \int_a^b F[\mathbf{x}(t)]\{\dot{\mathbf{x}}(t) \cdot \dot{\mathbf{x}}(t)\}^{1/2}\, dt. \tag{3.13.2}$$

In general the line integral depends on F, the two end points and the path between them. If, however, the integral around any simple closed curve vanishes, the value of the integral from A to B is independent of the path. To see this we take any two paths between A and B, say C_1 and C_2, and denote by C the closed curve formed by following C_1 from A to B and C_2 back from B to A. Now

$$\begin{aligned}\int_C F\, ds &= \left[\int_A^B F\, ds\right]_{C_1} + \left[\int_B^A F\, ds\right]_{C_2} \\ &= \left[\int_A^B F\, ds\right]_{C_1} - \left[\int_A^B F\, ds\right]_{C_2}\end{aligned} \tag{3.13.3}$$

and this vanishes by hypothesis, so that the integrals along the two different paths are equal.

If $\mathbf{a}(x_1, x_2, x_3)$ is any vector function of position, $\mathbf{a} \cdot \boldsymbol{\tau}$ is the projection of $\mathbf{a}$ tangent to the curve. The integral around a simple closed curve C of $\mathbf{a} \cdot \boldsymbol{\tau}$ is called the *circulation* of $\mathbf{a}$ around C,

$$\oint_C \mathbf{a} \cdot \boldsymbol{\tau}\, ds = \int a_i[x_1(s), x_2(s), x_3(s)] \cdot \tau_i(s)\, ds \tag{3.13.4}$$

Since $dx_i/ds = \tau_i$ this integral is sometimes written $\int a_i\, dx_i$ or $\int \mathbf{a} \cdot d\mathbf{x}$; we shall prefer however to write the integral with the unit tangent vector explicit.

Exercise 3.13.1. Evaluate the integral $\oint_C \mathbf{a} \cdot \boldsymbol{\tau}\, ds$ where C is the circle of radius a and center the origin lying in the plane $\mathbf{m} \cdot \mathbf{x} = 0$ and $\mathbf{a} = \boldsymbol{\omega} \wedge \mathbf{x}$. $\boldsymbol{\tau}$ is the unit tangent to C.

Exercise 3.13.2. Show that if C is a simple closed curve in the plane $O12$ then it encloses an area

$$\oint x_1 t_2\, ds = -\oint x_2 t_1\, ds = \tfrac{1}{2}\oint (x_1 t_2 - x_2 t_1)\, ds;$$

all the integrals being taken around C.

Exercise 3.13.3. C is any simple closed curve in space and $\mathbf{t}$ its tangent; C_3 is the projection of C on a plane $O12$ and C_3 is also a simple closed curve. Show that the area of C_3 is given by the same formulae as in the previous question.

3.14. Surface integrals

We shall need also to consider integrals over surfaces and should say something about their construction and the surfaces for which they can be constructed. A *closed surface* is one which lies within a bounded region of space and has an inside and an outside. We can pass from any inside point to another inside point by a curve which does not cross the surface and, similarly, from any outside point to any other. However, to pass from an inside point to an outside point the path must cross the surface. Familiar examples of closed surfaces are the sphere and surfaces that could be deformed into a sphere. A continuous surface which has no inside or outside, known as the Klein bottle, is shown in Fig. 3.2d. These pathological surfaces are the happy hunting ground of the topologist; they serve to preserve the engineer from becoming complacent in his assumed normality. If the normal to the surface varies continuously over a part of the surface that part is called *smooth*. Some closed surfaces (see for example Fig. 3.2a) are smooth everywhere, others are made up of a number of subregions which are smooth (Fig. 3.2b) and are called *piece-wise smooth*. A closed curve on a surface which can be continuously shrunk to a point is called *reducible*, as for example the equator of a sphere which can be continuously moved to any line of latitude until it shrinks to a pole. If all closed curves on a surface are reducible, the surface is called *simply connected*. The sphere is simply connected but the torus or anchor ring is not, for a closed curve such as is shown in Fig. 3.2c is not reducible.

If a surface is not closed it normally has a space curve as its boundary, as for example a hemisphere with the equator as boundary. It has two sides if it is impossible to go from a point on one side to a point on the other (in particular the point, which is just on the other side) along a continuous curve that does not cross the boundary. The surface is sometimes called the *cap* of the space curve. Again it is not necessary for a surface to have two sides; the Mobius strip (Fig. 3.2f) is the well-known example of this. Indeed a closed curve could be capped by either a one- or a two-sided surface for the boundary curve of the hemisphere (Fig. 3.2e) would be distorted into the boundary of the Mobius strip and the hemisphere would be a two-sided cap. These sophisticated considerations need not daunt us however, for given a simple closed curve we can imagine a soap film across it which can then be distorted into a piece-wise smooth cap.

Just as the line integral is a natural extension of the common integral over an interval, so the surface integral is an extension of the double integral. The double integral over a region R in the $O12$ plane of a function $F(x_1, x_2)$ is written

$$\iint_R F(x_1, x_2)\, dx_1\, dx_2$$

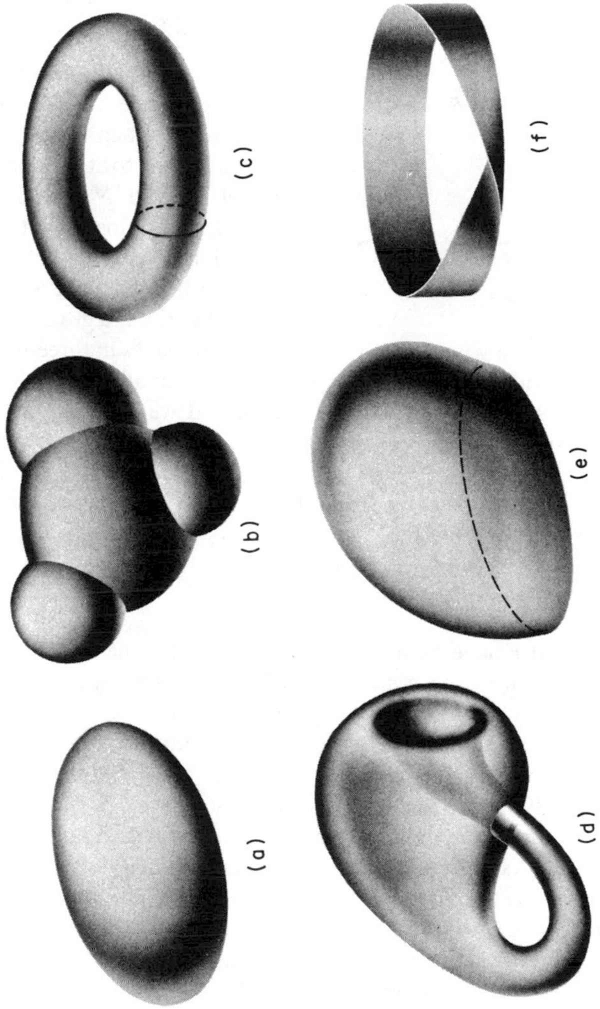

Fig. 3.2

and is constructed as follows. The region R is divided into a large number of small areas by a grid of lines $x_1 = a_0, a_1, \ldots, a_m$, $x_2 = b_0, b_1, \ldots, b_n$, which cover the region (see Fig. 3.3). If we consider any typical rectangle, say $a_{r-1} < x_1 \leqq a_r$, $b_{s-1} < x_2 \leqq b_s$, which is wholly or partly in R, we may select a point (α_r, β_s) in this and in R and evaluate F at this point. Then the sum over all the small areas

$$\Sigma\, F(\alpha_r, \beta_s)\, dS_{rs},$$

where dS_{rs} is the area of the part of the rectangle which is within R, is an approximation to the integral. We now let M and N increase without limit but always insist that the largest subdivision of area dS_{rs} must tend to zero, then if the sum tends to a limit that limit is the integral.

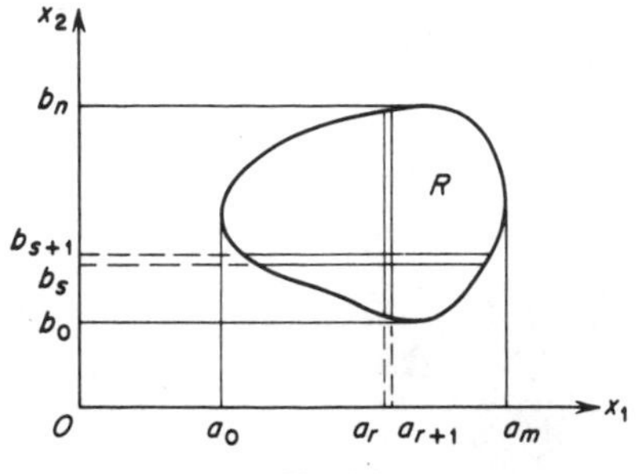

Fig. 3.3

Now if S is a piece-wise smooth surface with two sides in three-dimensional space, we can divide it up into a large number of small regions by a grid in much the same way as with a plane region R. If we are given a function F defined on the surface, it can be evaluated for some point of each subregion of the surface and the sum $\Sigma\, F\, dS_{rs}$ computed. Then, as the subdivisions increase in number and become finer, the limit that this may tend to is called the double integral of F over S

$$\iint\limits_S F\, dS.$$

The arguments of F have been left vague here. It may be that F is given as a function of position in space and we therefore evaluate it at a point on the surface, or it may be defined only on the surface itself.

The area of a curved surface is not an easy thing to define though more mystery than is necessary is often accorded it (see J. Serrin, Amer. Math. Mon. *68*, May, 1961, p. 435 for an elegant discussion). For many common surfaces however, we may relate the area of an element of the curved surface dS to the area of its projection on a coordinate plane (say $O12$) by $dA_{12} = n_3\, dS$ where n_3 is the third component of the normal. This is shown in Fig. 3.4 from which it is clear what must be done to calculate the areas of a cap. The cap S with boundary curve Γ has to have a projection on the $O12$ plane consisting of the simply connected region R with boundary C and for any point x_1, x_2 of R there must correspond only one point of S. If the normal to the surface $\mathbf{n}$ is defined everywhere on it, and $n_3 \neq 0$, then

$$dS = \frac{dA}{n_3}$$

and

$$\iint_S F \, dS = \iint_R \frac{F}{n_3} \, dx_1 \, dx_2. \tag{3.14.1}$$

Such a cap could be called 3-elementary.

A surface might fail to be 3-elementary either by having a sharp ridge at which **n** is discontinuous or by having more than one point of S correspond to each point of R or by having a whole region where $n_3 = 0$. Except in the last

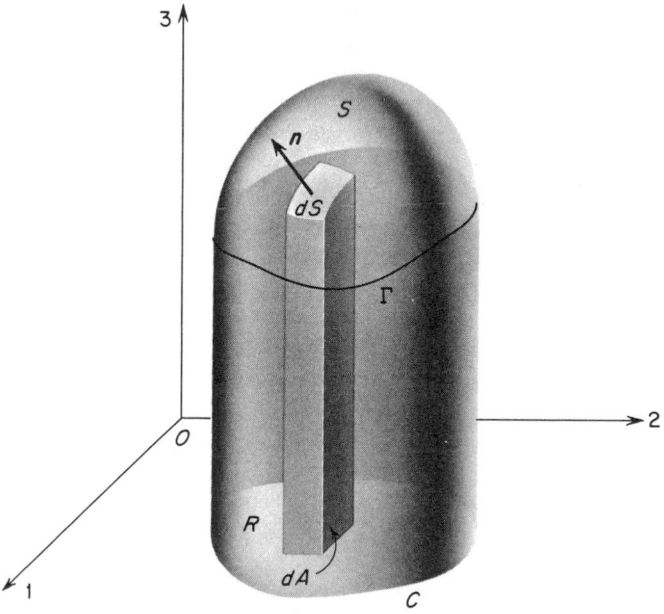

Fig. 3.4

case we can divide the surface up into parts that are 3-elementary, and evaluate the integral as the sum of the integrals over the several parts. If $n_3 = 0$ then the surface is either 1-elementary or 2-elementary meaning that we can project the area on to the $O23$ or the $O31$ planes in a similar way.

Exercise 3.14.1. S is the hemisphere

$$\mathbf{x} \cdot \mathbf{x} = a^2 \qquad \mathbf{m} \cdot \mathbf{x} \geqq 0,$$

where **m** is a fixed unit vector. Calculate $\iint_S \mathbf{a} \cdot \mathbf{n} \, dS$ for an arbitrary constant vector **a**.

Exercise 3.14.2. A surface can be given parametrically by three functions $x_i = g_i(u_1, u_2)$. Show that

$$\mathbf{n}\, dS = \left(\frac{\partial \mathbf{g}}{\partial u_1} \wedge \frac{\partial \mathbf{g}}{\partial u_2}\right) du_1\, du_2.$$

Deduce that area is independent of the choice of Cartesian coordinates.

3.15. Volume integrals

We shall also have occasion to integrate over the volume inside or outside of a closed surface. If a volume is such that a line parallel to $O\,3$ meets its bounding surface in two and only two points (say $x_3 = f_+(x_1, x_2)$, $x_3 = f_-(x_1, x_2)$, $f_+ \geqq f_-$) we may call it 3-elementary.* A sphere is the most obvious example of this. Similar definitions apply to the terms 1- and 2-elementary. If a volume can be divided into a number of smaller volumes each of which is 3-elementary, the volume may be called 3-composite. A region which is 1-, 2- and 3-composite will be called a composite volume.

An integration throughout the volume V is written

$$\iiint_V F(x_1, x_2, x_3)\, dV \tag{3.15.1}$$

and is, as before, the limit of a sum of the products of a very small subdivision of the volume and the function F evaluated somewhere within it. The limit is taken by letting the number of subdivisions increase without limit and the size of the largest tends to zero. If the volume V is 3-elementary and its projection on the $O12$ plane is a region R, then

$$\iiint_V F\, dV = \iint_R G\, dx_1\, dx_2$$

where (3.15.2)

$$G(x_1, x_2) = \int_{f_-(x_1, x_2)}^{f_+(x_1, x_2)} F(x_1, x_2, x_3)\, dx_3$$

This is shown in Fig. 3.5. The integral over a 3-composite volume may be calculated as the sum of such parts.

The double integral may be similarly reduced. We can call a simple closed curve C in the plane elementary if it can be traversed so that the slope of its tangent does not decrease. A composite curve is one that can be divided into elementary curves, as in Fig. 3.6. If R is bounded by an elementary closed

* This convenient nomenclature is taken from W. L. Ferrar, Integral Calculus, Oxford, 1958.

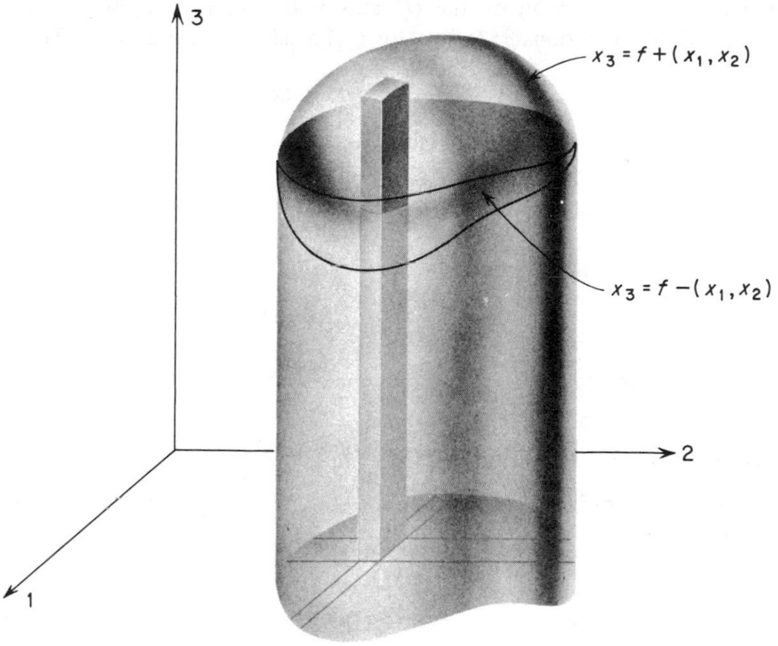

Fig. 3.5

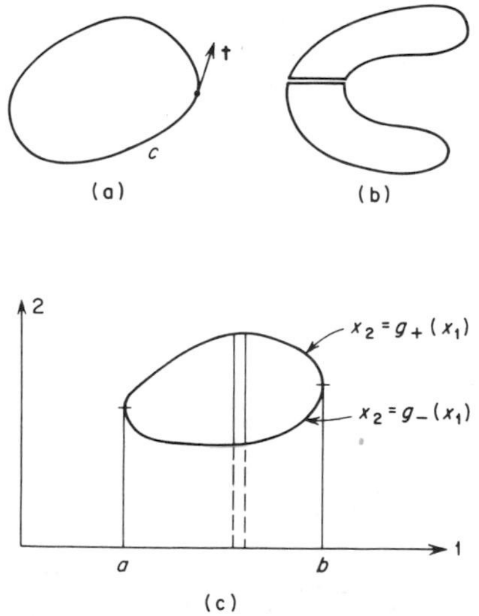

Fig. 3.6

curve C whose projection on the $O1$ axis is the interval (a, b), it can be specified by two functions $g_+(x_1)$ and $g_-(x_1)$ as shown in Fig. 3.6c. Then

$$\iint_R G\, dA = \int_a^b H(x_1)\, dx_1$$

where (3.15.3)

$$H(x_1) = \int_{g_-(x_1)}^{g_+(x_1)} G(x_1, x_2)\, dx_2.$$

Exercise 3.15.1. Evaluate

$$\iiint_V \frac{\partial^2 f(r)}{\partial x_i\, \partial x_i}\, dV$$

when V is the sphere $r = a$ and $r^2 = x_i x_i$.

3.16. Change of variable with multiple integrals

In Cartesian coordinates the element of volume dV is simply the volume of a rectangular parallelepiped of sides dx_1, dx_2, dx_3 and so

$$dV = dx_1\, dx_2\, dx_3. \tag{3.16.1}$$

Suppose, however, that it is convenient to describe the position by some other coordinates, say ξ_1, ξ_2, ξ_3. We may ask what volume is to be associated with the three small changes $d\xi_1$, $d\xi_2$, $d\xi_3$.

The change of coordinates must be given by specifying the Cartesian point $\mathbf{x}$ that is to correspond to a given set ξ_1, ξ_2, ξ_3, by

$$x_i = x_i(\xi_1, \xi_2, \xi_3). \tag{3.16.2}$$

Then by partial differentiation the small differences corresponding to a change $d\xi_j$ are

$$dx_i = \frac{\partial x_i}{\partial \xi_j}\, d\xi_j\, .$$

Let $d\mathbf{x}^{(j)}$ be the vectors with components $(\partial x_i/\partial \xi_j)\, d\xi_j$ for $j = 1$, 2, and 3. Then the volume element is

$$\begin{aligned} dV &= d\mathbf{x}^{(1)} \cdot (d\mathbf{x}^{(2)} \wedge d\mathbf{x}^{(3)}) \\ &= \epsilon_{ijk} \frac{\partial x_i}{\partial \xi_1} d\xi_1 \frac{\partial x_j}{\partial \xi_2} d\xi_2 \frac{\partial x_k}{\partial \xi_3} d\xi_3 \\ &= J\, d\xi_1\, d\xi_2\, d\xi_3 \end{aligned} \tag{3.16.3}$$

where

$$J = \frac{\partial(x_1, x_2, x_3)}{\partial(\xi_1, \xi_2, \xi_3)} = \epsilon_{ijk} \frac{\partial x_i}{\partial \xi_1} \frac{\partial x_j}{\partial \xi_2} \frac{\partial x_k}{\partial \xi_3}$$

is called the Jacobian of the transformation of variables.

Exercise 3.16.1. Show that the volume element in the frame $O\bar{1}\bar{2}\bar{3}$ of coordinates $\bar{x}_j = l_{ij}x_i$ is $d\bar{x}_1\, d\bar{x}_2\, d\bar{x}_3$.

Exercise 3.16.2. Obtain the volume elements in cylindrical and spherical polars by the Jacobian and check with a simple geometrical picture.

3.21. Vector fields

When the components of a vector or tensor depend on the coordinates we speak of a vector or tensor field. The flow of a fluid is a perfect realization of a vector field for at each point in the region of flow we have a certain velocity vector $\mathbf{v}(x_1, x_2, x_3)$. If the flow is unsteady then the velocity depends on the time as well as position, $\mathbf{v} = \mathbf{v}(x_1, x_2, x_3, t)$. When it is necessary to be specific we shall refer to this as a time dependent vector field. It is sometimes convenient to abbreviate these to $\mathbf{v}(\mathbf{x})$ and $\mathbf{v}(\mathbf{x}, t)$, or to use the index notation and write $v_i(\mathbf{x}, t)$, $A_{ij}(\mathbf{x}, t)$, etc.

Associated with any vector field $\mathbf{a}(\mathbf{x})$ are its trajectories, which is the name given to the family of curves everywhere tangent to the local vector $\mathbf{a}$. They are the solutions of the simultaneous equations

$$\frac{d\mathbf{x}}{ds} = \mathbf{a}(\mathbf{x}); \quad \text{that is,} \quad \frac{dx_i}{ds} = a_i(x_1, x_2, x_3). \tag{3.21.1}$$

where s is a parameter along the trajectory. (It will be the arc length if $\mathbf{a}$ is always a unit vector.) Though we have yet to define them, the streamlines of a steady flow are probably sufficiently familiar to be mentioned as the realization of these trajectories. For a time dependent vector field the trajectories will also be time dependent since they are solutions of

$$\frac{dx_i}{ds} = a_i(x_1, x_2, x_3, t). \tag{3.21.2}$$

If C is any closed curve in the vector field and we take the trajectories through all points of C, they describe a surface known as a *vector tube* of the field.

3.22. The vector operator ∇—gradient of a scalar

The symbol ∇ (enunciated as "del" or "nabla") is used for the symbolic vector operator whose i^{th} component is $\partial/\partial x_i$. Thus if ∇ operates on a scalar function of position φ it produces a vector $\nabla\varphi$ with components $\partial\varphi/\partial x_i$. We should of course establish that $\nabla\varphi$ is indeed a vector. In the coordinate frame $O\bar{1}\bar{2}\bar{3}$ the vector $\overline{\nabla\varphi}$ will have components $\partial\varphi/\partial\bar{x}_j$. However,

$$\frac{\partial\varphi}{\partial\bar{x}_j} = \frac{\partial\varphi}{\partial x_i}\frac{\partial x_i}{\partial\bar{x}_j} = l_{ij}\frac{\partial\varphi}{\partial x_i} \tag{3.22.1}$$

since $x_i = l_{ij}\bar{x}_j$, so that $\nabla\varphi$ is a vector.

In Cartesian coordinates the operation of partial differentiation with respect to three coordinates gives the components of a tensor of the next higher order. To show this, suppose that $A_{ij}(x_1, x_2, x_3)$ is a second order tensor field; then the 81 quantities $\partial^2 A_{ij}/\partial x_k\, \partial x_m$ are the components of a fourth order tensor. For since $x_i = l_{ij}\bar{x}_j$,

$$\frac{\partial x_i}{\partial \bar{x}_j} = l_{ij}, \qquad \frac{\partial^2 x_i}{\partial \bar{x}_j\, \partial \bar{x}_k} = 0. \tag{3.22.2}$$

If we write $A_{ij,km}$ for the second derivative $\partial^2 A_{ij}/\partial x_k\, \partial x_m$, then in the frame $O\bar{1}\bar{2}\bar{3}$

$$\begin{aligned}\bar{A}_{pq,rs} = \frac{\partial^2 \bar{A}_{pq}}{\partial \bar{x}_r\, \partial \bar{x}_s} &= \frac{\partial^2}{\partial \bar{x}_r\, \partial \bar{x}_s}(l_{ip}l_{jq}A_{ij}) \\ &= \frac{\partial x_k}{\partial \bar{x}_r}\frac{\partial}{\partial x_k}\left[\frac{\partial x_m}{\partial \bar{x}_s}\frac{\partial}{\partial x_m}(l_{ip}l_{jq}A_{ij})\right] \\ &= l_{ip}l_{jq}l_{kr}l_{ms}A_{ij,km}\end{aligned}$$

which shows that the $A_{ij,km}$ are components of a fourth order tensor.

The suffix notation $,i$ for the partial derivative with respect to x_i is a very convenient one and will be taken over for the generalization of this operation that must be made for non-Cartesian frames of reference. The notation "grad" for ∇ is often used and referred to as the gradient operator. Thus grad $\varphi = \nabla\varphi$ is the vector which is the gradient of the scalar, and grad $\mathbf{A}$ would be a tensor $A_{ij,k}$. ∇ is also sometimes written $\partial/\partial\mathbf{x}$ and can be expanded in the form

$$\nabla \equiv \mathbf{e}_{(i)}\, \partial/\partial x_i. \tag{3.22.3}$$

If $\varphi(x_1, x_2, x_3) = \varphi(\mathbf{x})$ is a scalar function of position and $dx_i = n_i\, dr$ a small displacement in the direction $\mathbf{n}$, then

$$\lim_{\delta r \to 0} \frac{\varphi(\mathbf{x} + \mathbf{n}\, dr) - \varphi(\mathbf{x})}{dr}$$

is the derivative in the direction $\mathbf{n}$ and is sometimes denoted by $\partial\varphi/\partial n$. By Taylor's theorem,

$$\varphi(\mathbf{x} + \mathbf{n}\, dr) = \varphi(x) + (\mathbf{n}\, dr) \cdot \nabla\varphi + 0(dr^2),$$

so

$$\frac{\partial \varphi}{\partial n} = \nabla\varphi \cdot \mathbf{n}. \tag{3.22.4}$$

Exercise 3.22.1. If $F(\rho) = \int^{\rho} f(\sigma)\, d\sigma$ is the indefinite integral of $f(\rho)$, show that $\nabla F(\rho) = f(\rho)\, \nabla\rho$.

It will be convenient to refer to the following formulae (2)–(4) later. They are all elementary but the exercise of establishing them is valuable.

Exercise 3.22.2. $\nabla(\varphi\psi) = \psi\, \nabla\varphi + \varphi\, \nabla\psi$.

Exercise 3.22.3. $\nabla(A_{jk}x_jx_k) = (A_{ij} + A_{ji})x_j\, \mathbf{e}_{(i)}$ if A is constant. (Summation on i).

Exercise 3.22.4. $\nabla f(r) = (f'(r)/r)\mathbf{x}$ where $r^2 = \mathbf{x} \cdot \mathbf{x}$.

Exercise 3.22.5. Show that if $\varphi(\mathbf{x}) = c$ is a surface, $\nabla\varphi$ is normal to the surface.

Exercise 3.22.6. If $A_{ij}x_ix_j = 1$ is a central quadric defined by a symmetric tensor A_{ij}, then $A_{ij}x_j$ transforms the radius vector $\mathbf{x}$ into a vector normal to the surface. The angle between the radius and the normal is given by

$$\tan^2\theta = (x_ix_i)(x_jA_{jk}A_{km}x_m) - (x_iA_{ij}x_j)^2$$
$$= (\mathbf{x} \cdot \mathbf{x})(\mathbf{x} \cdot \mathbf{A}^2 \cdot \mathbf{x}) - (\mathbf{x} \cdot \mathbf{A} \cdot \mathbf{x})^2.$$

3.23. The divergence of a vector field

The symbolic scalar or dot product of a vector and the operator ∇ is called the *divergence* of the vector field. Thus for any differentiable $\mathbf{a}(x_1, x_2, x_3)$, we write

$$\operatorname{div} \mathbf{a} = \nabla \cdot \mathbf{a} = a_{i,i} = \frac{\partial a_1}{\partial x_1} + \frac{\partial a_2}{\partial x_2} + \frac{\partial a_3}{\partial x_3}. \tag{3.23.1}$$

The divergence is a scalar since it is the contraction (or trace) of the second order tensor $a_{i,j}$.

Suppose that an elementary parallelepiped is set up with one corner P at x_1, x_2, x_3 and the diagonally opposite one Q at $x_1 + dx_1, x_2 + dx_2, x_3 + dx_3$ as shown in Fig. 3.7. The outward unit normal to the face through Q which is perpendicular to $O1$ is $\mathbf{e}_{(1)}$, whereas the outward normal to the parallel face through P is $-\mathbf{e}_{(1)}$. On the first of these faces

$$\mathbf{a} = \mathbf{a}(x_1 + dx_1, \xi_2, \xi_3)$$

where

$$x_2 \leqq \xi_2 \leqq x_2 + dx_2$$

and

$$x_3 \leqq \xi_3 \leqq x_3 + dx_3,$$

whereas on the second it is $\mathbf{a}(x_1, \xi_2, \xi_3)$.

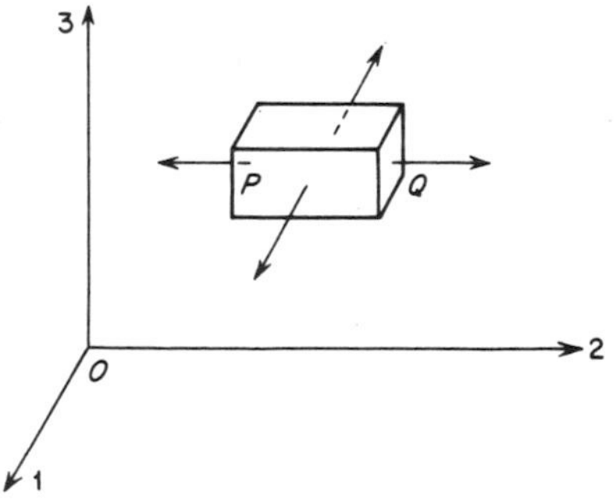

Fig. 3.7

Thus if $\mathbf{n}$ denotes the outward normal and dS is the area $dx_2\,dx_3$ of these faces, we have a contribution from them to the surface integral $\iint \mathbf{a} \cdot \mathbf{n}\, dS$ of

$$[a_1(x_1 + dx_1, \xi_2, \xi_3) - a_1(x_1, \xi_2, \xi_3)]\,dx_2\,dx_3$$
$$= \left[a_1(x_1, x_2, x_3) + \frac{\partial a_1}{\partial x_1}dx_1 - a_1(x_1, x_2, x_3)\right]dx_2\,dx_3 + 0(d^4)$$
$$= \frac{\partial a_1}{\partial x_1}dx_1\,dx_2\,dx_3 + 0(d^4),$$

where $0(d^4)$ denotes terms proportional to fourth powers of the dx. Similiar terms with $\partial a_2/\partial x_2$, $\partial a_3/\partial x_3$ will be given by the contributions of the other faces so that for the whole parallelepiped whose volume $dV = dx_1\,dx_2\,dx_3$ we have

$$\frac{1}{dV}\iint \mathbf{a}\cdot\mathbf{n}\,dS = \frac{\partial a_1}{\partial x_1} + \frac{\partial a_2}{\partial x_2} + \frac{\partial a_3}{\partial x_3} + 0(d). \tag{3.23.2}$$

If we let the volume shrink to zero we have

$$\lim_{dV\to 0}\frac{1}{dV}\iint \mathbf{a}\cdot\mathbf{n}\,dS = \nabla\cdot\mathbf{a}. \tag{3.23.3}$$

If $\mathbf{a}$ is thought of as a flux, then $\iint \mathbf{a}\cdot\mathbf{n}\,dS$ is the net flux out of the volume. A vector field whose divergence vanishes identically is called *solenoidal*. If the flux field of a certain property is solenoidal there is no generation of that property within the field, for all that flows into an infinitesimal element flows out again.

If $\mathbf{a}$ is the gradient of a scalar function $\nabla\varphi$, its divergence is called the *Laplacian* of φ

$$\nabla^2\varphi = \text{div grad }\varphi = \varphi_{,ii} = \frac{\partial^2\varphi}{\partial x_1^2} + \frac{\partial^2\varphi}{\partial x_2^2} + \frac{\partial^2\varphi}{\partial x_3^2}. \tag{3.23.4}$$

A function that satisfied Laplace's equation $\nabla^2\varphi = 0$ is called a *potential function.*

If A is a tensor, the notation div A is sometimes used for the vector $A_{ij,i}$, then div A' would be $A_{ij,j}$; we shall generally prefer the index notation for tensors.

Exercise 3.23.1. $\nabla\cdot(\varphi\mathbf{a}) = \nabla\varphi\cdot a + \varphi(\nabla\cdot a)$.

Exercise 3.23.2. $\nabla\cdot(\mathbf{a}\wedge\mathbf{b}) = (\nabla\wedge\mathbf{a})\cdot\mathbf{b} - \mathbf{a}\cdot(\nabla\wedge\mathbf{b})$.

Exercise 3.23.3. $\nabla^2 f(r) = f''(r) + 2f'(r)/r, \qquad r^2 = \mathbf{x}\cdot\mathbf{x}$.

Exercise 3.23.4. $\{\mathbf{x}^{(n)}\mathbf{a}\}$ denotes the symmetric tensor of order $n+1$

$$[a_i x_j x_k \ldots x_p + x_i a_j x_k \ldots x_p + x_i x_j a_k \ldots x_p + \ldots + x_i x_j x_k \ldots a_p]$$

whereas $\mathbf{a}\mathbf{x}^{(n+1)}$ is the tensor $a_i x_j x_k \ldots x_q$ of order $n+2$. Show that div $(\mathbf{a}\mathbf{x}^{(n+1)}) = \{\mathbf{x}^{(n)}\mathbf{a}\}$ if div $\mathbf{a} = 0$.

Exercise 3.23.5. Interpret $\nabla^2\varphi$ physically by thinking of it as $\nabla\cdot(\nabla\varphi)$.

Exercise 3.23.6. If $\varphi(x) = \varphi(x_1, x_2, x_3)$ is a potential function and $r^2 = \mathbf{x}\cdot\mathbf{x}$, show that

$$\varphi(x) + \frac{a}{r}\varphi\left(\frac{a^2x}{r^2}\right) - \frac{2}{ar}\int_0^a \lambda\varphi\left(\frac{\lambda^2 x}{r^2}\right)d\lambda$$

is also a potential function and that its normal derivative on the sphere $r = a$ vanishes. (Weiss, P. Proc. Camb. Phil. Soc., **40,** (1944), 249.)

3.24. The curl of a vector field

The symbolic vector or cross product of ∇ and a vector field **a** is called the curl of the vector field. It is the vector

$$\nabla \wedge a = \text{curl } \mathbf{a} = \epsilon_{ijk} a_{k,j} \mathbf{e}_{(i)},$$

with three components

$$\left(\frac{\partial a_3}{\partial x_2} - \frac{\partial a_2}{\partial x_3}\right), \quad \left(\frac{\partial a_1}{\partial x_3} - \frac{\partial a_3}{\partial x_1}\right), \quad \left(\frac{\partial a_2}{\partial x_1} - \frac{\partial a_1}{\partial x_2}\right). \tag{3.24.1}$$

It is connected, as we shall see, with the rotation of the field and is sometimes written rot **a** in older texts.* In dyadic notation curl $\mathbf{a} = (\text{grad } \mathbf{a})_\times$.

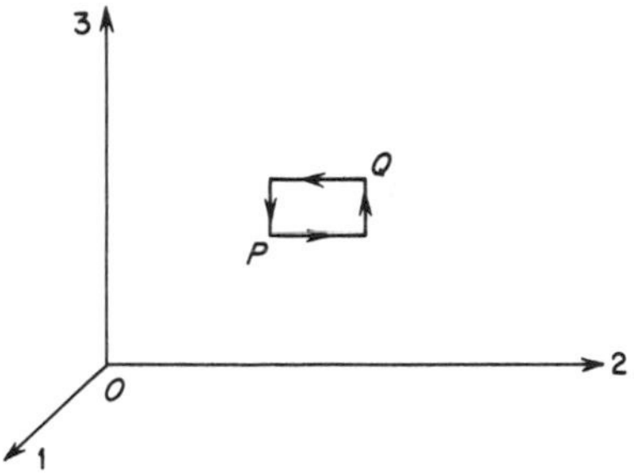

Fig. 3.8

Consider an elementary rectangle in the plane normal to $O1$ with one corner P at (x_1, x_2, x_3) and the diagonally opposite one Q at $(x_1, x_2 + dx_2, x_3 + dx_3)$, as shown in Fig. 3.8. We wish to calculate the line integral around this elementary circuit of $\mathbf{a} \cdot \mathbf{t}\, ds$, where $\mathbf{t}$ is the tangent. Now the line through P parallel to $O3$ has tangent $-\mathbf{e}_{(3)}$ and the parallel side through Q has tangent $\mathbf{e}_{(2)}$, and each is of length dx_3. Accordingly, they contribute to $\mathbf{a} \cdot \mathbf{t}\, ds$ an amount

$$\begin{aligned}[a_3(x_1, x_2 + dx_2, \xi_3) - a_3(x_1, x_2, \xi_3)]\, dx_3 \\ &= \left[a_3(x_1, x_2, \xi_3) + \frac{\partial a_3}{\partial x_2} dx_2 - a_3(x_1, x_2, \xi_3)\right] dx_3 + 0(d^3) \\ &= \frac{\partial a_3}{\partial x_2} dx_3\, dx_2 + 0(d^3).\end{aligned}$$

Similarly, from the other two sides, there is a contribution,

$$-\frac{\partial a_2}{\partial x_3} dx_3\, dx_2 + 0(d^3).$$

Thus writing $dA = dx_3 dx_2$, we have

$$\frac{1}{dA} \oint_1 \mathbf{a} \cdot \mathbf{t}\, ds = \left(\frac{\partial a_3}{\partial x_2} - \frac{\partial a_2}{\partial x_3}\right) + 0(d) \tag{3.24.3}$$

and in the limit

$$\lim_{dA \to 0} \frac{1}{dA} \oint_1 \mathbf{a} \cdot \mathbf{t}\, ds = \left(\frac{\partial a_3}{\partial x_2} - \frac{\partial a_2}{\partial x_3}\right). \tag{3.24.4}$$

* rot **a** is used by Toupin (Handbuch der Physik, III/1, Sect. 268) to denote the antisymmetric tensor of which curl **a** is the vector; this usage lends itself to generalization.

The suffix 1 has been put on the integral sign to show that the line integral is in a plane normal to $O1$, and Eq. (3.24.4) shows that the limit is the first component of the curl. An entirely similar treatment would give the other two components for line integrals around rectangles in planes perpendicular to the $O2$ and $O3$ axes.

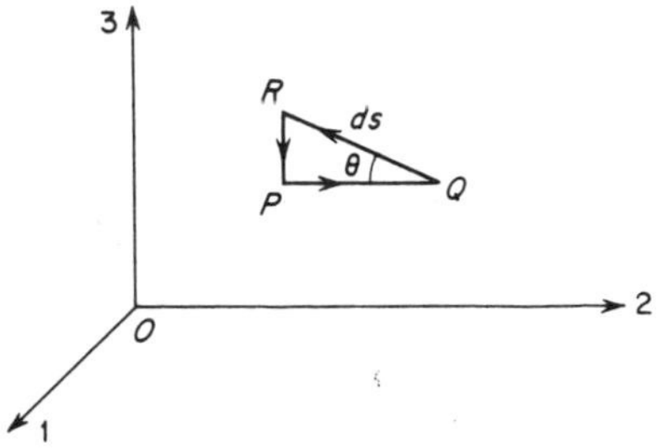

Fig. 3.9

We can treat an infinitesimal triangle (PQR of Fig. 3.9) in a similar way. If the length of PQ is $dx_2 = ds \cos\theta$, and the length of PR is $dx_3 = ds \sin\theta$, the area is

$$dA = \tfrac{1}{2}\, ds^2 \cos\theta \sin\theta.$$

The unit tangents around the triangle are $\mathbf{e}_{(2)}$, $(0, -\cos\theta, \sin\theta)$, $-\mathbf{e}_{(3)}$. Thus approximating each part of the line integral by the length of side multiplied by $\mathbf{a} \cdot \mathbf{t}$ evaluated at its midpoint,

$$\begin{aligned}
\oint \mathbf{a} \cdot \mathbf{t}\, ds &= a_2(x_1, x_2 + \tfrac{1}{2}dx_2, dx_3)\, dx_2 \\
&\quad + \{-\cos\theta\, a_2(x_1, x_2 + \tfrac{1}{2}\, dx_2, x_3 + \tfrac{1}{2}\, dx_3) \\
&\quad + \sin\theta\, a_3(x_1, x_2 + \tfrac{1}{2}\, dx_2, x_3 + \tfrac{1}{2}\, dx_3)\}\, ds \\
&\quad - a_3(x_1, x_2, x_3 + \tfrac{1}{2}\, dx_3)\, dx_3 \\
&= \{a_3(x_1, x_2 + \tfrac{1}{2}\, dx_2, x_3 + \tfrac{1}{2}\, dx_3) - a_3(x_1, x_2, x_3 + \tfrac{1}{2}\, dx_3)\}\, ds \sin\theta \\
&\quad - \{a_2(x_1, x_2 + \tfrac{1}{2}\, dx_2, x_3 + \tfrac{1}{2}\, dx_3) - a_2(x_1, x_2 + \tfrac{1}{2}\, dx_2, x_3)\}\, ds \cos\theta \\
&= \left(\frac{\partial a_3}{\partial x_2} - \frac{\partial a_2}{\partial x_3}\right) \tfrac{1}{2}\, ds^2 \cos\theta \sin\theta + 0(ds^3).
\end{aligned}$$

Again, as ds, and so dA, tends to zero,

$$\lim_{dA \to 0} \frac{1}{dA_1} \oint_1 a \cdot t\, ds = \epsilon_{1jk} \frac{\partial a_k}{\partial x_j}$$

and similar forms hold for triangles in planes normal to $O2$ and $O3$.

Consider now a fourth point S at $(x_1 + dx_1, x_2, x_3)$ so that QRS is a plane triangle of area dA whose unit normal is $\mathbf{n}$. Thus the areas of the triangles PQR, PRS, PSQ are

$$dA_1 = n_1 dA, \qquad dA_2 = n_2 dA, \qquad dA_3 = n_3 dA$$

respectively. However, the line integral around QRS can be taken to be the sum of those around PQR, PRS, PSQ since the parts PQ, PR, PS are traversed once in each direction and so cancel (see Fig. 3.10). Thus, if dA is very small,

$$\begin{aligned}\oint_{QRS} \mathbf{a} \cdot \mathbf{t}\, ds &= \oint_1 \mathbf{a} \cdot \mathbf{t}\, ds + \oint_2 \mathbf{a} \cdot \mathbf{t}\, ds + \oint_3 \mathbf{a} \cdot \mathbf{t}\, ds \\ &= \epsilon_{ijk}\, dA_i \frac{\partial a_k}{\partial x_j} \\ &= dA\, \epsilon_{ijk} n_i \frac{\partial a_k}{\partial x_j}.\end{aligned}$$

Then in the limit

$$\lim \left[\frac{1}{dA} \oint_{QRS} \mathbf{a} \cdot \mathbf{t}\, ds\right] = (\text{curl } \mathbf{a}) \cdot \mathbf{n}. \tag{3.24.5}$$

By working a little harder (Exercise 3.24.9), it can be shown that if any small curve in the plane with normal $\mathbf{n}$ shrinks on the point $\mathbf{x}$, the limit of the $\oint \mathbf{a} \cdot \mathbf{t}\, ds$ divided by the area is the projection of curl $\mathbf{a}$ on the normal, $\mathbf{n}$.

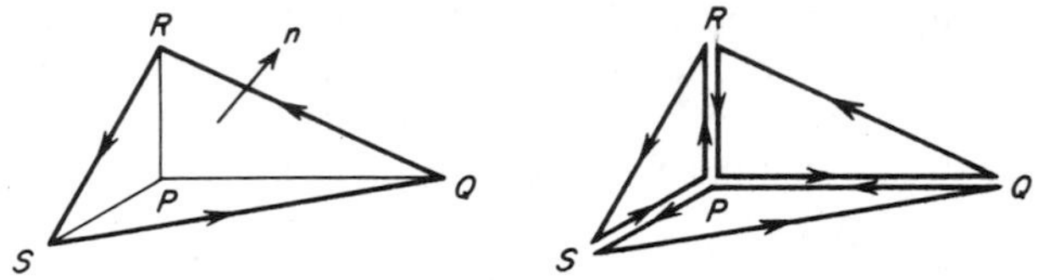

Fig. 3.10

A vector field $\mathbf{a}$ for which

$$\nabla \wedge \mathbf{a} = \text{curl } \mathbf{a} = 0 \tag{3.24.6}$$

is called *irrotational*, for evidently the circulation around any infinitesimal curve vanishes.

Exercise 3.24.1. $\nabla \cdot (\nabla \wedge \mathbf{a}) = 0.$

Exercise 3.24.2. $\nabla \wedge (\nabla \varphi) = 0.$

Exercise 3.24.3. $\nabla \wedge (\varphi \mathbf{a}) = (\nabla \varphi) \wedge \mathbf{a} + \varphi \nabla \wedge \mathbf{a}.$

Exercise 3.24.4. $\nabla \wedge (\mathbf{a} \wedge \mathbf{b}) = \mathbf{a}(\nabla \cdot \mathbf{b}) - \mathbf{b}(\nabla \cdot \mathbf{a}) + (\mathbf{b} \cdot \nabla)\mathbf{a} - (\mathbf{a} \cdot \nabla)\mathbf{b}.$

Exercise 3.24.5. $\nabla \wedge (\nabla \wedge \mathbf{a}) = \nabla(\nabla \cdot \mathbf{a}) - \nabla^2 \mathbf{a}.$

Exercise 3.24.6. $\nabla(\mathbf{a} \cdot \mathbf{b}) = (\mathbf{a} \cdot \nabla)\mathbf{b} + (\mathbf{b} \cdot \nabla)\mathbf{a} + \mathbf{a} \wedge (\nabla \wedge \mathbf{b}) + \mathbf{b} \wedge (\nabla \wedge \mathbf{a}).$

Exercise 3.24.7. $\mathbf{a} \wedge (\nabla \wedge \mathbf{a}) = \frac{1}{2}\nabla(\mathbf{a} \cdot \mathbf{a}) - (\mathbf{a} \cdot \nabla)\mathbf{a}.$

Exercise 3.24.8. If $\mathbf{v}(\mathbf{x})$ is the velocity of a rigid body due to a rotation $\boldsymbol{\omega}$, show that $\nabla \wedge \mathbf{v} = 2\boldsymbol{\omega}$.

Exercise 3.24.9. If C is a infinitesimal curve about $\mathbf{x}_0$ in a plane with normal $\mathbf{n}$, use Taylor's theorem to approximate the line integral

$$\oint_C \mathbf{a} \cdot \mathbf{t}\, ds = \oint [\mathbf{a}(\mathbf{x}_0) + \{(\mathbf{x} - \mathbf{x}_0) \cdot \nabla\}\mathbf{a}(\mathbf{x}_0) + \dots] \cdot \mathbf{t}\, ds$$

Hence, using Ex. 3.13.2, show that this integral is $\nabla \wedge \mathbf{a}(\mathbf{x}_0) \cdot \mathbf{n}\, dS$.

Exercise 3.24.10. If C_i is the projection on a plane normal to Oi of a simple closed curve C and is itself simple, show that the areas C_1, C_2, C_3 are components of the vector

$$\tfrac{1}{2} \oint_C (\mathbf{x} \wedge \mathbf{t})\, ds.$$

3.31. Green's theorem and some of its variants

One of the most valuable transformations of tensor analysis is that of Green* which relates a certain volume integral to an integral over the bounding surface. We shall give two demonstrations of it. The surface integral occurring is the integral of $\mathbf{a} \cdot \mathbf{n}$, where $\mathbf{n}$ is the outward normal to the surface. If we think of $\mathbf{a}$ as the flux of some physical property the integral of $\mathbf{a} \cdot \mathbf{n}$ over the whole surface is thus the total flux out of a closed volume. Green's theorem says that this total flux equals the integral of $\nabla \cdot \mathbf{a}$ throughout the enclosed volume.

Suppose V is a volume with a closed surface S and $\mathbf{a}$ any vector field defined in V and on S. Then if S is piece-wise smooth with outward normal $\mathbf{n}$ and $\mathbf{a}$ continuously differentiable,

$$\iiint_V \nabla \cdot \mathbf{a}\, dV = \iint_S \mathbf{a} \cdot \mathbf{n}\, dS. \tag{3.31.1}$$

We have shown for infinitesimal volumes that

$$\nabla \cdot \mathbf{a}\, dV = \mathbf{a} \cdot \mathbf{n}\, dS$$

and the integral on the left-hand side is by definition the limit of the sum of a number of infinitesimal volumes into which V can be divided. However, if it is so divided, the contributions of $\mathbf{a} \cdot \mathbf{n}\, dS$ from the touching faces of two adjacent elements of volume are equal in magnitude but opposite in sign since the outward normals point in opposite directions. Thus in a summation of $\mathbf{a} \cdot \mathbf{n}\, dS$, the only terms that survive are those on the outer surface S, and Eq. (3.31.1) follows.

Another form of Green's theorem is the following. Let V be a composite

* This is no place to enter the difficult subject of the proper attribution of names to theorems. This theorem was given in various forms by Lagrange (1762) Gauss (1813) and Ostrogradsky (1831). Its best known source, however, is Green's "Essay on the application of mathematical analysis to the theories of electricity and magnetism" (1828). The non-denominational name "Divergence theorem" is also used.

volume with piece-wise smooth boundary S and F a continuous function whose derivatives $F_{,i}$ are continuous in V. Then

$$\iiint_V F_{,i}\, dV = \iint_S F n_i\, dS. \tag{3.31.2}$$

Let us prove this for a volume which is 3-elementary and with $i = 3$; then by joining elementary volumes it will hold for a 3-composite volume and so by extension to a composite volume for $i = 1, 2,$ or 3. Figure 3.5 illustrates a 3-elementary volume and by Eq. (3.15.2) we can write

$$\iiint_V \frac{\partial F}{\partial x_3}\, dV = \iint_R G\, dx_1\, dx_2$$

where (3.31.3)

$$G = \int_{f_-}^{f_+} \frac{\partial F}{\partial x_3}\, dx_3 = F[x_1, x_2, f_+(x_1, x_2)] - F[x_1, x_2, f_-(x_1, x_2)]$$

and R is the projection of the volume V on the $O12$ plane. However, if we use a suffix $+$ or $-$ to denote the upper and lower surfaces, respectively, the element of area in R

$$dx_1\, dx_2 = n_{3+}\, dS_+ \quad \text{or} \quad -n_{3-}\, dS_- \tag{3.31.4}$$

since the elements of area are all positive but n_{3-} is negative. Thus, substituting in Eq. (3.31.3),

$$\iiint_V \frac{\partial F}{\partial x_3}\, dV = \iint F[x_1, x_2, f_+(x_1, x_2)] n_{3+}\, dS_+ + \iint F[x_1, x_2, f_-(x_1, x_2)] n_{3-}\, dS_-.$$

However, the right-hand side is now just the definition of the surface integral $Fn_3\, dS$, so the theorem is established for a 3-elementary region. A 3-composite volume can be divided up into a finite number of 3-elementary volumes in each of which Eq. (3.31.2) will hold. However, on adjacent faces of the elementary volumes the outward normals point in opposite directions so that the contributions to the total surface integral cancel. Thus in summing over all the elementary volumes the only contributions to the right-hand side of Eq. (3.31.2) that survive are those from the outer surface of the composite volume.

The theorem can be extended to a function F which is continuous in V but whose derivative $F_{,i}$ is only piece-wise continuous. The volume V may increase without limit or the integrand have singularities provided that the integrals converge. For further lightening of restrictions the reader is referred to the book by Kellogg mentioned in the Bibliography at the end of this chapter.

F may be a scalar or the component of a vector or tensor. If $F = a_i$ we have the form of Green's theorem originally given. If $F = \epsilon_{kij}a_j$, then $F_{,i} = \epsilon_{kij}a_{j,i}$ is the kth component of $\nabla \wedge \mathbf{a}$, whereas $Fn_i = \epsilon_{kij}n_i a_j$ is the kth component of $\mathbf{n} \wedge \mathbf{a}$. Thus

$$\iiint_V (\nabla \wedge \mathbf{a})\, dV = \iint_S (\mathbf{n} \wedge \mathbf{a})\, dS. \tag{3.31.5}$$

If $\mathbf{a} = \nabla\varphi$ we have

$$\iiint_V \nabla^2\varphi\, dV = \iint_S \nabla\varphi \cdot \mathbf{n}\, dS = \iint_S \frac{\partial\varphi}{\partial n}\, dS, \tag{3.31.6}$$

where $\partial\varphi/\partial n$ denotes the derivative in the direction of the outward normal. If $\mathbf{a} = \psi\nabla\varphi$, then

$$\iiint_V (\psi\nabla^2\varphi + \nabla\psi \cdot \nabla\varphi)\, dV = \iint_S \psi \frac{\partial\varphi}{\partial n}\, dS \tag{3.31.7}$$

whence

$$\iiint_V (\psi\nabla^2\varphi - \varphi\nabla^2\psi)\, dV = \iint_S \left(\psi \frac{\partial\varphi}{\partial n} - \varphi \frac{\partial\psi}{\partial n}\right) dS. \tag{3.31.8}$$

Exercise 3.31.1. Use Green's theorem to obtain the answer to Ex. 3.14.1 without integration.

Exercise 3.31.2. Show that

$$\iiint_V [(\nabla \wedge \mathbf{a}) \wedge \mathbf{b} + (\nabla \wedge \mathbf{b}) \wedge \mathbf{a} + \mathbf{a}(\nabla \cdot \mathbf{b}) + \mathbf{b}(\nabla \cdot \mathbf{a})]\, dV$$
$$= \iint_S [\mathbf{n} \cdot (\mathbf{ab} + \mathbf{ba}) - (\mathbf{a} \cdot \mathbf{b})\mathbf{n}]\, dS$$

Dyadic notation is used here, so $\mathbf{n} \cdot (\mathbf{ab} + \mathbf{ba})$ is the vector $n_i(a_i b_j + b_i a_j)$.

Exercise 3.31.3. By taking $\mathbf{a}$ to be independent of x_3 and $a_3 = 0$, show that if A is an area in the $O12$ plane bounded by a curve C, then

$$\iint_A (a_{1,1} + a_{2,2})\, dA = \oint_C (a_1 t_2 - a_2 t_1)\, ds$$

where $\mathbf{t}$ is the unit tangent vector to C.

Exercise 3.31.4. Deduce from the preceding question that

$$\oint_C (a_1 t_1 + a_2 t_2)\, ds = \iint_A (a_{2,1} - a_{1,2})\, dA.$$

Exercise 3.31.5. V is composed of two volumes V_1 and V_2 divided by a surface S'. The vector **a** is continuous in $V = V_1 + V_2$ and its derivatives are continuous in V_1 and V_2 separately and the normal derivative is continuous across S'. Show that Green's theorem holds good for any volume V^* within V.

3.32. Stokes' theorem

The previous theorem concerned the relation of an integral over a closed volume and its relation to an integral over the bounding surface. Stokes' theorem* relates the surface integral over a cap to a line integral around the bounding curve. The line integral appearing is that of $\mathbf{a} \cdot \mathbf{t}$, that is, the total circulation, and the theorem says that this is equal to the surface integral of the normal component of curl **a**. We shall again give two demonstrations.

We have shown earlier in Section 3.24 and Ex. 3.24 that for an infinitesimal triangular area the line integral

$$\oint_C \mathbf{a} \cdot \mathbf{t}\, ds = (\nabla \wedge \mathbf{a}) \cdot \mathbf{n}\, dS. \tag{3.32.1}$$

If S is the cap of a closed curve C, we can divide its surface into a large number of small triangles for each of which Eq. (3.32.1) is true. Then in summing the right-hand sides we shall have the surface integral over the whole cap, whereas on the left-hand sides contributions from adjacent sides of triangles will cancel since they will be traversed in opposite directions. The only remaining contributions from the line integrals will thus be those from the bounding curve C and

$$\oint_C \mathbf{a} \cdot \mathbf{t}\, ds = \iint_S (\nabla \wedge \mathbf{a}) \cdot \mathbf{n}\, dS. \tag{3.32.2}$$

The convention that the normal to the curve should be right-handed with respect to the direction of traversing the curve C is observed throughout.

More precisely stated, Stokes' theorem says that for any two-sided piecewise smooth surface S spanning the closed curve C and any continuous vector field **a** whose partial derivatives are continuous, Eq. (3.32.2) holds. We shall give another proof depending on Green's theorem. Suppose first that the cap S can be specified by a single function $x_3 = f(x_1, x_2)$ in the region R within the closed curve C', the projection of C on the plane $O12$. Consider the terms in a_1, namely,

$$\oint_C a_1 t_1\, ds = \iint_S \left(n_2 \frac{\partial a_1}{\partial x_3} - n_3 \frac{\partial a_1}{\partial x_2} \right) dS.$$

* It appears that the attribution of this theorem to Stokes is less appropriate than that of the previous one to Green. It is actually due to Kelvin though the usage of Stokes' name is too entrenched to be changed. See Truesdell, "Kinematics of Vorticity," footnote, p. 12.

Now on S, $x_3 = f(x_1, x_2)$ and

$$a_1 = a_1[x_1, x_2, f(x_1, x_2)] = g(x_1, x_2)$$

so

$$\frac{\partial g}{\partial x_2} = \frac{\partial a_1}{\partial x_2} + \frac{\partial a_1}{\partial x_3}\frac{\partial f}{\partial x_2}.$$

However, $n_2/n_3 = -\partial f/\partial x_2$ because the direction cosines of the normal are proportional to $\partial f/\partial x_1$, $\partial f/\partial x_2$, and -1, respectively. Hence,

$$\iint_S \left(n_2 \frac{\partial a_1}{\partial x_3} - n_3 \frac{\partial a_1}{\partial x_2}\right) dS = \iint_S -\frac{\partial g}{\partial x_2} n_3 \, dS$$
$$= -\iint_R \frac{\partial g}{\partial x_2} \, dA$$

by Eq. (3.15.1). However, by putting $a_1 = 0$, $a_2 = g$ in Ex. 3.31.3 which is a two-dimensional form of Green's theorem,

$$-\iint_R \frac{\partial g}{\partial x_2} \, dA = \oint_{C'} g t_1' \, ds' = \oint_{C'} a_1[x_1, x_2, f(x_1, x_2)] t_1' \, ds',$$

where t_1' and ds' are the component of the tangent and element of arc C' respectively. However, $t_1' \, ds' = t_1 \, ds$ so this last integral is really the line integral around C and the theorem is established for a_1. A similar proof goes through for a_2 and a_3 and for surfaces that can be decomposed into parts which are either 1, 2, or 3-elementary. A further discussion will be found in Kellogg's book.

Other forms of Stokes' theorem may be derived by inserting various vectors; in particular the components a_i may be a set of components from a tensor, say A_{ijk} with j, k fixed. Thus it is convenient to write

$$\iint_S \epsilon_{ijk} n_i F_{k,j} \, dS = \oint_C F_k t_k \, ds. \tag{3.32.3}$$

Exercise 3.32.1. Relate the results of Exercises 3.13.1 and 3.14.1 by Stokes' theorem.

Exercise 3.32.2. Show that the vanishing of the integral of $(\nabla \wedge \mathbf{a}) \cdot \mathbf{n}$ over a closed surface is a consequence both of Green's and Stokes' theorems.

Exercise 3.32.3. Show that

$$\iint_S (n_i a_{i,j} - a_{i,i} n_j) \, dS = \oint_C \epsilon_{ijk} a_i t_k \, ds.$$

Exercise 3.32.4. Show that

$$\oint_C (\nabla \wedge \mathbf{a}) \cdot \mathbf{t}\, ds = \iint_S \left[\frac{\partial}{\partial n}(\nabla \cdot \mathbf{a}) - \mathbf{n} \cdot (\nabla^2 \mathbf{a})\right] dS.$$

Exercise 3.32.5. The flux of **a** through the cap S is $\iint_S \mathbf{a} \cdot \mathbf{n}\, dS$. Suppose $\mathbf{a} = \mathbf{a}(x_1, x_2, x_3, t)$ and S moves in space its velocity being given by a vector field **v**. Consider the volume bounded by S_1 the position of S at time t and S_2 its position at time $t + dt$ and a bounding surface of vectors $\mathbf{v}\, dt$ for all points of the boundary C. Then applying Green's theorem to the volume, show that

$$\begin{aligned}\frac{d}{dt}\left[\iint \mathbf{a} \cdot \mathbf{n}\, dS\right] &= \lim \frac{1}{dt}\left[\iint_{S_2} \mathbf{a}(\mathbf{x}, t + dt) \cdot \mathbf{n}\, dS - \iint_{S_1} \mathbf{a}(\mathbf{x}, t) \cdot \mathbf{n}\, dS\right] \\ &= \iint_S \left[\frac{\partial \mathbf{a}}{\partial t} + (\nabla \cdot \mathbf{a})\mathbf{v}\right] \cdot \mathbf{n}\, dS + \oint_C \mathbf{a} \cdot (\mathbf{v} \wedge \mathbf{t})\, ds \\ &= \iint_S \left[\frac{\partial \mathbf{a}}{\partial t} + (\nabla \cdot \mathbf{a})\mathbf{v} + \nabla \wedge (\mathbf{a} \wedge \mathbf{v})\right] \cdot \mathbf{n}\, dS.\end{aligned}$$

Exercise 3.32.6. If V and **a** are as in Ex. 3.31.5 save that this time it is the tangential derivative that is continuous across S', show that Stokes' theorem still holds.

3.41. The classification and representation of vector fields

We have already noted two distinct types of vector field; namely, the solenoidal, for which $\nabla \cdot \mathbf{a} = 0$, and the irrotational, for which $\nabla \wedge \mathbf{a} = 0$. Such fields occur physically, and it is of interest to explore their properties a little further. We are particularly interested in relating the three components of **a** to certain scalar functions of position. For example, if **a** is the gradient of a scalar function φ, then it is certainly irrotational for $\nabla \wedge \nabla\varphi = 0$ identically. Thus if a problem calls for an irrotational vector field, it may be possible to turn it into a problem that requires finding only the function φ; certainly an easier matter than finding all three components of **a**. Before this can be done however it must be proved that all irrotational vector fields can be represented as the gradient of a scalar function, which is slightly more difficult than showing that the gradient of a scalar is irrotational.

To show the value of these representations let us anticipate the results of the next few sections and show that a vector field which is both irrotational and solenoidal is uniquely determined in a volume V if it is specified over S, the surface of V. If a vector field is irrotational we shall show that it can always be written as $\nabla\varphi$, where φ is determined up to an arbitrary constant.

If $\mathbf{a} = \nabla\varphi$ and $\mathbf{a}$ is solenoidal, $\nabla \cdot \mathbf{a} = \nabla^2\varphi = 0$; that is, φ is a potential function or solution of Laplace's equation. Now suppose it were possible for two different functions φ_1 and φ_2 to satisfy $\nabla^2\varphi = 0$ in V and take the same values on S. Then the difference between them $\varphi = \varphi_1 - \varphi_2$ would satisfy

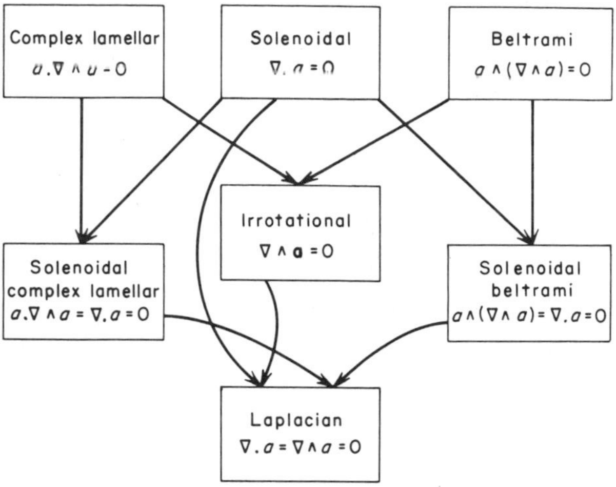

Fig. 3.11

$\nabla^2\varphi = 0$ in V but φ would be identically zero on S. Now put $\psi = \varphi$ in Eq. (3.31.7) so that

$$\iiint_V [\varphi\nabla^2\varphi + (\nabla\varphi)^2]\, dV = \iint_S \varphi \frac{\partial\varphi}{\partial n}\, dS. \tag{3.41.1}$$

However, since $\nabla^2\varphi = 0$ in V and $\varphi = 0$ in S, the first term on the left-hand side and the integral on the right are both identically zero. It follows that

$$\iiint_V (\nabla\varphi)^2\, dV = 0. \tag{3.41.2}$$

However, the integrand is everywhere positive and so the integral could not vanish unless $\nabla\varphi$ itself were everywhere zero. However, this means that φ is a constant, and since it is zero on the boundary this constant must be zero everywhere, that is, $\varphi_1 = \varphi_2$.

Apart from the solenoidal and irrotational, several other types of fields have been named. We will give a brief description of these though we shall not be able to go fully into the representation of them all.

The name *lamellar* is also applied to an irrotational vector field ($\nabla \wedge \mathbf{a} \equiv 0$) and it is a special case of the *complex lamellar* field for which $\mathbf{a} \cdot (\nabla \wedge \mathbf{a}) = 0$. The condition for a field to be complex lamellar is evidently that it should be orthogonal to its curl, which is less restrictive than the requirement that the curl should vanish. Another type is the *Beltrami* field for which the curl is parallel to the original vector, that is, $\mathbf{a} \wedge (\nabla \wedge \mathbf{a}) = 0$. The relations between these types are best shown in a diagram (Fig. 3.11). Arrows coming together at the same place denote the simultaneous possession of the character of the type from which they come. Thus, if a field is both a complex lamellar and a Beltrami field, it is irrotational (except possibly where $\mathbf{a} = 0$), for curl $\mathbf{a}$ is both orthogonal and parallel to $\mathbf{a}$ and so must be zero if $\mathbf{a} \neq 0$. A field which is both solenoidal and irrotational is sometimes called *Laplacian* since it is the gradient of a potential function. If curl $\mathbf{a}$ is not only parallel to $\mathbf{a}$ but is proportional to $\mathbf{a}$ with a constant that does not vary with position (that is, $\nabla \wedge \mathbf{a} = k\mathbf{a}$, $\nabla k = 0$), it is called *Trkalian*.

Exercise 3.41.1. Show that if the curl of a Beltrami field is itself a Beltrami field, then the field is Trkalian (Bjørgum).

Exercise 3.41.2. If $\boldsymbol{\tau} = \mathbf{a}/|\mathbf{a}|$ is the unit vector tangent to the field $\mathbf{a}$, show that

$$\boldsymbol{\tau} \cdot (\nabla \wedge \boldsymbol{\tau}) = \lim_{S \to 0} \frac{1}{S} \oint_C \boldsymbol{\tau} \cdot \mathbf{t}\, ds$$

where S is a cap of the curve C that shrinks on the point in such a way that the normal to S is $\boldsymbol{\tau}$. ($\mathbf{t}$ is the unit tangent of the curve C.)

3.42. Irrotational vector fields

We wish to show that if $\nabla \wedge \mathbf{a} = 0$ then there exists a scalar function φ such that

$$\mathbf{a} = \nabla\varphi \tag{3.42.1}$$

This function φ is often called the potential of $\mathbf{a}$ by analogy with force fields for which the force on a particle can be obtained as the gradient of the potential energy.

Since $\nabla \wedge \mathbf{a} = 0$ we know by Stokes' theorem that the circulation integral round any closed curve vanishes,

$$\oint_C \mathbf{a} \cdot \mathbf{t}\, ds = 0. \tag{3.42.2}$$

It follows (see Section 3.13) that the line integral from a fixed point (say the origin) to a point P, (x_1, x_2, x_3), is independent of the path. Let

$$\varphi(P) = \varphi(x_1, x_2, x_3) = \int_O^P \mathbf{a} \cdot \mathbf{t}\, ds, \tag{3.42.3}$$

then this is a definite scalar function depending only on the position of P. Consider a nearby point Q, $(x_1 + dx_1, x_2, x_3)$, then

$$\varphi(Q) - \varphi(P) = \int_O^Q \mathbf{a} \cdot \mathbf{t}\, ds - \int_O^P \mathbf{a} \cdot \mathbf{t}\, ds = \int_P^Q \mathbf{a} \cdot \mathbf{t}\, ds$$

since we are at liberty to take the path OQ through P. We may also choose the path from P to Q to be the straight line parallel to the axis $O1$, and then $\mathbf{t} = \mathbf{e}_{(1)}$. Thus,

$$\begin{aligned}\varphi(x_1 + dx_1, x_2, x_3) - \varphi(x_1, x_2, x_3) &= \int_{x_1}^{x_1+dx_1} a_1(\xi, x_2, x_3)\, d\xi \\ &= dx_1 a_1(x_1 + \theta\, dx_1, x_2, x_3)\end{aligned}$$

where $0 \leqq \theta \leqq 1$. In the limit as $dx_1 \to 0$ we have

$$\frac{\partial \varphi}{\partial x_1} = a_1(x_1, x_2, x_3). \tag{3.42.4}$$

This can be repeated with PQ parallel to the other two axes and establishes the representation (3.42.1).

Thus an irrotational vector field $\mathbf{a}$ can be characterized by any one of three equivalent properties. They are:

(i) curl $\mathbf{a} \equiv 0$,
(ii) $\oint \mathbf{a} \cdot \mathbf{t}\, ds \equiv 0$ for any closed curve C,
(iii) $\mathbf{a} \equiv \nabla\varphi$, where φ is a scalar point function.

If the partial derivatives are only piece-wise continuous, the equivalence has to be framed with slightly more care, but for continuously differentiable vector fields it can be baldly stated. Some writers regard (ii) as the more fundamental physical property and define irrotationality from this equation.

Since $\nabla\varphi$ is in the direction of the normal to a family of surfaces $\varphi(x_1, x_2, x_3) =$ constant, the irrotationality of the vector field implies that there is a family of surfaces everywhere normal to the trajectories of the vector field. Actually $\mathbf{a} = \nabla\varphi$ says a little more than this for it would be sufficient that $\mathbf{a} = \lambda\nabla\varphi$, with λ not necessarily constant, for the normal to be in the direction of $\mathbf{a}$. Suppose that $\mathbf{a} \neq \nabla\varphi$ but a function μ can be found so that $\mu\mathbf{a} = \nabla\varphi$. Then

$$\nabla \wedge (\mu\mathbf{a}) = \nabla\mu \wedge \mathbf{a} + \mu\nabla \wedge \mathbf{a} = \nabla \wedge \nabla\varphi \equiv 0. \tag{3.42.5}$$

Now $\mathbf{a} \cdot (\nabla\mu \wedge \mathbf{a})$ is identically zero, so that forming the scalar product of $\mathbf{a}$ with Eq. (3.42.5)

$$\mu\mathbf{a} \cdot (\nabla \wedge \mathbf{a}) = 0.$$

Thus if $\mu \neq 0$ it is necessary that

$$\mathbf{a} \cdot (\nabla \wedge \mathbf{a}) = 0 \tag{3.42.6}$$

if the integrating factor μ is to exist. It follows that for such a family of surfaces to exist the vector field must be complex lamellar. Conversely it can be shown (Ex. 3.42) that if the field is complex lamellar, an integrating factor exists. The family of surfaces $\varphi =$ constant is called the *normal congruence* of the vector field $\mathbf{a}$, and it is necessary and sufficient for a field to be complex lamellar if it is to have a normal congruence.

Exercise 3.42.1. If $\mathbf{a} \cdot \nabla \wedge \mathbf{a} = 0$ and μ is any function of x, then

$$(\mu\mathbf{a}) \cdot \nabla \wedge (\mu\mathbf{a}) = 0.$$

Exercise 3.42.2. If $a_1^2 + a_2^2 \neq 0$, show that the existence of an integrating factor μ, such that $\mu(a_1\,dx_1 + a_2\,dx_2) = \nabla\varphi$, follows from the existence of solutions of the ordinary differential equation $dx_2/dx_1 = -a_1/a_2$.

Exercise 3.42.3. By first taking x_3 to be constant, show that there exists a function $A(x_1, x_2, x_3)$ and an integrating factor μ, such that $\mathbf{a} \cdot d\mathbf{x} = dA + B\,dx_3$ where $B = \mu a_3 - \partial A/\partial x_3$. Show further that

$$\mu\mathbf{a} \cdot (\nabla \wedge \mu\mathbf{a}) = \frac{\partial A}{\partial x_1}\frac{\partial B}{\partial x_2} - \frac{\partial B}{\partial x_1}\frac{\partial A}{\partial x_2} = \frac{\partial(A, B)}{\partial(x_1, x_2)}$$

Exercise 3.42.4. The vanishing of the Jacobian $\partial(A, B)/\partial(x_1, x_2)$ implies a relation between A and B which is independent of x_1 and x_2. By combining the results of the three preceding exercises, show that $\mathbf{a} \cdot (\nabla \wedge \mathbf{a}) = 0$ is a sufficient condition for the existence of an integrating factor λ such that $\lambda\mathbf{a} = \nabla\varphi$.

3.43. Solenoidal vector fields

A solenoidal vector field is one for which

$$\nabla \cdot \mathbf{a} = 0. \tag{3.43.1}$$

By Green's theorem this is equivalent to saying that

$$\iint_S \mathbf{a} \cdot \mathbf{n}\,dS = 0 \tag{3.43.2}$$

for any closed surface S. We shall now show that $\mathbf{a}$ can be represented as the curl of a vector field $\psi\nabla\chi$, depending on the two scalar functions ψ and χ.

Consider the system of differential equations

$$\frac{dx_1}{a_1(x_1, x_2, x_3)} = \frac{dx_2}{a_2(x_1, x_2, x_3)} = \frac{dx_3}{a_3(x_1, x_2, x_3)}. \tag{3.43.3}$$

If the a_i are continuously differentiable, these equations have two independent solutions which may be written

$$f_1(x_1, x_2, x_3) = c_1 \quad \text{and} \quad f_2(x_1, x_2, x_3) = c_2.$$

Since $\mathbf{a}$ is tangent to either of these surfaces, it is orthogonal to both their normals and so

$$\mathbf{a} = \lambda\nabla f_1 \wedge \nabla f_2 \tag{3.43.4}$$

for some scalar function λ. Now let $f_3(x_1, x_2, x_3)$ be a third function such that

$$(\nabla f_1 \wedge \nabla f_2) \cdot \nabla f_3 \neq 0; \tag{3.43.5}$$

that is, the normal to the surfaces $f_3 = c_3$ is not in the plane of the other two normals. Then we can take f_1, f_2, and f_3 as coordinates of the point $\mathbf{x}$, and by the rules of partial differentiation

$$\frac{\partial}{\partial x_i} = \frac{\partial f_j}{\partial x_i}\frac{\partial}{\partial f_j}.$$

Thus

$$\begin{aligned}
\nabla \cdot \mathbf{a} = \frac{\partial a_i}{\partial x_i} &= \frac{\partial}{\partial x_i}\left(\lambda \epsilon_{ijk} \frac{\partial f_1}{\partial x_j}\frac{\partial f_2}{\partial x_k}\right) \\
&= \frac{\partial \lambda}{\partial f_p}\epsilon_{ijk}\frac{\partial f_p}{\partial x_i}\frac{\partial f_1}{\partial x_j}\frac{\partial f_2}{\partial x_k} \\
&= \frac{\partial \lambda}{\partial f_3}\epsilon_{ijk}\frac{\partial f_3}{\partial x_i}\frac{\partial f_1}{\partial x_j}\frac{\partial f_2}{\partial x_k}
\end{aligned} \tag{3.43.6}$$

since all other terms vanish. However, if $\nabla \cdot \mathbf{a} = 0$ comparison with Eq. (3.43.5) shows that

$$\frac{\partial \lambda}{\partial f_3} = 0$$

or λ is a function of f_1 and f_2 only. Let

$$\psi = \int \lambda \, df_1$$

where the integration is carried out with f_2 constant. Then

$$\nabla \psi = \lambda \nabla f_1 + \frac{\partial \psi}{\partial f_2}\nabla f_2$$

and

$$\nabla \psi \wedge \nabla f_2 = \lambda \nabla f_1 \wedge \nabla f_2. \tag{3.43.7}$$

If, therefore, we let $\chi = f_2$, we have, by Eq. (3.43.4) and this last equation,

$$\mathbf{a} = \nabla \psi \wedge \nabla \chi = \nabla \wedge (\psi \nabla \chi), \tag{3.43.8}$$

which is the required representation. Notice that

$$\nabla \cdot (\nabla \psi \wedge \nabla \chi) \equiv 0$$

so that the representation can also be given in the form

$$\mathbf{a} = \nabla \wedge \boldsymbol{\alpha}, \qquad \nabla \cdot \boldsymbol{\alpha} = 0, \tag{3.43.9}$$

where $\boldsymbol{\alpha}$ is a vector function of position. $\boldsymbol{\alpha}$ is not unique for we could add to $\boldsymbol{\alpha}$ any irrotational vector field $\boldsymbol{\beta}$ and, since $\nabla \wedge \boldsymbol{\beta} = 0$, we would have

$$\nabla \wedge (\boldsymbol{\alpha} + \boldsymbol{\beta}) = \nabla \wedge \boldsymbol{\alpha} = \mathbf{a}.$$

A concrete example will help to make this clear. Suppose

$$a_1 = 0, \qquad a_2 = \frac{-x_3}{r}, \qquad a_3 = \frac{x_2}{r},$$

where

$$r^2 = x_i x_i.$$

Then Eq. (3.43.3) are

$$\frac{dx_1}{0} = \frac{dx_2}{-x_3/r} = \frac{dx_3}{x_2/r} \tag{3.43.10}$$

of which we immediately have two integrals

$$f_1 = x_1, \qquad f_2 = \rho = \{x_2^2 + x_3^2\}^{1/2}$$

Now $\nabla f_1 = \mathbf{e}_{(1)}$ and $\nabla f_2 = (x_2 \mathbf{e}_{(2)} + x_3 \mathbf{e}_{(3)})/\rho$ so

$$\nabla f_1 \wedge \nabla f_2 = (-x_3 \mathbf{e}_{(2)} + x_2 \mathbf{e}_{(3)})/\rho.$$

Hence $\lambda = \rho/r$ so that

$$\psi = \int \lambda \, df_1 = \int \frac{\rho \, dx_1}{\{\rho^2 + x_1^2\}^{1/2}} = \rho \ln (x_1 + r)$$

and we can write

$$\mathbf{a} = \nabla \wedge [\rho \ln (x_1 + r) \nabla \rho].$$

Notice that this representation is not unique, for another pair of functions satisfying Eq. (3.43.10) would be

$$f_1 = x_1, \qquad f_2 = r = \{x_1^2 + x_2^2 + x_3^2\}^{1/2},$$

and now $\nabla f_1 \wedge \nabla f_2 = (-x_3 \mathbf{e}_{(2)} + x_2 \mathbf{e}_{(3)})/r$, so that a simpler pair of functions is $\psi = x_1$, $\chi = r$.

The continuously differentiable solenoidal vector field thus has three equivalent characteristics:

(i) $\nabla \cdot \mathbf{a} = 0$,
(ii) $\iint \mathbf{a} \cdot \mathbf{n} \, dS = 0$ for any closed surface S,
(iii) $\mathbf{a} = \nabla \wedge \boldsymbol{\alpha}$, where $\boldsymbol{\alpha}$ is itself solenoidal.

Exercise 3.43.1. Show that $f(r)\boldsymbol{\omega} \wedge \mathbf{x}$, where $\boldsymbol{\omega}$ is a constant vector and $r^2 = \mathbf{x} \cdot \mathbf{x}$, is a solenoidal vector field. Construct its representation as $\nabla \wedge (\psi \nabla \chi)$ and $\nabla \wedge \boldsymbol{\alpha}$.

Exercise 3.43.2. Show that the strength of a vector tube in a solenoidal field is the same for all cross-sections.

3.44. Helmholtz' representation

The results that have been obtained for the representation of solenoidal and irrotational vector fields can be combined to give a representation of an arbitrary continuously differentiable vector field. Thus for any finite,

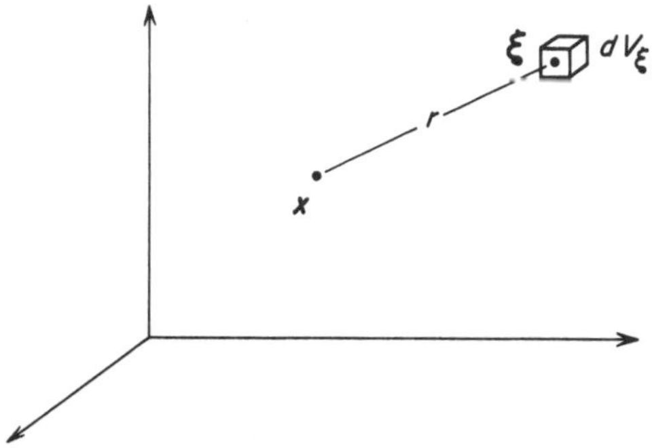

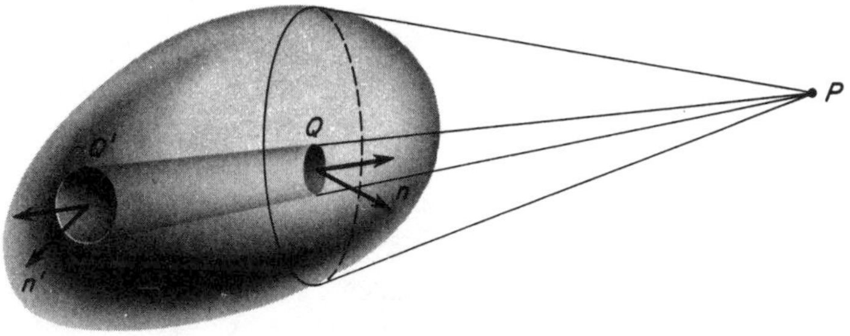

Fig. 3.12

continuous vector field which vanishes at infinity we may always find three scalar functions φ, ψ, and χ such that

$$\mathbf{a} = \nabla\varphi + \nabla \wedge (\psi\nabla\chi). \tag{3.44.1}$$

Equivalently we may find a scalar function φ and a solenoidal vector field $\boldsymbol{\alpha}$ such that

$$\mathbf{a} = \nabla\varphi + \nabla \wedge \boldsymbol{\alpha}. \tag{3.44.2}$$

Thus the vector field is decomposed into an irrotational and a solenoidal part.

To prove this theorem we need a formula for the solution of Poisson's equation

$$\nabla^2\varphi = -f(x_1, x_2, x_3). \tag{3.44.3}$$

This is provided by the integral

$$\varphi(\mathbf{x}) = \iiint \frac{f(\boldsymbol{\xi})}{4\pi r}\, dV_\xi. \tag{3.44.4}$$

The integral is over the whole of space, for if f is only defined in a certain region, we may set it equal to zero outside and if it is defined everywhere we require that it tend to zero towards infinity. Fig. 3.12 illustrates the integral; r is the distance between the point $\mathbf{x}$ and the element of integtation at $\boldsymbol{\xi}$ so

$$r^2 = |\mathbf{x} - \boldsymbol{\xi}|^2 = (\mathbf{x} - \boldsymbol{\xi}) \cdot (\mathbf{x} - \boldsymbol{\xi}).$$

Consider the gradient of φ, the vector

$$\nabla\varphi = -\iiint \frac{f(\boldsymbol{\xi})}{4\pi r^3} (\mathbf{x} - \boldsymbol{\xi})\, dV_\xi.$$

If we integrate $\nabla^2\varphi$ over some finite volume V we have by Green's theorem

$$\begin{aligned}\iiint_V \nabla^2\varphi\, dV_x &= \iint_S \nabla\varphi \cdot \mathbf{n}\, dS_x \\ &= -\iint_S dS_x \iiint \frac{f(\boldsymbol{\xi})}{4\pi r^3} (\mathbf{x} - \boldsymbol{\xi}) \cdot \mathbf{n}\, dV_\xi \\ &= -\iiint f(\boldsymbol{\xi})\, dV_\xi \iint_S \frac{(\mathbf{x} - \boldsymbol{\xi}) \cdot \mathbf{n}}{4\pi r^3}\, dS_x,\end{aligned}$$

by changing the order of integration. However, $(\mathbf{x} - \boldsymbol{\xi})/r$ is a unit vector in the direction PQ [Fig. 3.12b], so $\mathbf{n} \cdot (\mathbf{x} - \boldsymbol{\xi})\, dS/r^3$ is just the solid angle subtended at P by the element of area at Q. Now if P is outside of V, the integral of this solid angle is zero for the contributions from the elements at Q and Q' are equal and opposite. However, if P is within V, then the integral is the total solid angle, namely, 4π. It follows that

$$\iint_S \frac{(\mathbf{x} - \boldsymbol{\xi}) \cdot \mathbf{n}}{4\pi r^3}\, dS = \begin{cases} 1 & \text{if } \boldsymbol{\xi} \text{ is in } V, \\ 0 & \text{if } \boldsymbol{\xi} \text{ is outside } V. \end{cases} \tag{3.44.5}$$

Thus the last integral over the whole space of $\boldsymbol{\xi}$ has zero integrand outside of V and so may be regarded as the integral over V only. Then

$$\iiint_V \nabla^2\varphi\, dV = -\iiint_V f(\mathbf{x})\, dV.$$

Since the volume V was arbitrary this equation can only be satisfied if $\nabla^2\varphi = -f$ everywhere, and we see that Eq. (3.44.4) does give the solution.

Returning now to the arbitrary vector field $\mathbf{a}$ we see that if it can be written in the form $\nabla\varphi + \nabla \wedge (\psi\nabla\chi)$ then we have $\nabla \cdot \mathbf{a} = \nabla^2\varphi$. Now this equation may be solved by the formula we have just constructed and φ given by

$$\varphi = -\iiint \frac{(\nabla \cdot \mathbf{a})\, dV_\xi}{4\pi r}. \tag{3.44.6}$$

Now $\mathbf{a} - \nabla\varphi$ is a solenoidal vector for $\nabla \cdot (\mathbf{a} - \nabla\varphi) = 0$. Hence by the methods of the preceding paragraph we can construct functions ψ and χ or a solenoidal vector $\boldsymbol{\alpha}$ such that

$$\mathbf{a} - \nabla\varphi = \nabla \wedge (\psi\nabla\chi) = \nabla \wedge \boldsymbol{\alpha}$$

which was to be proved.

Exercise 3.44.1. By the vector form of Poisson's equation we mean $\nabla^2\boldsymbol{\alpha} = \mathbf{g}$ where each component has to satisfy Poisson's equation. By considering $\nabla \wedge \mathbf{a}$ show that the vector $\boldsymbol{\alpha}$ of Eq. (3.44.2) is given by

$$\boldsymbol{\alpha} = \iiint \frac{(\nabla \wedge \mathbf{a})\, dV_\xi}{4\pi r}$$

and complete the representation in the form of Eq. (3.44.2).

3.45. Other representations

We can also represent an arbitrary vector field as the super-position of irrotational and complex lamellar fields,

$$\mathbf{a} = \nabla\varphi + \psi\nabla\chi. \tag{3.45.1}$$

for $\nabla \wedge \mathbf{a}$ is certainly solenoidal, and so by Section 3.43 may be represented as $\nabla \wedge (\psi\nabla\chi)$. However, $\nabla \wedge (\mathbf{a} - \psi\nabla\chi) = 0$, so that $\mathbf{a} - \psi\nabla\chi$ is an irrotational vector field and therefore can be represented as the gradient of a scalar φ. φ, ψ, and χ in this representation are known as Monge's potentials.

$$\begin{aligned}\mathbf{a}\cdot(\nabla \wedge \mathbf{a}) &= (\nabla\varphi + \psi\nabla\chi)\cdot(\nabla\psi \wedge \nabla\chi)\\ &= \nabla\varphi\cdot(\nabla\psi \wedge \nabla\chi)\\ &= \frac{\partial(\varphi, \psi, \chi)}{\partial(x_1, x_2, x_3)}.\end{aligned} \tag{3.45.2}$$

A field for which $\mathbf{a} \wedge (\nabla \wedge \mathbf{a}) = 0$ has been called a Beltrami field. This equation means that $\mathbf{a}$ and curl $\mathbf{a}$ are parallel and so

$$\nabla \wedge \mathbf{a} = \Omega\mathbf{a} \tag{3.45.3}$$

where Ω is called the *abnormality* of the field. $\Omega = 0$ is the condition for the vector field to have a normal congruence of surfaces and this is the reason for the term. If $\mathbf{a}$ is represented in the form of Eq. (3.45.1) and is a Beltrami field, then

$$\begin{aligned}\mathbf{a} \wedge (\nabla \wedge \mathbf{a}) &= (\nabla\varphi + \psi\nabla\chi) \wedge (\nabla\psi \wedge \nabla\chi)\\ &= \nabla\psi\{\nabla\chi\cdot(\nabla\varphi + \psi\nabla\chi)\} - \nabla\chi\{\nabla\psi\cdot(\nabla\varphi + \psi\nabla\chi)\}\\ &= 0.\end{aligned}$$

Now if the quantities in the brackets $\{\}$ are not zero, this means that $\nabla\psi$ and $\nabla\chi$ are parallel, and, therefore,

$$\nabla\psi \wedge \nabla\chi = \nabla \wedge \mathbf{a} = 0.$$

Thus, except where $\nabla \wedge \mathbf{a} = 0$, we must have

$$\begin{aligned}\nabla\chi\cdot(\nabla\varphi + \psi\nabla\chi) &= \nabla\chi\cdot\mathbf{a} = 0,\\ \nabla\psi\cdot(\nabla\varphi + \psi\nabla\chi) &= \nabla\psi\cdot\mathbf{a} = 0.\end{aligned} \tag{3.45.4}$$

It follows from the first of these that

$$\psi = \frac{-\nabla\chi \cdot \nabla\varphi}{(\nabla\chi)^2} \tag{3.45.5}$$

so that

$$\mathbf{a} = \nabla\varphi - \frac{\nabla\chi \cdot \nabla\varphi}{(\nabla\chi)^2}\nabla\chi$$

and from the second,

$$\nabla\left(\frac{\nabla\chi \cdot \nabla\varphi}{(\nabla\chi)^2}\right) \cdot \left[\nabla\varphi - \frac{\nabla\chi \cdot \nabla\varphi}{(\nabla\chi)^2}\nabla\chi\right] = 0$$

A Trkalian field is one for which $\Omega = k$, a constant. Since in this case $\mathbf{a} = (\nabla \wedge \mathbf{a})/k$ it follows that $\nabla \cdot \mathbf{a} = 0$ and so a Trkalian field is a solenoidal Beltrami field. There are, however, solenoidal Beltrami fields which are not Trkalian. Since

$$\nabla \wedge \mathbf{a} = k\mathbf{a}$$

$$\nabla \wedge (\nabla \wedge \mathbf{a}) = k\nabla \wedge \mathbf{a} = k^2\mathbf{a}$$

However, by Ex. 3.24.5 and $\nabla \cdot \mathbf{a} = 0$,

$$\nabla^2\mathbf{a} + k^2\mathbf{a} = 0. \tag{3.45.6}$$

Thus a Trkalian vector field satisfies the Helmholtz equation. Bjørgum and Godal have shown that if h is any solution of the wave equation

$$\nabla^2 h + k^2 h = 0$$

then

$$\mathbf{a} = k\,\nabla h \wedge \mathbf{e} + \mathbf{e}k^2 h + (\mathbf{e} \cdot \nabla)\nabla h, \tag{3.45.7}$$

where $\mathbf{e}$ is an arbitrary constant unit vector, is the representation of a Trkalian vector field. The interest in these fields arises from the importance of Beltrami motions in which $\mathbf{a}$ is the velocity vector. These have been extensively studied by Truesdell and Bjørgum.

Another form of decomposition with some similarity to the Helmholtz theorem has recently been given. Any vector field can be represented as

$$\mathbf{a} = (\mathbf{x} \wedge \nabla)\varphi + (\mathbf{x} \wedge \nabla) \wedge \boldsymbol{\alpha} + \boldsymbol{\alpha} \tag{3.45.8}$$

where $\boldsymbol{\alpha}$ is a vector satisfying

$$(\mathbf{x} \wedge \nabla) \cdot \boldsymbol{\alpha} = 0,$$

and

$$(\mathbf{x} \wedge \nabla) \wedge (\mathbf{x} \wedge \nabla)\varphi = -(\mathbf{x} \wedge \nabla)\varphi.$$

The proof of this is beyond our scope here, but is to be found in the paper by J. S. Lomont and H. E. Moses. (Communications on Pure and Applied Mathematics, **14** (1961), pp. 69–76.)

Exercise 3.45.1. Show that for the Beltrami field

$$\Omega = \frac{\nabla\varphi \cdot (\nabla\psi \wedge \nabla\chi)}{\{(\nabla\varphi)^2 - \psi^2(\nabla\chi)^2\}} . \qquad \text{(Bjørgum)}$$

Exercise 3.45.2. Show that if $\nabla \wedge \mathbf{a} = \Omega\mathbf{a}$, then

$$\Omega = \frac{(\nabla \wedge \mathbf{a}) \cdot \nabla \wedge (\nabla \wedge \mathbf{a})}{(\nabla \wedge \mathbf{a})^2} . \qquad \text{(Truesdell)}$$

Exercise 3.45.3. Show that successive curls of a Trkalian field are Trkalian with the same constant k.

Exercise 3.45.4. If $\mathbf{a} = \nabla\varphi + \psi\nabla\chi$, show that

$$(\mathbf{a} \cdot \nabla)\mathbf{a} = \tfrac{1}{2}\nabla(\nabla\varphi + \psi\nabla\chi)^2 + [\nabla\psi \cdot (\nabla\varphi + \psi\nabla\chi)]\nabla\chi - [\nabla\chi \cdot (\nabla\varphi + \psi\nabla\chi)]\nabla\psi.$$

BIBLIOGRAPHY

3.1. An elementary discussion of space curves such as we have given is common to most books on vectors and tensors. A more extended treatment may be found in

Eisenhart, L. P. An introduction to differential geometry. Princeton: Princeton University Press, 1940.

A good introduction to multiple integrals is provided by

Ferrar, W. L., Integral calculus. Oxford: Oxford University Press, 1958.

and in the many available advanced calculus texts.

3.2. A valuable elementary discussion of the points that must be examined to give completely rigorous definitions is given in Chapter 4 of

Newell, H. E., Jr, Vector analysis. New York: McGraw-Hill, 1955.

3.3. A full discussion of the types of surface for which Green's and Stokes' theorem can be established will be found in Chap. IV of

Kellogg, O. D., Foundations of potential theory. Berlin: Springer-Verlag, 1929.

3.4. Truesdell, C. A., The kinematics of vorticity. Bloomington: Indiana University Press, 1954

summarises the characterization of vector fields and gives an unusually ample bibliography.

Some general discussion and references, as well as a full treatment of Beltrami fields, are to be found in

Bjørgum, O., "On Beltrami vector fields and flows. Part I. A comparative study of some basic types of vector fields." Universitetet i Bergen Arbok (1951). Naturvitenskapelig rekke Nr. 1.

and

Bjørgum, O. and T. Godal, "On Beltrami vector fields and flows. Part II. The case when Ω is constant in space." Universitetet i Bergen Arbok (1952). Naturvitenskapelig rekke Nr. 13.

The Kinematics of Fluid Motion

4

4.11. Particle paths

Kinematics is the description of motion per se. It takes no account of how the motion is brought about or of the forces involved for these are in the realm of dynamics. Consequently the results of kinematical studies apply to all types of fluid and are the ground work on which the dynamical results are constructed.

The basic mathematical idea of a fluid motion is that it can be described by a *point transformation*. At some instant we look at the fluid and remark that a certain "particle" is at a position $\boldsymbol{\xi}$ and at a later time the same particle is at position $\mathbf{x}$. Without loss of generality, we can take the first instant to be the time $t = 0$ and if the later instant is time t we say that $\mathbf{x}$ is a function of t and the initial position $\boldsymbol{\xi}$,

$$\mathbf{x} = \mathbf{x}(\boldsymbol{\xi}, t) \quad \text{or} \quad x_i = x_i(\xi_1, \xi_2, \xi_3, t). \tag{4.11.1}$$

Of course we have immediately violated the concepts of the kinetic theory of fluids for in this theory the particles are the molecules and these are in random motion. In fact we have replaced the molecular picture by that of a continuum whose velocity at any point is the average velocity of the molecules in a suitable neighborhood of the point. As we have noted in Chapter 1, the definition of average needs some care in this context, but this idealization which endows the elementary portions of the fluid with a permanence denied them by molecular theory is the key to the classical treatment of fluid motion.

The initial coordinates $\boldsymbol{\xi}$ of a particle will be referred to as the *material coordinates* of the particle and, when convenient, the particle itself may be called the particle $\boldsymbol{\xi}$. The terms *convected* and *Lagrangian* coordinates are also used. The former is a sensible term since the material coordinate system is convected with the fluid; the latter is both a misnomer* and, lacking descriptive quality, is often forgotten or confused by the student. The *spatial coordinates* $\mathbf{x}$ of the particle may be referred to as its *position* or *place*. It will be assumed that the motion is continuous, single valued and that Eq. (4.11.1) can be inverted to give the initial position or material coordinates of the particle which is at any position $\mathbf{x}$ at time t; that is,

$$\boldsymbol{\xi} = \boldsymbol{\xi}(\mathbf{x}, t) \quad \text{or} \quad \xi_i = \xi_i(x_1, x_2, x_3, t) \tag{4.11.2}$$

are also continuous and single valued. Physically this means that a continuous arc of particles does not break up during the motion or that the particles in the neighborhood of a given particle continue in its neighborhood during the motion. The single valuedness of the equations means that a particle cannot split up and occupy two places nor can two distinct particles occupy the same place. Assumptions must also be made about the continuity of derivatives. It is usual (see, for example, Serrin, Handbuch der Physik Bd. VIII/1, p. 129) to assume continuity up to the third order derivatives. Exception to these requirements may be allowed on a finite number of singular surfaces, lines or points, as for example when a fluid divides around an obstacle. It is shown in Appendix B that a necessary and sufficient condition for the inverse functions to exist is that the Jacobian

$$J = \frac{\partial(x_1, x_2, x_3)}{\partial(\xi_1, \xi_2, \xi_3)} \tag{4.11.3}$$

should not vanish.

The transformation (4.11.1) may be looked at as the parametric equation of a curve in space with t as parameter. The curve goes through the point $\boldsymbol{\xi}$, corresponding to the parameter $t = 0$, and these curves are called the *particle paths*. Any property of the fluid may be followed along the particle path. For example, we might be given the density in the neighborhood of a particle as a function $\rho(\boldsymbol{\xi}, t)$, meaning that for any prescribed particle $\boldsymbol{\xi}$ we have the density as a function of time, that is, the density that an observer riding on the particle would see. (Position itself is a "property" in this general sense so that the equations of the particle path are of this form.) This *material*

* The term Eulerian is also applied to the spatial coordinates x. Truesdell, in a footnote of customary erudition (Kinematics of vorticity, p. 30), has traced the origin of this incorrect usage.

description of the change of some property, say $\mathscr{F}(\boldsymbol{\xi}, t)$, can be changed into a *spatial* description $\mathscr{F}(\mathbf{x}, t)$ by Eq. (4.11.2),

$$\mathscr{F}(\mathbf{x}, t) = \mathscr{F}[\boldsymbol{\xi}(\mathbf{x}, t), t]. \tag{4.11.4}$$

Physically this says that the value of the property at position $\mathbf{x}$ and time t is the value appropriate to the particle which is at $\mathbf{x}$ at time t. Conversely, the material description can be derived from the spatial one by Eqs. (4.11.1)

$$\mathscr{F}(\boldsymbol{\xi}, t) = \mathscr{F}[\mathbf{x}(\boldsymbol{\xi}, t), t], \tag{4.11.5}$$

meaning that the value as seen by the particle at time t is the value at the position it occupies at that time.

Associated with these two descriptions are two derivatives with respect to time. We shall denote them by

$$\frac{\partial}{\partial t} \equiv \left(\frac{\partial}{\partial t}\right)_x \equiv \text{derivative with respect to time keeping } \mathbf{x} \text{ constant,} \tag{4.11.6}$$

and

$$\frac{d}{dt} \equiv \left(\frac{\partial}{\partial t}\right)_\xi \equiv \text{derivative with respect to time keeping } \boldsymbol{\xi} \text{ constant.} \tag{4.11.7}$$

Thus $\partial\mathscr{F}/\partial t$ is the rate of change of $\mathscr{F}$ as observed at a fixed point $\mathbf{x}$, whereas $d\mathscr{F}/dt$ is the rate of change as observed when moving with the particle. The latter we call the *material derivative*.* In particular the material derivative of the position of a particle is its velocity. Thus, putting $\mathscr{F} = x_i$, we have

$$v_i = \frac{dx_i}{dt} = \frac{\partial}{\partial t} x_i(\xi_1, \xi_2, \xi_3, t) \tag{4.11.8}$$

or

$$\mathbf{v} = \frac{d\mathbf{x}}{dt}.$$

This allows us to establish a connection between the two derivatives, for

$$\begin{aligned} \frac{d\mathscr{F}}{dt} &= \frac{\partial}{\partial t}\mathscr{F}(\boldsymbol{\xi}, t) = \frac{\partial}{\partial t}\mathscr{F}[\mathbf{x}(\boldsymbol{\xi}, t), t] \\ &= \frac{\partial\mathscr{F}}{\partial x_i}\left(\frac{\partial x_i}{\partial t}\right)_\xi + \left(\frac{\partial\mathscr{F}}{\partial t}\right)_x \\ &= v_i\frac{\partial\mathscr{F}}{\partial x_i} + \frac{\partial\mathscr{F}}{\partial t}. \end{aligned} \tag{4.11.9}$$

* It is also called the convected or convective derivative and often denoted by D/Dt.

It is sometimes convenient to write this as

$$\frac{d\mathscr{F}}{dt} = \frac{\partial \mathscr{F}}{\partial t} + (\mathbf{v} \cdot \nabla)\mathscr{F}. \tag{4.11.10}$$

Exercise 4.11.1. It is not always necessary to use the initial position as material coordinate. Consider the equations for the particle paths in Gerstner waves

$$x_1 = a + (e^{-bk}/k) \sin k(a + ct),$$
$$x_2 = -b - (e^{-bk}/k) \cos k(a + ct),$$
$$x_3 = \text{constant}.$$

Relate the constants a and b to the initial position and show that the particle paths are circles. Find the velocity vector and show that $d\,|\mathbf{v}|\,/dt = 0$.

Exercise 4.11.2. Show that the Jacobian Eq. (4.11.3) is 1 for the Gerstner wave.

Exercise 4.11.3. Show that $f(\mathbf{x}, t) = 0$ is a surface of the same material particles if and only if

$$\frac{\partial f}{\partial t} + \frac{\partial f}{\partial x_i} v_i = 0.$$

Exercise 4.11.4. If $f(\mathbf{x}, t)$ is not a material surface but moves with a speed $\mathbf{u}$ different from the stream speed $\mathbf{v}$, show that

$$(\mathbf{v} - \mathbf{u}) \cdot \mathbf{n} = \frac{df/dt}{|\nabla f|}$$

where $\mathbf{n}$ is the normal to the surface.

4.12. Streamlines

From the material description $\mathbf{x}(\boldsymbol{\xi}, t)$ of the flow we have derived a vector field

$$\mathbf{v} = d\mathbf{x}/dt = \mathbf{v}(x_1, x_2, x_3, t). \tag{4.12.1}$$

The flow is called *steady* if the velocity components are independent of time. The trajectories of the velocity field are called *streamlines*; they are the solutions of the three simultaneous equations

$$\frac{d\mathbf{x}}{ds} = \mathbf{v} \quad \text{or} \quad \frac{dx_i}{ds} = v_i(x_1, x_2, x_3, t) \tag{4.12.2}$$

where s is a parameter along the streamline. This parameter s is not to be confused with the time, for in Eq. (4.12.2) t is held fixed while the equations are integrated, and the resulting curves are the streamlines *at the instant t*. These may vary from instant to instant and in general will not coincide with the particle paths.

To obtain the particle paths from the velocity field we have to follow the motion of each particle. This means we have to solve the differential equations

$$\frac{dx_i}{dt} = v_i(x_1, x_2, x_3, t)$$

subject to $x_i = \xi_i$ at $t = 0$.

If the functions v_i do not depend on t, then the parameter s along the streamlines may be taken to be t and clearly the streamlines and particle paths will coincide. Exercise 4.12.1 shows that streamlines and particle paths may coincide for an unsteady motion.

The acceleration or rate of change of velocity is defined as

$$\mathbf{a} = \frac{d\mathbf{v}}{dt} = \frac{\partial \mathbf{v}}{\partial t} + (\mathbf{v} \cdot \nabla)\mathbf{v}. \tag{4.12.3}$$

Notice that in steady flow this does not vanish but reduces to

$$\mathbf{a} = (\mathbf{v} \cdot \nabla)\mathbf{v}. \tag{4.12.4}$$

The higher rates of change are sometimes used and defined by repeated material differentiation; thus the $(n - 1)^{\text{th}}$ acceleration* is

$$\mathbf{v}^{(n)} = d\mathbf{v}^{(n-1)}/dt. \tag{4.12.5}$$

If C is a closed curve in region of flow, the streamlines through every point of C generate a surface known as a *stream tube*. Let S be a surface with C as bounding curve, then

$$\iint_S \mathbf{v} \cdot \mathbf{n}\, dS$$

is known as the *strength of the stream tube* at its cross-section S.

Exercise 4.12.1. Show that the streamlines and particle paths coincide for the flow $v_i = x_i/(1 + t)$.

* This is the notation of Rivlin and Ericksen, "Stress deformation relations for isotropic materials," J. Rat. Mech. and Anal. **4** (1955), p. 329.

Exercise 4.12.2. Show that if $v_i/|\mathbf{v}|$ is independent of t, then streamlines and particle paths will coincide.

Exercise 4.12.3. Find the streamlines and particle paths for

$$v_i = \frac{x_i}{1 + a_i t}$$

where a_i are positive constants. (There is no summation on i here.) Describe the paths and streamlines if $a_1 = 2a_2 = 1$, $a_3 = 0$.

4.13. Streaklines

The name streakline is applied to the curve traced out by a plume of smoke or dye which is continuously injected at a fixed point but does not diffuse. Thus at time t the streakline through a fixed point $\mathbf{y}$ is a curve going from $\mathbf{y}$ to $\mathbf{x}(\mathbf{y}, t)$, the position reached by the particle which was at $\mathbf{y}$ at time $t = 0$. A particle is on the streakline if it passed the fixed point $\mathbf{y}$ at some time between 0 and t. If this time were s, then the material coordinates of the particle would be given by Eq. (4.11.2) $\boldsymbol{\xi} = \boldsymbol{\xi}(\mathbf{y}, s)$. However, at time t this particle is at $\mathbf{x} = \mathbf{x}(\boldsymbol{\xi}, t)$ so that the equation of the streakline at time t is given by

$$\mathbf{x} = \mathbf{x}[\boldsymbol{\xi}(\mathbf{y}, s), t], \tag{4.13.1}$$

where the parameter s along it lies in the interval $0 \leqq s \leqq t$. If we regard the motion as having been proceeding for all time, then the origin of time is arbitrary and s can take negative values $-\infty < s \leqq t$.

These concepts may be illustrated by the simple plane flow

$$v_1 = x_1/(1 + t), \qquad v_2 = x_2, \qquad v_3 = 0.$$

Here the streamlines at time t are the solutions of

$$\frac{dx_1}{ds} = \frac{x_1}{1 + t}, \qquad \frac{dx_2}{ds} = x_2, \qquad \frac{dx_3}{ds} = 0. \tag{4.13.2}$$

Thus keeping t constant the streamline through a is

$$x_1 = a_1 e^{s/(1+t)}, \qquad x_2 = a_2 e^s, \qquad x_3 = a_3$$

which is a curve in the plane $x_3 = a_3$

$$\frac{x_2}{a_2} = \left(\frac{x_1}{a_1}\right)^{(1+t)} \tag{4.13.3}$$

The streamlines are shown for increasing t in Fig. 4.1a.

The particle paths are solutions of

$$\frac{dx_1}{dt} = \frac{x_1}{1+t}, \qquad \frac{dx_2}{dt} = x_2, \qquad \frac{dx_3}{dt} = 0. \tag{4.13.4}$$

These are $x_1 = \xi_1(1 + t)$, $x_2 = \xi_2 e^t$, $x_3 = \xi_3$ or the curves in the plane $x_3 = \xi_3$

$$x_2 = \xi_2 \, e^{(x_1 - \xi_1)/\xi_1} \tag{4.13.5}$$

They are shown for several initial positions in Fig. 4.1b.

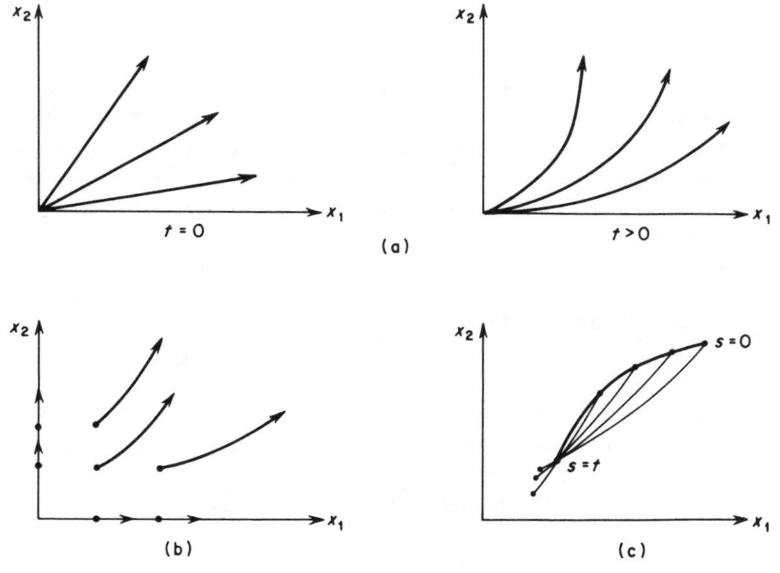

Fig. 4.1

For the inverse relations defining the particle at **y**, at times we have

$$\xi_1 = \frac{y_1}{1+s}, \qquad \xi_2 = y_2 e^{-s}, \qquad \xi_3 = y_3. \tag{4.13.6}$$

Hence, the streakline is given by

$$x_1 = y_1 \frac{1+t}{1+s}, \qquad x_2 = y_2 e^{t-s}, \qquad x_3 = y_3. \tag{4.13.7}$$

This, with some of the particle paths that contribute to it, is shown in Fig. 4.1c. (For other examples of streamlines, streaklines and particle paths see Truesdell and Toupin, Handbuch der Physik III/1, Berlin, Springer-Verlag, 1960, pages 331–336, where further references will be found.)

4.21. Dilatation

We have noticed in Section 3.16 that if the coordinate system is changed from coordinates $\boldsymbol{\xi}$ to coordinates $\mathbf{x}$, then the element of volume changes by the formula

$$dV = \frac{\partial(x_1, x_2, x_3)}{\partial(\xi_1, \xi_2, \xi_3)}\, d\xi_1\, d\xi_2\, d\xi_3 = J\, dV_0. \tag{4.21.1}$$

If we think of $\boldsymbol{\xi}$ as the material coordinates, they are Cartesian coordinates at $t = 0$, so that $d\xi_1\, d\xi_2\, d\xi_3$ is the volume dV_0 of an elementary parallelepiped. Consider this elementary parallelepiped about a given point $\boldsymbol{\xi}$ at the initial instant. By the motion this parallelepiped is moved and distorted but because the motion is continuous it cannot break up and so at time t is some neighborhood of the point $\mathbf{x} = \mathbf{x}(\boldsymbol{\xi}, t)$. By Eq. (4.21.1), its volume is $dV = J\, dV_0$ and hence

$$J = \frac{dV}{dV_0} = \text{ratio of an elementary material volume to its initial volume.} \tag{4.21.2}$$

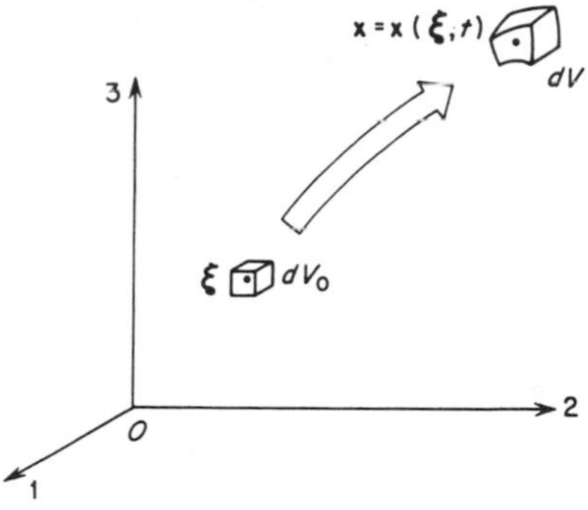

Fig. 4.2

It is called the *dilatation* or *expansion*. The assumption that the Eq. (4.11.1) can be inverted to give Eq. (4.11.2), and vice versa, is equivalent to requiring that neither J nor J^{-1} vanish. Thus,

$$0 < J < \infty. \tag{4.21.3}$$

We can now ask how the dilatation changes as we follow the motion. To answer this we calculate the material derivative dJ/dt. However,

$$J = \begin{vmatrix} \frac{\partial x_1}{\partial \xi_1} & \frac{\partial x_1}{\partial \xi_2} & \frac{\partial x_1}{\partial \xi_3} \\ \frac{\partial x_2}{\partial \xi_1} & \frac{\partial x_2}{\partial \xi_2} & \frac{\partial x_2}{\partial \xi_3} \\ \frac{\partial x_3}{\partial \xi_1} & \frac{\partial x_3}{\partial \xi_2} & \frac{\partial x_3}{\partial \xi_3} \end{vmatrix}. \tag{4.21.4}$$

Now

$$\frac{d}{dt}\left(\frac{\partial x_i}{\partial \xi_j}\right) = \frac{\partial}{\partial \xi_j}\frac{dx_i}{dt} = \frac{\partial v_i}{\partial \xi_j} \tag{4.21.4}$$

for d/dt is differentiation with $\boldsymbol{\xi}$ constant so that the order can be interchanged. Now if we regard v_i as a function of x_1, x_2, and x_3,

$$\frac{\partial v_i}{\partial \xi_j} = \frac{\partial v_i}{\partial x_1}\frac{\partial x_1}{\partial \xi_j} + \frac{\partial v_i}{\partial x_2}\frac{\partial x_2}{\partial \xi_j} + \frac{\partial v_i}{\partial x_3}\frac{\partial x_3}{\partial \xi_j} = \frac{\partial v_i}{\partial x_k}\frac{\partial x_k}{\partial \xi_j}. \tag{4.21.5}$$

In Appendix A it is shown that the derivative of the determinant is the sum of three terms in each of which only one row is differentiated. Thus for dJ/dt we should have the sum of three determinants of which the first would be

$$\begin{vmatrix} \dfrac{\partial v_1}{\partial \xi_1} & \dfrac{\partial v_1}{\partial \xi_2} & \dfrac{\partial v_1}{\partial \xi_3} \\ \dfrac{\partial x_2}{\partial \xi_1} & \dfrac{\partial x_2}{\partial \xi_2} & \dfrac{\partial x_2}{\partial \xi_3} \\ \dfrac{\partial x_3}{\partial \xi_1} & \dfrac{\partial x_3}{\partial \xi_2} & \dfrac{\partial x_3}{\partial \xi_3} \end{vmatrix} = \begin{vmatrix} \dfrac{\partial v_1}{\partial x_k}\dfrac{\partial x_k}{\partial \xi_1} & \dfrac{\partial v_1}{\partial x_k}\dfrac{\partial x_k}{\partial \xi_2} & \dfrac{\partial v_1}{\partial x_k}\dfrac{\partial x_k}{\partial \xi_3} \\ \dfrac{\partial x_2}{\partial \xi_1} & \dfrac{\partial x_2}{\partial \xi_2} & \dfrac{\partial x_2}{\partial \xi_3} \\ \dfrac{\partial x_3}{\partial \xi_1} & \dfrac{\partial x_3}{\partial \xi_2} & \dfrac{\partial x_3}{\partial \xi_3} \end{vmatrix}.$$

In expanding this determinant by the first row we see that only the first term ($k = 1$) of the elements in the first row survive, for with $k = 2$ or 3 the coefficient of $\partial v_1/\partial x_k$ is a determinant with two rows the same and so vanishes. The value of this determinant is thus $(\partial v_1/\partial x_1)J$. Considering also the other two terms in the differentiation we see that

$$\frac{dJ}{dt} = \left(\frac{\partial v_1}{\partial x_1} + \frac{\partial v_2}{\partial x_2} + \frac{\partial v_3}{\partial x_3}\right)J$$

or

$$d(\ln J)/dt = \operatorname{div} \mathbf{v}. \tag{4.21.5}$$

We thus have an important physical meaning for the divergence of the velocity field. It is the relative rate of change of the dilatation following a particle path. It is evident, that for an incompressible fluid motion,

$$\nabla \cdot \mathbf{v} = 0. \tag{4.21.6}$$

Exercise 4.21.1. Calculate J for (a) Gerstner waves and (b) the flow of Ex. 4.12.3 and confirm Eq. (4.21.5).

Exercise 4.21.2. Show that in steady flow $d^2J/dt^2 = J\nabla \cdot \{(\nabla \cdot \mathbf{v})\mathbf{v}\}$ and find a similar expression for higher derivatives.

4.22. Reynolds' transport theorem

An important kinematical theorem can be derived from the so-called Euler expansion formula, Eq. (4.21.5). It is due to Reynolds and concerns the rate of change not of an infinitesimal element of volume but of any volume integral. Let $\mathscr{F}(\mathbf{x}, t)$ be any function and $V(t)$ be a closed volume moving with the fluid, that is, consisting of the same fluid particles. Then

$$F(t) = \iiint_{V(t)} \mathscr{F}(\mathbf{x}, t)\, dV \tag{4.22.1}$$

is a function of t that can be calculated. We are interested in its material derivative dF/dt. Now the integral is over the varying volume $V(t)$ so we cannot take the differentiation through the integral sign. If, however, the integration were with respect to a volume in $\boldsymbol{\xi}$-space it would be possible to interchange differentiation and integration since d/dt is differentiation with respect to time keeping $\boldsymbol{\xi}$ constant. However, the transformation $\mathbf{x} = \mathbf{x}(\boldsymbol{\xi}, t)$, $dV = J\, dV_0$ allows us to do just this, for $V(t)$ has been defined as a moving material volume and so has come from some fixed volume V_0 at time $t = 0$. Thus

$$\begin{aligned}
\frac{d}{dt}\iiint_{V(t)} \mathscr{F}(\mathbf{x}, t)\, dV &= \frac{d}{dt}\iiint_{V_0} \mathscr{F}[\mathbf{x}(\boldsymbol{\xi}, t), t] J\, dV_0 \\
&= \iiint_{V_0} \left(\frac{d\mathscr{F}}{dt} J + \mathscr{F}\frac{dJ}{dt}\right) dV_0 \\
&= \iiint_{V_0} \left\{\frac{d\mathscr{F}}{dt} + \mathscr{F}(\nabla\cdot\mathbf{v})\right\} J\, dV_0 \\
&= \iiint_{V(t)} \left\{\frac{d\mathscr{F}}{dt} + \mathscr{F}(\nabla\cdot\mathbf{v})\right\} dV.
\end{aligned} \tag{4.22.2}$$

Since $d/dt = (\partial/\partial t) + \mathbf{v}\cdot\nabla$ we can throw this formula into a number of different shapes. Substituting for the material derivative and collecting the gradient terms gives

$$\frac{d}{dt}\iiint_{V(t)} \mathscr{F}(\mathbf{x}, t)\, dV = \iiint_{V(t)} \left\{\frac{\partial\mathscr{F}}{\partial t} + \nabla\cdot(\mathscr{F}\mathbf{v})\right\} dV \tag{4.22.3}$$

Now applying Green's theorem to the second integral we have

$$\frac{d}{dt}\iiint_{V(t)} \mathscr{F}(\mathbf{x}, t)\, dV = \iiint_{V(t)} \frac{\partial\mathscr{F}}{\partial t}\, dV + \iint_{S(t)} \mathscr{F}\mathbf{v}\cdot\mathbf{n}\, dS, \tag{4.22.4}$$

where $S(t)$ is the surface of $V(t)$. This admits of an immediate physical picture for it says that the rate of change of the integral of $\mathscr{F}$ within the moving volume is the integral of the rate of change at a point plus the net flow of $\mathscr{F}$ over the surface. $\mathscr{F}$ can be any scalar or tensor component, so that this is a kinematical result of wide application.

Exercise 4.22.1. Show that if $\nabla \cdot \mathbf{v} = 0$ the motion proceeds without change of volume. (Such a motion is called *isochoric*. An incompressible fluid cannot but be in isochoric motion, but isochoric motions of a compressible fluid are also possible.)

Exercise 4.22.2. Show that in isochoric motion the strength of any stream tube is constant.

Exercise 4.22.3. Show that the moment

$$\iiint_V \{\mathbf{v} \cdot \mathbf{x}^{(n)}\}\, dV = \iint_{S_2} (\mathbf{v} \cdot \mathbf{n})\mathbf{x}_2^{(n+1)}\, dS - \iint_{S_1} (\mathbf{v} \cdot \mathbf{n})\mathbf{x}_1^{(n+1)}\, dS.$$

V is the volume between any two sections S_1 and S_2 of a stream tube and the motion is isochoric. (See Ex. 3.23.4 for notation and hint.) What is the value of the moment if the stream tube should close on itself?

Exercise 4.22.4. V is composed of two volumes V_1 and V_2 divided by an internal surface Σ and S is the external surface of V. V is a material volume but as Σ moves with arbitrary velocity $\mathbf{u}$ and across it $\mathscr{F}$ suffers a discontinuity, $\mathscr{F}_1$ and $\mathscr{F}_2$ being its values on either side. If $\boldsymbol{\nu}$ is the normal to Σ in the direction from V_1 to V_2 show that Eq. (4.22.4) may be generalised to

$$\frac{d}{dt}\iiint_V \mathscr{F}\, dV = \iiint_V \frac{\partial \mathscr{F}}{\partial t}\, dV + \iint_S \mathscr{F}\mathbf{v} \cdot \mathbf{n}\, dS + \iint_\Sigma (\mathscr{F}_1 - \mathscr{F}_2)\mathbf{u} \cdot \boldsymbol{\nu}\, dS.$$

(Truesdell and Toupin).

Exercise 4.22.5. The surface element dS with normal $\mathbf{n}$ corresponding to two displacements $d\mathbf{x}$ and $d\mathbf{y}$ is given by $\mathbf{n}\, dS = d\mathbf{x} \wedge d\mathbf{y}$. By making these displacements correspond to differences in material coordinates $d\boldsymbol{\xi}$ and $d\boldsymbol{\eta}$ that define an area element $\boldsymbol{\nu}\, d\sigma$, show that

$$n_i \frac{\partial x_i}{\partial \xi_p}\, dS = J\nu_p\, d\sigma.$$

By differentiating this materially with respect to time, show that

$$\frac{d}{dt}(n_i\, dS) = \frac{\partial v_j}{\partial x_j} n_i\, dS - \frac{\partial v_j}{\partial x_i} n_j\, dS.$$

Exercise 4.22.6. Obtain a transport theorem for surface integrals in the form

$$\frac{d}{dt}\iint_S \mathbf{a} \cdot \mathbf{n}\, dS = \iint_S \left[\frac{d\mathbf{a}}{dt} + \mathbf{a}(\nabla \cdot \mathbf{v}) - (\mathbf{a} \cdot \nabla)\mathbf{v}\right] \cdot \mathbf{n}\, dS,$$

and reconcile this with Ex. 3.32.5.

Exercise 4.22.7. Show that if

$$\frac{\partial \mathbf{a}}{\partial t} + \mathbf{v}(\nabla \cdot \mathbf{a}) + \nabla \wedge (\mathbf{a} \wedge \mathbf{v}) = 0,$$

the strength of any vector tube in the field **a** at a material cross-section remains constant.

4.3. Conservation of mass and the equation of continuity

Although the idea of mass is not a kinematical one it is convenient to introduce it here and to obtain the continuity equation. Let $\rho(\mathbf{x}, t)$ be the mass per unit volume of a homogeneous fluid at position $\mathbf{x}$ and time t. Then the mass of any finite volume V is

$$m = \iiint_V \rho(\mathbf{x}, t)\, dV. \tag{4.3.1}$$

If V is a material volume, that is, if it is composed of the same particles, and there are no sources or sinks in the medium we take it as a principle that the mass does not change. By inserting $\mathscr{F} = \rho$ in Eq. (4.22.2) we have

$$\frac{dm}{dt} = \iiint_V \left\{\frac{d\rho}{dt} + \rho(\nabla \cdot \mathbf{v})\right\} dV = 0. \tag{4.3.2}$$

Now this is true for an arbitrary volume and hence the integrand itself must vanish everywhere. Suppose it did not vanish at some point P, but were positive there. Then since the integrand is continuous it would have to be positive for some neighborhood of P and we might take V to be entirely within this neighborhood, and for this V the integral would not vanish. It follows that

$$\frac{d\rho}{dt} + \rho(\nabla \cdot \mathbf{v}) = \frac{\partial \rho}{\partial t} + \nabla \cdot (\rho \mathbf{v}) = 0 \tag{4.3.3}$$

which is the equation of continuity.

Combining the equation of continuity with Reynolds' transport theorem for a function $\mathscr{F} = \rho F$ we have

$$\begin{aligned}\frac{d}{dt}\iiint_{V(t)} \rho F\, dV &= \iiint_{V(t)} \left\{\frac{d}{dt}(\rho F) + \rho F(\nabla \cdot \mathbf{v})\right\} dV \\ &= \iiint_{V(t)} \left\{\rho \frac{dF}{dt} + F\left(\frac{d\rho}{dt} + \rho \nabla \cdot \mathbf{v}\right)\right\} dV \\ &= \iiint_{V(t)} \rho \frac{dF}{dt}\, dV \end{aligned} \tag{4.3.4}$$

since the second term vanishes by (4.3.3).

A fluid for which the density ρ is constant is called *incompressible.* In this case the equation of continuity becomes

$$\nabla \cdot \mathbf{v} = 0 \tag{4.3.5}$$

and the motion is isochoric or the velocity field solenoidal.

Exercise 4.3.1. Show that if $\rho_0(\boldsymbol{\xi})$ is the distribution of density of the fluid at time $t = 0$ and $\nabla(\nabla \cdot \mathbf{v}) = 0$, then the distribution at time t is

$$\rho(\mathbf{x}, t) = \rho_0[\boldsymbol{\xi}(\mathbf{x}, t)] \exp -\int_0^t (\nabla \cdot \mathbf{v})\, dt.$$

Exercise 4.3.2. Show that, for the motion of Ex. 4.12.3,

$$\rho x_1 x_2 x_3 = \rho_0 \xi_1 \xi_2 \xi_3.$$

4.41. Deformation and rate of strain

Consider two nearby points P and Q with material coordinates $\boldsymbol{\xi}$ and $\boldsymbol{\xi} + d\boldsymbol{\xi}$. At time t they are to be found at $\mathbf{x}(\boldsymbol{\xi}, t)$ and $\mathbf{x}(\boldsymbol{\xi} + d\boldsymbol{\xi}, t)$. Now

$$x_i(\boldsymbol{\xi} + d\boldsymbol{\xi}, t) = x_i(\boldsymbol{\xi}, t) + \frac{\partial x_i}{\partial \xi_j} d\xi_j + 0(d^2), \tag{4.41.1}$$

where $0(d^2)$ represents terms of order $d\boldsymbol{\xi}^2$ and higher which will be neglected from this point onwards. Thus the small displacement vector $d\boldsymbol{\xi}$ has now become

$$d\mathbf{x} = \mathbf{x}(\boldsymbol{\xi} + d\boldsymbol{\xi}, t) - \mathbf{x}(\boldsymbol{\xi}, t)$$

where

$$dx_i = \frac{\partial x_i}{\partial \xi_j} d\xi_j. \tag{4.41.2}$$

It is clear from the quotient rule (since $d\boldsymbol{\xi}$ is arbitrary) that the nine quantities $\partial x_i/\partial \xi_j$ are the components of a tensor. It may be called the displacement gradient tensor and is basic to the theory of elasticity. For fluid motion, its material derivative is of more direct application and we will concentrate on this.

If $\mathbf{v} = d\mathbf{x}/dt$ is the velocity, the relative velocity of two particles $\boldsymbol{\xi}$ and $\boldsymbol{\xi} + d\boldsymbol{\xi}$ has components

$$dv_i = \frac{\partial v_i}{\partial \xi_k} d\xi_k = \frac{d}{dt}\left(\frac{\partial x_i}{\partial \xi_j}\right) d\xi_j. \tag{4.41.3}$$

However, by inverting the relation of Eq. (4.41.2), we have

$$dv_i = \frac{\partial v_i}{\partial \xi_k}\frac{\partial \xi_k}{\partial x_j} dx_j = \frac{\partial v_i}{\partial x_j} dx_j \tag{4.41.4}$$

expressing the relative velocity in terms of current relative position. Again it is evident that the $(\partial v_i/\partial x_j)$ are components of a tensor, the *velocity gradient tensor*, for which we need to obtain a sound physical feeling.

We first observe that if the motion is a rigid body translation with velocity $\mathbf{u}$,

$$\mathbf{x} = \boldsymbol{\xi} + \mathbf{u}t \tag{4.41.5}$$

and the velocity gradient tensor vanishes identically. Secondly, the velocity gradient tensor can be written as the sum of symmetric and antisymmetric parts,

$$\begin{aligned}\frac{\partial v_i}{\partial x_j} &= \frac{1}{2}\left(\frac{\partial v_i}{\partial x_j} + \frac{\partial v_j}{\partial x_i}\right) + \frac{1}{2}\left(\frac{\partial v_i}{\partial x_j} - \frac{\partial v_j}{\partial x_i}\right) \\ &= e_{ij} + \Omega_{ij}.\end{aligned} \tag{4.41.5}$$

Now we have seen (Section 2.45) that a relative velocity dv_i related to the relative position dx_j by an antisymmetric tensor Ω_{ij}: therefore $dv_i = \Omega_{ij}\, dx_j$, represents a rigid body rotation with angular velocity $\boldsymbol{\omega} = -\text{vec}\,\boldsymbol{\Omega}$. In this case

$$\omega_i = -\frac{1}{2}\epsilon_{ijk}\Omega_{jk} = \frac{1}{2}\epsilon_{ijk}\frac{\partial v_k}{\partial x_j} \tag{4.41.6}$$

or

$$\boldsymbol{\omega} = \tfrac{1}{2}\,\text{curl}\,\mathbf{v}.$$

(Cf. Ex. 3.24.8.) Thus the antisymmetric part of the velocity gradient tensor corresponds to a rigid body rotation, and, if the motion is a rigid one (composed of a translation plus a rotation), the symmetric part of the velocity gradient tensor will vanish. For this reason the tensor e_{ij} is called the *deformation* or *rate of strain* tensor and its vanishing is necessary and sufficient for the motion to be without deformation, that is, rigid.

Exercise 4.41.1. If $e_{ij} \equiv 0$ show that

$$v_i = \epsilon_{ijk}\omega_j x_k + u_i$$

where ω_i and u_i are constants.

4.42. Physical interpretation of the deformation tensor

To interpret the tensor e_{ij} we shall see how a small element is changing during the motion. The length of the line segment from P to Q is ds, where

$$ds^2 = dx_i\, dx_i = \frac{\partial x_i}{\partial \xi_j}\frac{\partial x_i}{\partial \xi_k}\, d\xi_j\, d\xi_k. \tag{4.42.1}$$

Now P and Q are the material particles $\boldsymbol{\xi}$ and $\boldsymbol{\xi} + d\boldsymbol{\xi}$ so that $d\xi_j$ and $d\xi_k$ do not change during the motion. Thus

$$\frac{d}{dt}(ds^2) = \left(\frac{\partial v_i}{\partial \xi_j}\frac{\partial x_i}{\partial \xi_k} + \frac{\partial x_i}{\partial \xi_j}\frac{\partial v_i}{\partial \xi_k}\right) d\xi_j\, d\xi_k = 2\frac{\partial v_i}{\partial \xi_j}\frac{\partial x_i}{\partial \xi_k} d\xi_j\, d\xi_k$$

by symmetry. However,

$$\frac{\partial v_i}{\partial \xi_j} d\xi_j = \frac{\partial v_i}{\partial x_j} dx_j \quad \text{and} \quad \frac{\partial x_i}{\partial \xi_k} d\xi_k = dx_i$$

Thus

$$\frac{1}{2}\frac{d}{dt}(ds^2) = (ds)\frac{d}{dt}(ds) = \frac{\partial v_i}{\partial x_j} dx_j\, dx_i = e_{ij}\, dx_i\, dx_j \tag{4.42.2}$$

by symmetry, or

$$\frac{1}{(ds)}\frac{d}{dt}(ds) = e_{ij}\frac{dx_i}{ds}\frac{dx_j}{ds}. \tag{4.42.3}$$

Now dx_i/ds is the ith component of a unit vector in the direction of the segment PQ, so that this equation says that the rate of change of the length of the segment as a fraction of its length is related to its direction through the deformation tensor.

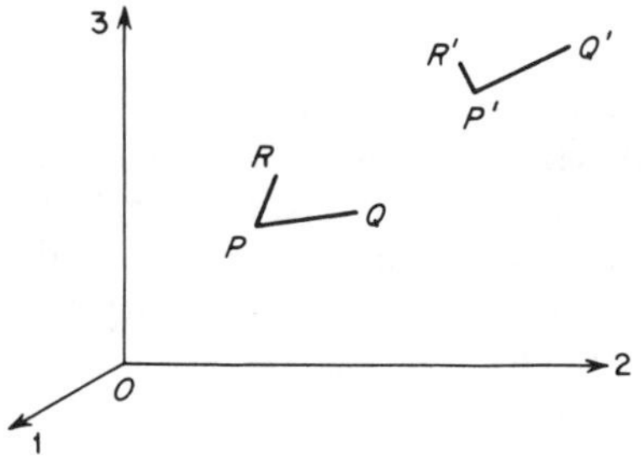

Fig. 4.3

In particular, if PQ is parallel to the coordinate axis $O1$ we have $d\mathbf{x}/ds = \mathbf{e}_{(1)}$ and

$$\frac{1}{dx_1}\frac{d}{dt}(dx_1) = e_{11}. \tag{4.42.4}$$

Thus e_{11} is the rate of longitudinal strain of an element parallel to the $O1$ axis. Similar interpretations apply to e_{22} and e_{33}.

Again consider two segments PQ and PR, where R is the particle $\boldsymbol{\xi} + d\boldsymbol{\xi}'$. If θ is the angle between them and ds' is the length of PR,

$$ds\, ds' \cos\theta = dx_i\, dx_i'.$$

Differentiating with respect to time we have

$$\frac{d}{dt}[ds\, ds' \cos\theta] = dv_i\, dx_i' + dx_i\, dv_i'$$

$$= \frac{\partial v_i}{\partial x_j} dx_j\, dx_i' + dx_i \frac{\partial v_i}{\partial x_j} dx_j'$$

since $dv_i' = (\partial v_i/\partial x_j)\, dx_j'$. The i and j are dummy suffixes so we may interchange them in the first term on the right, then performing the differentiation we have

$$\cos\theta\left\{\frac{1}{(ds)}\frac{d}{dt}(ds) + \frac{1}{(ds')}\frac{d}{dt}(ds')\right\} - \sin\theta\,\frac{d\theta}{dt}$$

$$= \left(\frac{\partial v_j}{\partial x_i} + \frac{\partial v_i}{\partial x_j}\right)\frac{dx_j}{ds}\frac{dx_i'}{ds'} = 2e_{ij}\frac{dx_i'}{ds'}\frac{dx_j}{ds}$$

Now suppose that dx' is parallel to the axis $O1$ and dx to the axis $O2$, so that $(dx_i'/ds') = \delta_{i1}$ and $(dx_j/ds) = \delta_{j2}$ and $\theta_{12} = \pi/2$. Then

$$-\frac{d\theta_{12}}{dt} = 2e_{12}. \tag{4.21.5}$$

Thus e_{12} is to be interpreted as one-half the rate of decrease of the angle between two segments originally parallel to the $O1$ and $O2$ axes respectively. Similar interpretations are appropriate to e_{23} and e_{31}.

The fact that the deformation tensor is linear in the velocity field has an important consequence. Since we may superimpose two velocity fields to form a third, it follows that the deformation tensor of this is the sum of the deformation tensors of the fields from which it is composed. Thus a flow with $v_1 = \lambda_1 x_1$, $v_2 = v_3 = 0$ would have only one nonvanishing component of the rate of strain tensor, $e_{11} = \lambda_1$. This represents a *pure stretching* in the $O1$ direction with no deformation of an element perpendicular. Again, if $v_i = \lambda_i x_i$ (no summation on i), we have a deformation which is the superposition of three stretchings parallel to the three axes. However, if $v_1 = f(x_2)$, $v_2 = v_3 = 0$ so that the only nonzero component of the deformation tensor is $e_{12} = \frac{1}{2}f'(x_2)$, the motion is one of *pure shear* in which elements parallel to the coordinate axes are not stretched at all. Note however that in pure stretching an element not parallel or perpendicular to the direction of stretching will suffer rotation. Likewise in pure shear an element not normal to or in the plane of shear will suffer stretching.

Exercise 4.42.1. Follow through the ideas of this section for the plane stagnation flow $v_1 = x_1$, $v_2 = -x_2$, $v_3 = 0$. Show that if θ is the angle between an infinitesimal material segment and the axis $O1$, then the rate of change of log tan θ is constant along a particle path.

Exercise 4.42.2. Find an expression for the rate of change of the angle between a material line segment and a fixed direction and analyse it.

4.43. Principal axes of deformation

The quadratic form (4.42.3) may be written

$$\frac{d}{dt} \ln (ds) = e_{ij} l_i l_j \tag{4.43.1}$$

where $\mathbf{l}$ is a unit vector in the direction PQ. From our knowledge of symmetric second order tensors we know that there are three mutually perpendicular directions along which this expression has stationary values (see Appendix A.12). Moreover we know from Section 2.5 and Appendix A.11 that we can find a rotation of coordinates to a frame $O\bar{1}\bar{2}\bar{3}$ such that the component $\bar{e}_{ij}$ in this frame of reference are zero if $i \neq j$. If $d_{(1)}$, $d_{(2)}$, $d_{(3)}$ are the values of $\bar{e}_{11}$, $\bar{e}_{22}$, and $\bar{e}_{33}$, they are roots of the cubic

$$\det (e_{ij} - d\delta_{ij}) = \Psi - d\Phi + d^2\Theta - d^3 = 0 \tag{4.43.2}$$

(cf. Eq. 2.5.4). The three directions $O\bar{1}$, $O\bar{2}$, $O\bar{3}$ are called the *principal axes* of stretching or rate of strain and $d_{(1)}$, $d_{(2)}$, $d_{(3)}$ the *principal rates of strain.* The three scalars Θ, Φ, Ψ are the *invariants of the deformation tensor* and the first of them we have already encountered as the dilatation. In fact

$$\Theta = e_{11} + e_{22} + e_{33} = \frac{\partial v_1}{\partial x_1} + \frac{\partial v_2}{\partial x_2} + \frac{\partial v_3}{\partial x_3}$$

$$\Phi = e_{22}e_{33} - e_{23}e_{32} + e_{33}e_{11} - e_{31}e_{13} + e_{11}e_{22} - e_{12}e_{21}$$

$$\Psi = \det e_{ij} = \epsilon_{ijk} e_{1i} e_{2j} e_{3k}.$$

We know that Θ is to be interpreted as the fractional rate of change of an infinitesimal volume. Dishington (Physics of Fluids, **3**, 1960, p. 482) has given a physical interpretation to the other invariants. He finds that

$$\lim_{V \to 0} \frac{1}{V} \frac{d^2V}{dt^2} = (\nabla \cdot \mathbf{a})_0 + 2\Phi$$

where $(\nabla \cdot \mathbf{a})_0$ is the divergence of the acceleration as measured by an observer moving and rotating with the element.* Thus 2Φ is the part contributed to the fractional acceleration of volume by the steady velocity of the surface. In a similar way 6Ψ is found to be the contribution to the limit of $(d^3V/dt^3)/V$.

A picture of the deformation may be formed by considering what happens to a small sphere of radius dr during a short interval of time ds. Suppose that we have chosen the coordinate system so that the axes are parallel to the principal axes of stretching at the point $\mathbf{x}$. The particles on a sphere of

* This interpretation is also implicit in the formulae given by Truesdell for $\nabla \cdot \mathbf{a}$ (Kinematics of Vorticity, p. 79).

radius dr and center $\mathbf{x}$ are $\mathbf{x} + \mathbf{l}\, dr$, where $\mathbf{l}$ is a unit vector and they have material coordinates $\boldsymbol{\xi} + d\boldsymbol{\xi}$ where

$$l_i\, dr = \left(\frac{\partial x_i}{\partial \xi_j}\right)_t d\xi_j. \tag{4.43.3}$$

In the interval from t to $t + dt$ the center moves from $\mathbf{x}(\boldsymbol{\xi}, t)$ to $\mathbf{x}(\boldsymbol{\xi}, t + dt)$. If $d\mathbf{y}$ is the position of a particle that was on the surface of the sphere relative to the new position of the center,

$$\begin{aligned} dy_i &= x_i(\boldsymbol{\xi} + d\boldsymbol{\xi}, t + dt) - x_i(\boldsymbol{\xi}, t + dt) \\ &= \left(\frac{\partial x_i}{\partial \xi_k}\right)_{t+dt} d\xi_k \end{aligned} \tag{4.43.4}$$

However, here $\partial x_i/\partial \xi_k$ is evaluated at $\boldsymbol{\xi}$ and $t + dt$, that is,

$$\begin{aligned} \left(\frac{\partial x_i}{\partial \xi_k}\right)_{t+dt} &= \left(\frac{\partial x_i}{\partial \xi_k}\right)_t + dt\, \frac{d}{dt}\left(\frac{\partial x_i}{\partial \xi_k}\right) \\ &= \left(\frac{\partial x_i}{\partial \xi_k}\right)_t + \left(\frac{\partial v_i}{\partial \xi_k}\right) dt. \end{aligned}$$

Substituting this value back into Eq. (4.43.4) and using the relation (4.43.3) in the form

$$d\xi_k = \left(\frac{\partial \xi_k}{\partial x_i}\right)_t l_i\, dr,$$

we have

$$\begin{aligned} dy_i &= l_i\, dr + \frac{\partial v_i}{\partial \xi_k}\frac{\partial \xi_k}{\partial x_j}\, l_j\, dr\, dt \\ &= \left(\delta_{ij} + \frac{\partial v_i}{\partial x_j}\, dt\right) l_j\, dr = A_{ij} l_j\, dr \end{aligned} \tag{4.43.5}$$

Now since the coordinate axes were chosen parallel to the principal axes of deformation $e_{ij} = 0$ for $i \neq j$,

$$A_{ii} = 1 + e_{ii}\, dt, \qquad i = j \qquad A_{ij} = \frac{1}{2}\left(\frac{\partial v_i}{\partial x_j} - \frac{\partial v_j}{\partial x_i}\right) = \Omega_{ij}, \quad i \neq j.$$

The off-diagonal terms thus represent the rigid body rotation as before and the remaining terms are purely diagonal giving, in the absence of rotation,

$$dy_i = (1 + e_{ii}\, dt) l_i\, dr \qquad \text{(no summation)}.$$

Since $\mathbf{l}$ is a unit vector,

$$1 = l_i l_i = \frac{dy_i\, dy_i}{(1 + e_{ii}\, dt)^2\, dr^2},$$

and this is an infinitesimal ellipsoid whose axes are coincident with the principal axes of stretching and of lengths $(1 + e_{ii}\, dt)\, dr$, $i = 1, 2, 3$. Thus in the complete deformation a small sphere is distorted into an ellipsoid and rotated, as shown in Fig. 4.4.

This insight into the character of deformation is expressed by the so-called *Cauchy-Stokes decomposition theorem*, which Truesdell formulates as follows: *an arbitrary instantaneous state of motion may be resolved at each*

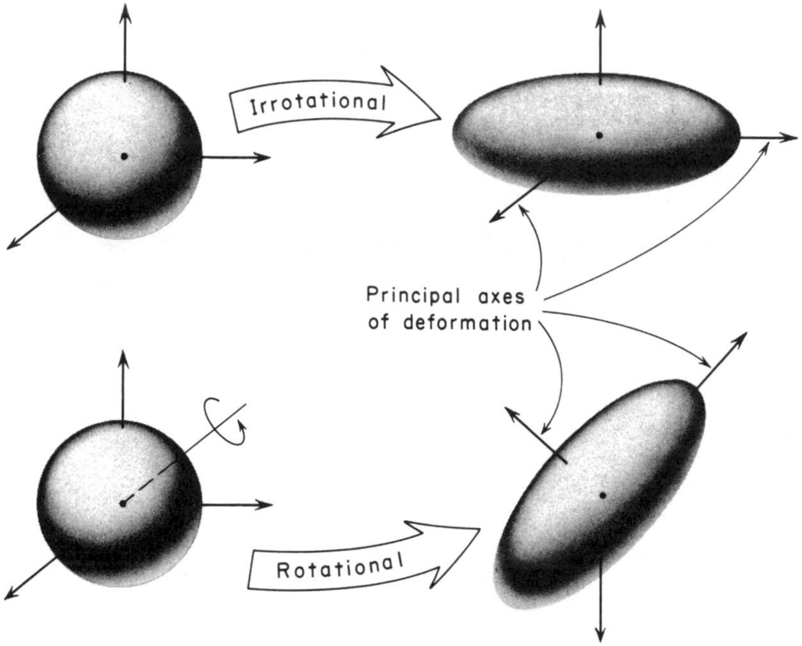

Fig. 4.4

point into a uniform translation, a dilatation along three mutually perpendicular axes, and a rigid rotation of these axes.

Exercise 4.43.1. Show that $e_{ij}e_{ij} = \Theta^2 - 2\Phi$ and deduce that $\Phi \leqq 0$ for an isochoric motion.

Exercise 4.43.2. The stretching is called spherical if all the principal rates of strain are equal. Show that in this case

$$\Psi'^2 = (\tfrac{1}{3}\Phi)^3 = (\tfrac{1}{3}\Theta)^6.$$

Exercise 4.43.3. If $v_1 = f(r)(x_2/r)$, $v_2 = -f(r)(x_1/r)$, $v_3 = 0$, $r^2 = x_1^2 + x_2^2$, the motion is a steady one with circular streamlines. Show that the deformation tensor is

$$e_{11} = -e_{22} = F \sin 2\theta \qquad e_{12} = -F \cos 2\theta$$

where $\tan\theta = x_2/x_1$ and $F = \frac{1}{2}\{f'(r) - f(r)/r\}$. Show that the principal rates of strain are equal and opposite and find the principal axes.

4.5. Vorticity, vortex lines, and tubes

We have seen that the antisymmetric part of the rate of strain tensor represents the local rotation, in fact $\text{vec}\,\boldsymbol{\Omega} = \frac{1}{2}\,\text{curl}\,\mathbf{v}$. The curl of the velocity is known as the *vorticity*,

$$\mathbf{w} = \nabla \wedge \mathbf{v}. \tag{4.5.1}$$

For an irrotational flow the vorticity vanishes everywhere. The trajectories of the vortex field are called *vortex lines* and the surface generated by the vortex lines through a closed curve C is a *vortex tube*. Since the strength of a vector tube of the field $\mathbf{a}$ at its section S has been defined as

$$\iint_S \mathbf{a} \cdot \mathbf{n}\, dS$$

we see that the strength of a vortex tube equals the circulation round the closed curve C which bounds the cross-section S, for

$$\iint_S (\nabla \wedge \mathbf{v}) \cdot \mathbf{n}\, dS = \oint_C \mathbf{v} \cdot \mathbf{t}\, ds \tag{4.5.2}$$

by Stokes' theorem.

The kinematics of vorticity has been elegantly and extensively treated by Truesdell in a monograph of that title (Indiana University Press, 1954) and in the appropriate sections of Bd. III/1 of the Handbuch der Physik. Here we have no intention of covering or even attempting to survey the full scope of the theory; it will be sufficient if the reader's buoyancy is increased enough for him to enjoy swimming in these waters. Truesdell collects four interpretations of the vorticity some of which we have already encountered. First, since the curl can be defined in terms of the circulation around an infinitesimal curve, the component of vorticity in a given direction is the circulation around a small circuit in a plane normal to that direction. Second, for a rigid body rotation the angular velocity is $\frac{1}{2}\,\text{curl}\,\mathbf{v}$. Now the principal axes are unchanged by the deformation part of the rate of strain tensor and hence $\frac{1}{2}\mathbf{w}$ is the angular velocity of the principal axes at a point with respect to a fixed coordinate system. Third, it may be shown that $\frac{1}{2}w_1$ is the mean value of the angular velocity of two line segments through the point parallel to the $O2$ and $O3$ axes. The fourth interpretation is related to the last and identifies $\frac{1}{2}w_1$ as the mean value of all the rates of rotation about an axis parallel to $O1$ of line segments in a plane normal to $O1$. The last two interpretations can of course be generalized to relate $\frac{1}{2}\mathbf{w} \cdot \mathbf{n}$ to the mean rates of rotation of segments in a plane with normal $\mathbf{n}$.

We observe that the strength of a vortex tube at any cross-section is the same, for $\mathbf{w}$ is a solenoidal vector. One characterisation of the solenoidal field is the vanishing of $\iint \mathbf{w}\cdot\mathbf{n}\,dS$ over any closed surface. Take this closed surface to be a stream tube with sections S_1 and S_2 whose boundary curves are C_1 and C_2. If $\mathbf{n}$ is the outward normal, it will be the positive normal (in the sense of being right-handed for a given circuit of C) for one of S_1 and S_2 and negative for the other. Since $\mathbf{w}\cdot\mathbf{n}$ vanishes identically over the surface of the vortex tube,

$$\iint_{S_1} \mathbf{w}\cdot\mathbf{n}\,dS = \iint_{S_2} \mathbf{w}\cdot(-\mathbf{n})\,dS \tag{4.5.3}$$

and the strength is constant.

Taking the curl of the formula (Ex. 4.5.1) for the acceleration

$$\begin{aligned}\nabla\wedge\mathbf{a} &= \partial\mathbf{w}/\partial t + \nabla\wedge(\mathbf{w}\wedge\mathbf{v})\\ &= d\mathbf{w}/dt - (\mathbf{w}\cdot\nabla)\mathbf{v} + \mathbf{w}(\nabla\cdot\mathbf{v}),\end{aligned} \tag{4.5.4}$$

where we have made use of Ex. 3.24.4. Thus, using the continuity equation,

$$\begin{aligned}\frac{d}{dt}\left(\frac{\mathbf{w}}{\rho}\right) &= \frac{1}{\rho}\frac{d\mathbf{w}}{dt} - \frac{\mathbf{w}}{\rho^2}\frac{d\rho}{dt}\\ &= \frac{1}{\rho}\{\nabla\wedge\mathbf{a} + (\mathbf{w}\cdot\nabla)\mathbf{v} - \mathbf{w}(\nabla\cdot\mathbf{v})\} + \frac{\mathbf{w}}{\rho}(\nabla\cdot\mathbf{v})\\ &= \left(\frac{\mathbf{w}}{\rho}\cdot\nabla\right)\mathbf{v} + \frac{1}{\rho}\nabla\wedge\mathbf{a}.\end{aligned} \tag{4.5.5}$$

If the acceleration is irrotational this equation may be solved. Setting $w_j = \rho c_i(\partial x_j/\partial\xi_i)$, where the c_i are components of a new vector $\mathbf{c}$, we have

$$\begin{aligned}\frac{d}{dt}\left(\frac{w_j}{\rho}\right) &= \frac{dc_i}{dt}\frac{\partial x_j}{\partial\xi_i} + c_i\frac{\partial v_j}{\partial\xi_i}\\ &= c_i\frac{\partial x_k}{\partial\xi_i}\frac{\partial v_j}{\partial x_k} = c_i\frac{\partial v_j}{\partial\xi_i}.\end{aligned}$$

Since J, the determinant of the coefficients of the dc_i/dt, is not zero, this gives

$$\frac{dc_i}{dt} = 0$$

or $\mathbf{c} = \mathbf{c}(\xi_1, \xi_2, \xi_3)$, and

$$\mathbf{w} = \rho\left[c_i(\boldsymbol{\xi})\frac{\partial}{\partial\xi_i}\right]\mathbf{x}. \tag{4.5.6}$$

If $\mathbf{w}_0$ and ρ_0 are the initial values of $\mathbf{w}$ and ρ of a particle,

$$(\mathbf{w}_0)_i = \rho_0 c_i(\boldsymbol{\xi})$$

and hence

$$\frac{\mathbf{w}}{\rho} = \left[\frac{(\mathbf{w}_0)_i}{\rho_0}\frac{\partial}{\partial\xi_i}\right]\mathbf{x}. \tag{4.5.7}$$

If initially an element $d\boldsymbol{\xi}$ is in a vortex line, $d\boldsymbol{\xi} = \mathbf{w}_0\, d\sigma$. However, under the motion, this element becomes

$$d\mathbf{x} = d\xi_i \frac{\partial \mathbf{x}}{\partial \xi_i} = (\mathbf{w}_0)_i \frac{\partial \mathbf{x}}{\partial \xi_i}\, d\sigma = \frac{\rho_0}{\rho}\, \mathbf{w}\, d\sigma;$$

in other words a material element tangent to a vortex line remains tangent to it. It follows that, *if* $\nabla \wedge \mathbf{a} = 0$ *then vortex lines are material lines.* It can be shown that this is also true under the broader condition $\mathbf{w} \wedge (\nabla \wedge \mathbf{a}) = 0$.

We shall return to this subject in Chapter 6, but must now pass on to some dynamical considerations.

Exercise 4.5.1. Show that the acceleration $\mathbf{a}$ is given by

$$\mathbf{a} = \partial \mathbf{v}/\partial t + \nabla(\tfrac{1}{2}\, |\mathbf{v}|^2) + \mathbf{w} \wedge \mathbf{v}.$$

Exercise 4.5.2. Establish thc third interpretation given above.

Exercise 4.5.3. Show that the abnormality of the velocity field is the ratio of the component of vorticity in the direction of motion to the speed. (Abnormality is defined in Section 3.45) (Truesdell).

Exercise 4.5.4. Show that $v_{i,j}v_{j,i} = e_{ij}e_{ij} - \frac{1}{2}w_i w_i$ and hence that

$$\nabla \cdot \mathbf{a} = d\Theta/dt + \Theta^2 - 2\Phi - \tfrac{1}{2}\, |\mathbf{w}|^2 \qquad \text{(Truesdell)}$$

Exercise 4.5.5. Show that the strength of a vortex tube remains constant if

$$\frac{\partial \mathbf{w}}{\partial t} + \nabla \wedge (\mathbf{w} \wedge \mathbf{v}) = 0.$$

(Cf. Ex. 4.22.7.)

BIBLIOGRAPHY

The basic material on kinematics goes back to the seventeenth century. Full references can be found in the works of Truesdell and a valuable survey has been given by him in his introduction to volume (2) *12* of L. Euleri Opera Omnia (Lausannae MCMLIV), "Rational Fluid Mechanics," 1687–1765.

4.1. The most extensive exposition is to be found in chapter B of

Truesdell, C. A., and R. Toupin, The classical field thcories, in Handbuch der Physik III/1, Ed. S. Flugge. 1960.

The "peculiar and characteristic glory of three-dimensional kinematics" is the subject of an earlier treatise

Truesdell, C. A., The kinematics of vorticity. Bloomington, Ind: Indiana University Press, 1954.

The basic material is given in

Serrin, J., Mathematical principles of classical fluid mechanics, in Handbuch der Physik VIII/1, Ed. S. Flugge, 1959.

We have followed his exposition in treating the vortex line in the last section.

Stress in Fluids

5

The object of this chapter is to show the tensorial character of stress and exhibit some of the ways in which it may be related to strain. We have already referred to stress as being a force per unit area and so having two directions (those of the force and the normal to the area) associated with it. This gives reason to suspect that stress can be represented by a tensor, but to establish this we follow a very elegant line of reasoning laid down by Cauchy in 1823. On the principle of the conservation of momentum we can then establish certain properties of the stress tensor. The relation between the stress tensor and the deformation tensor is known as the *mechanical constitutive equation* for the material and the remainder of the chapter will treat some elementary examples of these defining relations.

5.11. Cauchy's stress principle and the conservation of momentum

The forces acting on an element of a continuous medium may be of two kinds. *External or body forces*, such as gravitation or electromagnetic forces, can be regarded as reaching into the medium and acting throughout the volume. *Internal or contact forces* are to be regarded as acting on an element of volume through its bounding surface. If the element of volume has an external bounding surface, the forces there may be specified, as, for example, when a constant pressure is applied over a free surface. If the element is internal, the resultant force is that exerted by the material outside the surface upon that inside. Let **n** be the unit outward normal at a point of the surface S

and $\mathbf{t}_{(\mathbf{n})}$ the force per unit area exerted there by the material outside S. Then Cauchy's principle asserts that $\mathbf{t}_{(\mathbf{n})}$ is a function of the position $\mathbf{x}$, the time t, and the orientation $\mathbf{n}$ of the surface element. Thus the total internal force exerted on the volume V through its bounding surface S is

$$\iint_S \mathbf{t}_{(\mathbf{n})}\, dS. \tag{5.11.1}$$

If $\mathbf{f}$ is the external force per unit mass (for example if $O3$ is vertical, gravitation will exert a force $-g\mathbf{e}_{(3)}$ per unit mass or $-\rho g\mathbf{e}_{(3)}$ per unit volume), the total external force will be

$$\iiint_S \rho\mathbf{f}\, dV. \tag{5.11.2}$$

The principle of the conservation of linear momentum asserts that the sum of these two forces equals the rate of change of linear momentum of the volume; that is,

$$\frac{d}{dt}\iiint_V \rho\mathbf{v}\, dV = \iiint_V \rho\mathbf{f}\, dV + \iint_S \mathbf{t}_{(\mathbf{n})}\, dS. \tag{5.11.3}$$

This is an integral form of the equations of motion which can be changed when we know more about the nature of $\mathbf{t}_{(\mathbf{n})}$. If we assume that all torques arise from macroscopic forces, then not only momentum but also its moment are expressible in terms of $\mathbf{f}$ and $\mathbf{t}_{(\mathbf{n})}$ and

$$\frac{d}{dt}\iiint_V \rho(\mathbf{x}\wedge\mathbf{v})\, dV = \iiint_V \rho(\mathbf{x}\wedge\mathbf{f})\, dV + \iint_S (\mathbf{x}\wedge\mathbf{t}_{(\mathbf{n})})\, dS. \tag{5.11.4}$$

This is the case with many ordinary fluids, but a fluid with a strongly polar character is capable of transmitting stress torques and being subjected to body torques. We shall consider this briefly in Section 5.13.

From the form of these integral relations we can deduce an important relation. Suppose V is a volume of given shape with characteristic dimension d. Then the volume of V will be proportional to d^3 and the area of S to d^2, with the proportionality constants depending only on the shape. Now let V shrink on a point but preserve its shape, then the first two integrals in Eq. (5.11.3) will decrease as d^3 but the last will be as d^2. It follows that

$$\lim_{d\to 0}\frac{1}{d^2}\iint_S \mathbf{t}_{(\mathbf{n})}\, dS = 0 \tag{5.11.5}$$

or, *the stresses are locally in equilibrium.*

Exercise 5.11.1. Show that $\mathbf{t}_{(-n)} = -\mathbf{t}_{(n)}$.

Exercise 5.11.2. Establish a result for the moments of the stresses as the volume shrinks on a point.

5.12. The stress tensor

To elucidate the nature of the stress system at a point P we consider a small tetrahedron with three of its faces parallel to the coordinate planes through P and the fourth with normal $\mathbf{n}$ (see Fig. 5.1). If dA is the area of the slant face, the areas of the faces perpendicular to the coordinate axis Pi is $dA_i = n_i\, dA$. The outward normals to these faces are $-\mathbf{e}_{(i)}$ and we may denote the stress vector over these faces by $-\mathbf{t}_{(i)}$. ($\mathbf{t}_{(i)}$ denotes the stress vector when $+\mathbf{e}_{(i)}$ is

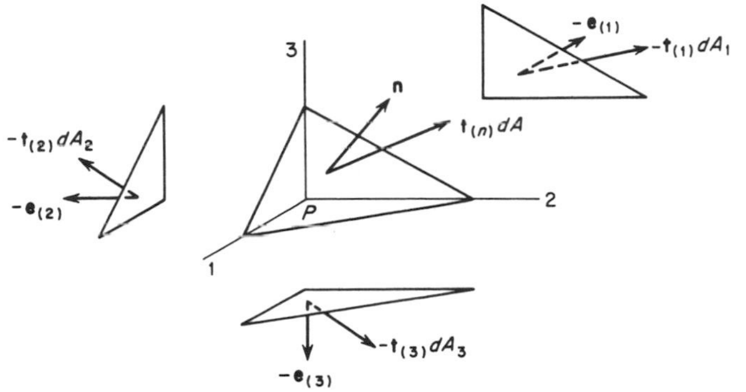

Fig. 5.1

the outward normal.) Then applying the principle of local equilibrium to the stress forces when the tetrahedron is very small we have

$$\mathbf{t}_{(n)}\, dA - \mathbf{t}_{(1)}\, dA_1 - \mathbf{t}_{(2)}\, dA_2 - \mathbf{t}_{(3)}\, dA_3$$
$$= (\mathbf{t}_{(n)} - \mathbf{t}_{(1)}n_1 - \mathbf{t}_{(2)}n_2 - \mathbf{t}_{(3)}n_3)\, dA = 0. \quad (5.12.1)$$

Now let T_{ji} denote the ith component of $\mathbf{t}_{(j)}$ and $t_{(n)i}$ the ith component of $\mathbf{t}_{(n)}$ so that this equation can be written

$$t_{(n)i} = T_{ji}n_j. \quad (5.12.2)$$

However, $\mathbf{t}_{(n)}$ is a vector and $\mathbf{n}$ is a unit vector quite independent of the T_{ji} so that by the quotient rule the T_{ji} are components of a second order tensor $\mathbf{T}$. In dyadic notation we might write

$$\mathbf{t}_{(n)} = \mathbf{n} \cdot \mathbf{T}. \quad (5.12.3)$$

This tells us that the system of stresses in a fluid is not so complicated as to demand a whole table of the functions $\mathbf{t}_{(n)}(\mathbf{x}, \mathbf{n})$ at any given instant, but that it depends rather simply on $\mathbf{n}$ through the nine quantities $T_{ji}(\mathbf{x})$. Moreover, because these are components of a tensor, any equation we derive with them will be true under any rotation of the coordinate axes.

Inserting Eq. (5.12.2) in Eq. (5.11.3) and using Green's theorem we have

$$\frac{d}{dt}\iiint_V \rho v_i \, dV = \iiint_V \rho \frac{dv_i}{dt} \, dV = \iiint_V \rho f_i \, dV + \iint_S T_{ji} n_j \, dS$$
$$= \iiint_V [\rho f_i + T_{ji,j}] \, dV.$$

However, since V is an arbitrary volume this equation is only satisfied if

$$\rho \frac{dv_i}{dt} = \rho f_i + T_{ji,j} \tag{5.12.4}$$

or

$$\rho \mathbf{a} = \rho \frac{d\mathbf{v}}{dt} = \rho \mathbf{f} + \nabla \cdot \mathbf{T}, \tag{5.12.5}$$

where $\mathbf{a} = d\mathbf{v}/dt$ is the acceleration. This is Cauchy's equation of motion. It holds for any continuum no matter how the stress tensor $\mathbf{T}$ is connected with the rate of strain.

Exercise 5.12.1. Show that Cauchy's equation of motion (5.12.4) can be written

$$\frac{\partial}{\partial t}(\rho v_i) = \rho f_i + (T_{ji} - \rho v_j v_i)_{,j}$$

and interpret this physically.

Exercise 5.12.2. Show that if $\mathscr{F}$ is any function of position and time,

$$\iint_S \mathscr{F} T_{ij} n_j \, dS = \iiint_V \left[T_{ij} \mathscr{F}_{,j} + \rho \mathscr{F} \left(\frac{dv_i}{dt} - f_i \right) \right] dV$$

(Theorem of stress means.)

5.13. The symmetry of the stress tensor

If the fluid is such that the torques within it arise only as the moments of direct forces we shall call it nonpolar. A polar fluid is one that is capable of transmitting stress couples and being subjected to body torques, as in polyatomic and certain non-Newtonian fluids. For the nonpolar fluid we can make the assumption either that angular momentum is conserved or that the stress tensor is symmetric. We will make the first assumption and deduce the symmetry and then discuss the more general situation that obtains for the polar case.

Since $\mathbf{v} \wedge \mathbf{v} = 0$, $d(\mathbf{x} \wedge \mathbf{v})/dt = \mathbf{x} \wedge \mathbf{a}$, applying the transport theorem (4.3.4) to the equation for moment of momentum (5.11.4) we have

$$\frac{d}{dt} \iiint_V \rho(\mathbf{x} \wedge \mathbf{v})\, dV = \iiint_V \rho(\mathbf{x} \wedge \mathbf{a})\, dV$$
$$= \iiint_V \rho(\mathbf{x} \wedge \mathbf{f})\, dV + \iint_S (\mathbf{x} \wedge \mathbf{t}_{(n)})\, dS. \qquad (5.13.1)$$

This last integral has as its ith component

$$\iint_S \epsilon_{ijk} x_j T_{kp} n_p \, dS = \iiint_V \epsilon_{ijk}(x_j T_{pk})_{,p}\, dV \qquad (5.13.2)$$

by Green's theorem. However, this integrand is

$$\epsilon_{ijk} x_j T_{pk,p} + \epsilon_{ijk} T_{jk},$$

since $x_{j,p} = \delta_{jp}$, and this is the ith component of $\mathbf{x} \wedge (\nabla \cdot \mathbf{T}) + \mathbf{T}_\times$, where $T_\times$ is the vector $\epsilon_{ijk} T_{jk}$. Substituting back into Eq. (5.13.1) and rearranging gives

$$\iiint_V \mathbf{x} \wedge (\rho\mathbf{a} - \rho\mathbf{f} - \nabla \cdot \mathbf{T})\, dV = \iiint_V \mathbf{T}_\times \, dV. \qquad (5.13.3)$$

However, the left-hand side vanishes identically by Cauchy's equation (5.12.5), hence the right-hand side vanishes for an arbitrary volume and so

$$\mathbf{T}_\times \equiv 0. \qquad (5.13.4)$$

However, the components of $\mathbf{T}_\times$ are $(T_{23} - T_{32})$, $(T_{31} - T_{13})$, and $(T_{12} - T_{21})$ and the vanishing of these implies

$$T_{ij} = T_{ji}, \qquad (5.13.5)$$

so that $\mathbf{T}$ is symmetric.

In the case of a polar fluid we must introduce a body torque per unit mass, $\rho\mathbf{g}$, in addition to the body force $\rho\mathbf{f}$, and a couple stress $\mathbf{c}_{(n)}$ in addition to the normal stress $\mathbf{t}_{(\mathbf{n})}$. Then just as $\mathbf{t}_{(\mathbf{n})}$ can be written as $\mathbf{n} \cdot \mathbf{T}$ so $\mathbf{c}_{(\mathbf{n})}$ can be expressed in the form $\mathbf{n} \cdot \mathbf{C}$. The angular momentum must also be conceived to consist of two parts, the moment of linear momentum $\rho\mathbf{x} \wedge \mathbf{v}$ and an intrinsic angular momentum $\rho\mathbf{l}$. Then a balance of total angular momentum gives

$$\frac{d}{dt} \iiint_V \rho(\mathbf{l} + \mathbf{x} \wedge \mathbf{v}) = \iiint_V (\rho\mathbf{g} + \mathbf{x} \wedge \rho\mathbf{f})\, dV + \iint_S (\mathbf{c}_{(\mathbf{n})} + \mathbf{x} \wedge \mathbf{t}_{(\mathbf{n})})\, dS$$
$$= \iiint_V [\rho\mathbf{g} + \mathbf{x} \wedge \rho\mathbf{f} + \nabla \cdot \mathbf{C} + \mathbf{x} \wedge (\nabla \cdot \mathbf{T}) + \mathbf{T}_\times]\, dV. \qquad (5.13.6)$$

This is an equation for the total angular momentum which may be written

$$\rho \frac{d}{dt}[\mathbf{l} + \mathbf{x} \wedge \mathbf{v}] = \rho\mathbf{g} + \mathbf{x} \wedge \rho\mathbf{f} + \nabla \cdot \mathbf{C} + \mathbf{x} \wedge (\nabla \cdot \mathbf{T}) + \mathbf{T}_{\times}. \quad (5.13.7)$$

From the vector product of $\mathbf{x}$ and Cauchy's equation we have

$$\rho\mathbf{x} \wedge \mathbf{a} = \rho \frac{d}{dt}(\mathbf{x} \wedge \mathbf{v}) = \mathbf{x} \wedge \rho\mathbf{f} + \mathbf{x} \wedge (\nabla \cdot \mathbf{T}) \quad (5.13.8)$$

and subtracting we see that

$$\rho \frac{d\mathbf{l}}{dt} = \rho\mathbf{g} + \nabla \cdot \mathbf{C} + \mathbf{T}_{\times}. \quad (5.13.9)$$

Thus the antisymmetric part of the stress tensor contributes to the rate of increase of the internal angular momentum. When the tensor is not symmetric, the external moment of momentum is not conserved in the usual sense for if we integrate the ith component of (5.13.8) throughout the volume V, we have

$$\begin{aligned}\frac{d}{dt}\iiint_V \epsilon_{ijk}x_j v_k \, dV &= \iiint_V \epsilon_{ijk}x_j f_k \, dV + \iiint_V \epsilon_{ijk}x_j T_{pk,p} \\ &= \iiint_V \epsilon_{ijk}x_j f_k \, dV + \iiint_V \epsilon_{ijk}(x_j T_{pk})_{,p} \, dV - \iiint_V \epsilon_{ijk}T_{jk} \, dV \\ &= \iiint_V \epsilon_{ijk}x_j f_k \, dV + \iint_S \epsilon_{ijk}x_j T_{pk}n_p \, dS - \iiint_V \epsilon_{ijk}T_{jk} \, dV.\end{aligned}$$

In dyadic notation this is

$$\frac{d}{dt}\iiint_V \rho(\mathbf{x} \wedge \mathbf{v}) \, dV = \iiint_V (\mathbf{x} \wedge \rho\mathbf{f}) \, dV + \iint_S \mathbf{x} \wedge \mathbf{t}_{(\mathbf{n})} \, dS - \iiint_V T_{\times} \, dV \quad (5.13.10)$$

which shows that there is a loss of external angular momentum per unit volume of $\mathbf{T}_{\times}$ which shows up as a gain in internal angular momentum in Eq. (5.13.9).

Exercise 5.13.1. Apply the result of Ex. 5.11.2 to an elementary parallelepiped to prove the symmetry of the stress tensor.

Exercise 5.13.2. Show that the symmetry of the stress tensor is equivalent to *Cauchy's reciprocal theorem:* Each of two stresses at a point has an equal projection on the normal to the surface on which the other acts.

5.14. Hydrostatic pressure

If the stress system is such that an element of area always experiences a stress normal to itself and this stress is independent of the orientation, the stress is called *hydrostatic*. All fluids at rest exhibit this stress behavior. It implies that $\mathbf{n} \cdot \mathbf{T}$ is always proportional to $\mathbf{n}$ and that the constant of proportionality is independent of $\mathbf{n}$. Let us write this constant $-p$, then

$$n_i T_{ij} = -p n_j. \tag{5.14.1}$$

However, this equation means that any vector is a characteristic vector of $\mathbf{T}$ which must therefore be spherical. Thus

$$T_{ij} = -p\delta_{ij} \tag{5.14.2}$$

for a state of hydrostatic stress.

For a compressible fluid at rest, p may be identified with the pressure of classical thermodynamics. On the assumption that there is local thermodynamic equilibrium even when the fluid is in motion this concept of stress may be retained. For an incompressible fluid the thermodynamic, or more correctly thermostatic, pressure cannot be defined except as the limit of pressure in a sequence of compressible fluids. We shall see later that it has to be taken as an independent dynamical variable.

The stress tensor may always be written

$$T_{ij} = -p\delta_{ij} + P_{ij},$$

and P_{ij} is called the viscous stress tensor. The mean of the three stresses T_{11}, T_{22}, and T_{33} is

$$\tfrac{1}{3} T_{ii} = -p + \tfrac{1}{3} P_{ii}. \tag{5.14.2}$$

In the case of hydrostatic stress, where P_{ij} vanishes, this mean stress equals the thermostatic pressure. We shall show later that for the incompressible Newtonian fluid this is also true, but a distinction must be made in general between the mean stress and the pressure. A perfect fluid is one for which P_{ij} vanishes identically.

Exercise 5.14.1. Show that if the external force field is irrotational then the acceleration is irrotational for a perfect fluid for which p is a function of ρ.

Exercise 5.14.2. Show that an irrotational external force field is equivalent to an assigned hydrostatic pressure.

5.15. Principal axes of stress and the notion of isotropy

The diagonal terms T_{11}, T_{22}, T_{33} of the stress tensor are sometimes called the *direct stresses* and the terms T_{12}, T_{21}, T_{31}, T_{13}, T_{23}, T_{32} the *shear stresses*. When there are no external or stress couples, the stress tensor is symmetric and we can invoke the known properties of symmetric tensors. In particular, there are three principal directions and referred to coordinates parallel to these, the shear stresses vanish. The remaining direct stresses are called the *principal stresses* and the axes *the principal axes of stress*. The mean pressure $-p$ is one third of the trace of the stress tensor and so is the mean of the principal stresses.

All the properties of the canonical representation of the symmetric second order tensor can be applied to the representation of the stress tensor, and one or two of these are given as examples at the end of this section. The essential point to recognize is that in a coordinate system with axes parallel to the principal axes all the stresses are direct stresses.

An *isotropic* fluid is such that a simple direct stress acting in it does not produce a shearing deformation. This is an entirely reasonable view to take for isotropy means that there is no internal sense of direction within the fluid so that a direct stress, say

$$T_{11} \neq 0, \qquad T_{ij} = 0 \qquad i, j \neq 1,$$

should not produce any differential motion in planes parallel to its line of action, in this case the axis $O1$. Another way of expressing the absence of any internally preferred direction is to say that the functional relation between stress and deformation must be independent of the orientation of the coordinate system. We shall show in the next section that this implies that the principal axes of stress and deformation coincide.

Exercise 5.15.1. Show that T_{ij} and P_{ij} have the same principal axes.

Exercise 5.15.2. If the stress tensor is symmetric, $T_{ij}x_ix_j = 1$ is called the Cauchy stress quadric. Show that the normal component of the stress on any plane is inversely proportional to the square of the distance from the center of the quadric to its surface in the direction of the normal to the plane.

Exercise 5.15.3. If the cone $T_{ij}x_ix_j = 0$ is real, show that it divides the space into two regions such that planes with normals in one suffer a compressive stress and those with normals in the other a tension.

5.21. The Stokesian fluid

In deriving the constitutive equations of a nonelastic fluid we shall follow closely the elegant presentation of Serrin (Handbuch der Physik VIII/1, p. 230 et seq.). The fundamental ideas are due to Stokes so that it is appropriate to call a fluid satisfying his hypotheses a Stokesian fluid. Whether or not a given fluid has the constitutive equations we shall derive must be established by experiment; suffice it to say that a large class of real fluids appear to be Stokesian.

We therefore start from the following assumptions:

I. The stress tensor T_{ij} is a continuous function of the deformation tensor e_{ij} and the local thermodynamic state, but independent of other kinematical quantities.

II. The fluid is homogeneous, that is, T_{ij} does not depend explicitly on $\mathbf{x}$.

III. The fluid is isotropic, that is, there is no preferred direction.

IV. When there is no deformation ($e_{ij} = 0$) the stress is hydrostatic, ($T_{ij} = -p\delta_{ij}$).

The first assumption implies that the relation between stress and rate of strain is independent of the rigid body rotation of an element given by the antisymmetric kinematical tensor Ω_{ij}. The thermodynamic variables, for example, pressure and temperature, will be carried along throughout this discussion without specific mention except where it is necessary for emphasis. We are concerned with a homogeneous portion of fluid so we assume in the second place that the stress tensor depends only on position through the variation of e_{ij} and the thermodynamic variables with position. The third assumption is that of isotropy and we shall first show that this implies that the principal directions of the two tensors coincide. To express this as an equation we write $T_{ij} = f_{ij}(e_{pq})$, then if there is no preferred direction, $\bar{T}_{ij}$ is the same function f_{ij} of $\bar{e}_{pq}$ as T_{ij} is of e_{pq}. Thus

$$\bar{T}_{ij} = f_{ij}(\bar{e}_{pq}). \tag{5.21.1}$$

The fourth assumption is that the tensor $P_{ij} = T_{ij} + p\delta_{ij}$ vanishes when there is no motion. P_{ij} is called the viscous stress tensor.

The Stokesian fluid is essentially nonelastic. We shall discuss later (Chapter 8) the case where both viscous and elastic behavior is present.

5.22. Constitutive equations of the Stokesian fluid

If l_{ij} is the set of direction cosines of a rotation of the axes from the system $O123$ to $O\bar{1}\bar{2}\bar{3}$, we have, since T_{ij} and e_{pq} are tensors,

$$\bar{T}_{ij} = l_{mi}l_{nj}T_{mn}, \qquad \bar{e}_{pq} = l_{rp}l_{sq}e_{rs}. \tag{5.22.1}$$

Hence Eq. (5.21.1) becomes

$$l_{mi}l_{nj}f_{mn}(e_{pq}) = f_{ij}(l_{rp}l_{sq}e_{rs}). \tag{5.22.2}$$

Now suppose that the coordinate system has been chosen so that it coincides with the principal axes of e_{ij}. Then we can take $e_{rr} = d_r$, $r = 1, 2$, or 3 and $e_{rs} = 0$, $r \neq s$. Then f_{mn} is a function of d_1, d_2, and d_3. If we take the rotation specified by $l_{11} = 1, l_{22} = l_{33} = -1$, and $l_{ij} = 0, i \neq j$, then the $\bar{e}_{rr} = e_{rr} = d_r$ are unchanged. However, $f_{13} = l_{m1}l_{n3}f_{mn} = -f_{13}$ and so both are zero. Similarly, $f_{12} = f_{21} = f_{31} = 0$ and a similar transformation with $l_{11} = l_{22} = -1$, $l_{33} = 1$ shows that $f_{23} = f_{32} = 0$. Thus in the principal coordinate system of the deformation tensor, the stress tensor has diagonal form and it therefore has the same principal axes. We may therefore write $P_{ii} = p_i$, $P_{ij} = 0, i \neq j$ and

$$p_i = f_i(d_1, d_2, d_3) \tag{5.22.3}$$

where, by the fourth assumption, the f_i must vanish when the d_i vanish.

We now ask what is the most general form that the f_i can have. Since a permutation of d_1, d_2, d_3 can be affected by an orthogonal transformation, such a permutation must permute the functions f_i in the same way. For example the diagonal tensor (d_1, d_2, d_3) is transformed to (d_3, d_1, d_2) by $l_{12} = l_{23} = l_{31} = 1$, other l's zero. Such a relation is given by

$$p_i = \alpha + \beta d_i + \gamma d_i^2, \tag{5.22.4}$$

where α, β, and γ can be functions of the three invariants Θ, Φ, Ψ since these are unaffected by a permutation. Moreover we do not need to assume any higher powers than the square of d_i, for these can be expressed as functions of d_i, d_i^2 and the three invariants. For example,

$$\begin{aligned} d^3 &= \Theta d^2 - \Phi d + \Psi, \\ d^4 &= \Theta d^3 - \Phi d^2 + \Psi d \\ &= (\Theta^2 - \Phi)d^2 - (\Theta\Phi - \Psi)d + \Theta\Psi, \end{aligned}$$

and so on. If the d_i are all different, the three equations (5.22.4) can be solved for α, β, and γ. If two of the d_i are the same, the corresponding p_i will be the same and there will really only be two distinct equations from which α and β may be determined to give a relation $p_i = \alpha + \beta d_i$. If all the d_i are the same, then all the p_i are the same and we have $p_i = \alpha(\Theta, \Phi, \Psi)$.

All cases are thus subsumed under the general formula (5.22.4) which is written for the principal coordinate system. If we transform back to any other system, the functions α, β, and γ must be the same or the requirement of isotropy will not be met. Thus in general

$$P_{ij} = \alpha\delta_{ij} + \beta e_{ij} + \gamma e_{ik}e_{kj} \tag{5.22.5}$$

or

$$T_{ij} = (-p + \alpha)\,\delta_{ij} + \beta e_{ij} + \gamma e_{ik}e_{kj}. \tag{5.22.6}$$

p depends only on the thermodynamic state but α, β, and γ depend as well on the invariants of the rate of strain tensor. This gives ample scope for the fitting of exceedingly complex relations, but the tensorial character is prescribed by the assumptions.

If the fluid is compressible, the thermodynamic pressure is a well-defined quantity and we should take p equal to this. Then, by the fourth assumption, $\alpha = 0$ when $e_{ij} = 0$. If the fluid is incompressible, the thermodynamic pressure is not defined and pressure has to be taken as one of the fundamental dynamical variables. We are at liberty to do this in the simplest possible way so that without losing any generality we can absorb α into the pressure p and write

$$T_{ij} = -p\delta_{ij} + \beta e_{ij} + \gamma e_{ik}e_{kj}, \tag{5.22.7}$$

which insures that T reduces to the hydrostatic form when the deformation vanishes.

Exercise 5.22.1. Show that for an incompressible fluid with the constitutive equation (5.22.7), the mean of the three direct stresses is

$$\begin{aligned} \tfrac{1}{3} \text{ trace } T_{ij} &= -p + \tfrac{1}{3}\gamma e_{ij}e_{ij} \\ &= -p - \tfrac{2}{3}\gamma\Phi. \end{aligned}$$

Exercise 5.22.2. The so-called power law of non-Newtonian fluids asserts that in certain circumstances the shear stress is proportional to a power of the shear strain. Give two different constitutive equations that reduce to

$$P_{12} = [f'(x_2)]^p$$

in the circumstance of Fig. 5.2.

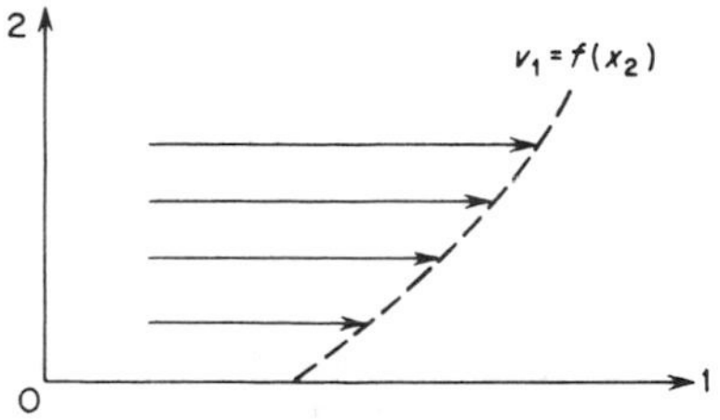

Fig. 5.2

Exercise 5.22.3. A fluid flows along a circular tube with axis parallel to $O1$ its velocity being $v_1 = f(r)$, $v_2 = v_3 = 0$, $r^2 = x_2^2 + x_3^2$. Show that

$$T_{ij} = (-p + \alpha')\begin{bmatrix} 1 & 0 & 0 \\ 0 & 1 & 0 \\ 0 & 0 & 1 \end{bmatrix} + \beta'\begin{bmatrix} 0 & x_2/r & x_3/r \\ x_2/r & 0 & 0 \\ x_3/r & 0 & 0 \end{bmatrix}$$
$$+ \gamma'\begin{bmatrix} 1 & 0 & 0 \\ 0 & (x_2/r)^2 & (x_2x_3/r^2) \\ 0 & (x_2x_3/r^2) & (x_3/r)^2 \end{bmatrix}$$

where α', β', and γ' are functions of $f'(r)$. Hence, calculate the total drag on the coaxial cylindrical surface of radius r and unit length.

Exercise 5.22.4. Treat similarly the case of a fluid between two rotating cylinders where the velocity is wholly circumferential.

5.23. The Newtonian fluid

The Newtonian fluid is a linear Stokesian fluid, that is, the stress components depend linearly on the rates of deformation. Moreover, since the viscous stress tensor must vanish with vanishing d_i, we must have

$$p_i = a_{ij}d_j. \tag{5.23.1}$$

However, we have observed that the assumption of isotropy implies that any permutation of the d's must effect the same permutation of the p's. Writing out these equations in full gives:

$$\begin{aligned} p_1 &= a_{11}d_1 + a_{12}d_2 + a_{13}d_3, \\ p_2 &= a_{21}d_1 + a_{22}d_2 + a_{23}d_3, \\ p_3 &= a_{31}d_1 + a_{32}d_2 + a_{33}d_3. \end{aligned}$$

Now permute the d_1, d_2, d_3 to d_3, d_1, d_2 and rearrange to obtain

$$\begin{aligned} a_{12}d_1 + a_{13}d_2 + a_{11}d_3 &= p_3 \\ a_{22}d_1 + a_{23}d_2 + a_{21}d_3 &= p_1 \\ a_{32}d_1 + a_{33}d_2 + a_{31}d_3 &= p_2. \end{aligned}$$

The right-hand side of each equation has been obtained by making the same permutation on the p's. Now we compare these two sets of equations; for example,

$$p_1 = a_{11}d_1 + a_{12}d_2 + a_{13}d_3 = a_{22}d_1 + a_{23}d_2 + a_{21}d_3,$$

gives $a_{11} = a_{22}$, $a_{12} = a_{23}$, $a_{13} = a_{21}$. Doing this for all, and for the set we could derive by permuting d_1, d_2, d_3 to d_2, d_3, d_1, we find

$$\begin{aligned} a_{11} &= a_{22} = a_{33} \\ a_{12} &= a_{21} = a_{23} = a_{32} = a_{31} = a_{13} \end{aligned}$$

Let the common value of the second row be λ and of the first $\lambda + 2\mu$, where these are for the moment numbers whose physical meaning has to be obtained. Then

$$\begin{aligned} p_i &= \lambda(d_1 + d_2 + d_3) + 2\mu d_i \\ &= \lambda\Theta + 2\mu d_i. \end{aligned} \tag{5.23.2}$$

Transforming to a general coordinate system

$$P_{ij} = \lambda\Theta\delta_{ij} + 2\mu e_{ij} \tag{5.23.3}$$

or

$$T_{ij} = (-p + \lambda\Theta)\delta_{ij} + 2\mu e_{ij}. \tag{5.23.4}$$

We have given this proof in extenso since it is a kind of link between the following two demonstrations. In the first of these we observe that the result follows immediately from imposing linearity on the general relation (5.22.6) for this requires that $\gamma = 0$, β be a constant, and α be proportional to Θ since it must vanish with Θ. Again, since P_{ij} is to be a linear combination of e_{pq}, it can be expressed as a tensor product

$$P_{ij} = A_{ijpq} e_{pq}.$$

Now A_{ijpq} must be symmetric in i and j and must be an isotropic fourth order tensor. However, by Section 2.7 we know that the most general form of A is

$$A_{ijpq} = \lambda \delta_{ij} \delta_{pq} + \mu(\delta_{ip}\delta_{jq} + \delta_{iq}\delta_{jp}) + \nu(\delta_{ip}\delta_{jq} - \delta_{iq}\delta_{jp}).$$

The symmetry requirement is met by putting $\nu = 0$ so that

$$\begin{aligned} P_{ij} &= \lambda \delta_{ij} e_{pp} + \mu(e_{ij} + e_{ji}) \\ &= \lambda \Theta \delta_{ij} + 2\mu e_{ij}. \end{aligned}$$

The first demonstration is a link between the latter two in that the same kind of arguments are used there as were used in establishing the form of the general isotropic tensor. However, since we had already dismissed the antisymmetric part, the argument was distinctly simpler.

5.24. Interpretation of the constants λ and μ

Consider the shear flow given by

$$v_1 = f(x_2), \qquad v_2 = v_3 = 0. \tag{5.24.1}$$

For this we have all the e_{ij} zero except

$$e_{12} = e_{21} = \tfrac{1}{2} f'(x_2). \tag{5.24.2}$$

Thus

$$P_{12} = P_{21} = \mu f'(x_2) \tag{5.24.3}$$

and all the other viscous stresses are zero. This is shown in Fig. 5.2 and it is evident that μ is the proportionality constant relating the shear stress to the velocity gradient. This is the common definition of the viscosity, or more precisely the coefficient of shear viscosity, of a fluid.

For an incompressible fluid we have seen that the pressure is the mean of the principal stresses since this is

$$\tfrac{1}{3} T_{ii} = -p + (\lambda + \tfrac{2}{3}\mu)\Theta$$

and $\Theta = 0$. For a compressible fluid we should take the pressure p as the thermodynamic pressure to be consistent with our ideas of equilibrium. Thus if we call $-\bar{p}$ the mean of the principal stresses,

$$\begin{aligned} \bar{p} - p &= -(\lambda + \tfrac{2}{3}\mu)\Theta = -(\lambda + \tfrac{2}{3}\mu)\nabla \cdot \mathbf{v} \\ &= \left(\lambda + \frac{2}{3}\mu\right) \frac{1}{\rho} \frac{d\rho}{dt}. \end{aligned} \tag{5.24.4}$$

Since p, the thermodynamic pressure, is in principle known from the equation of state, $\bar{p} - p$ is a measurable quantity. Equation (5.24.4) shows that it is proportional to $(d \ln \rho)/dt$ and the constant of proportionality is known as the *coefficient of bulk viscosity*. It is difficult to measure, however, since relatively large rates of change of ρ must be used and the assumption of linearity is then dubious. Stokes assumed that $p = \bar{p}$ and on this ground claimed that

$$\lambda + \tfrac{2}{3}\mu = 0 \qquad (5.24.5)$$

supporting this from an argument from the kinetic theory of gases. While this assumption seems to be reasonable for monatomic gases, it is certainly not true for polyatomic gases or liquids. However, the precise value of λ becomes unimportant for motions that are nearly isochoric or fluids nearly incompressible. For a fuller discussion see the bibliography at the end of this chapter.

BIBLIOGRAPHY

5.11. For a fuller analysis of the distinctions of force see

Truesdell, C. A. and R. Toupin, The classical field theories, in Handbuch der Physik Bd. III/1, 1960, pp. 536 et seq. and references given there.

5.13. Asymmetric stress tensors arise from the molecular theory of gases, see

Dahler, J. S., J. Chem. Phys. **30**, (1959), 1447–1475,

and

Grad, H., Comm. Pure Appl. Math. **5** (1952), 455–494.

The matter is treated by Truesdell in the above reference sects. 60–61, 205. See also

Dahler, J. S. and L. E. Scriven, Angular momentum of continua, Nature, **192**, No. 4797 (Oct. 7, 1961), 36–37.

5.21. Serrin, J., The derivation of stress deformation relations for a Stokesian fluid.

also his article in Handbuch der Physik VIII/1, sections 58–65.

5.24. A summary of the history of Stokes relation $3\lambda + 2\mu = 0$ is given by

Truesdell, C. A., Rat. Mech. and Anal. **1** (1952) pp. 228-231, where many other references are given.

The importance of the second viscosity in sound propagation is discussed by

Landau, L. D. and E. M. Lifshitz, Fluid mechanics, pp. 304–9, Reading, Mass: Addison-Wesley, 1959.

Equations of Motion and Energy in Cartesian Coordinates

6

6.11. Equations of motion of a Newtonian fluid

We now have all the material to assemble equations of fluid motion, for to Cauchy's equation of motion (5.12.4) we may add the constitutive Eq. (5.23.4). We shall do this first for the Newtonian fluid and obtain the equations known as the Navier-Stokes equations.

Cauchy's equation of motion is

$$\rho a_i = \rho \frac{dv_i}{dt} = \rho f_i + T_{ij,j} \tag{6.11.1}$$

for the symmetric stress tensor, which is related to the rate of strain tensor by

$$T_{ij} = (-p + \lambda\Theta)\delta_{ij} + 2\mu e_{ij}. \tag{6.11.2}$$

From the last equation we shall evidently need

$$\begin{aligned} e_{ij,j} &= \frac{1}{2}\frac{\partial}{\partial x_j}\left(\frac{\partial v_i}{\partial x_j} + \frac{\partial v_j}{\partial x_i}\right) = \frac{1}{2}\frac{\partial^2 v_i}{\partial x_j\,\partial x_j} + \frac{1}{2}\frac{\partial}{\partial x_i}\frac{\partial v_j}{\partial x_j} \\ &= \frac{1}{2}\nabla^2 v_i + \frac{1}{2}\frac{\partial}{\partial x_i}(\nabla \cdot \mathbf{v}). \end{aligned}$$

Thus

$$T_{ij,j} = -\frac{\partial p}{\partial x_i} + (\lambda + \mu)\frac{\partial}{\partial x_i}(\nabla \cdot \mathbf{v}) + \mu\nabla^2 v_i \tag{6.11.3}$$

which substituted in Eq. (6.11.1) gives

$$\rho\frac{dv_i}{dt} - \rho f_i - \frac{\partial p}{\partial x_i} + (\lambda + \mu)\frac{\partial}{\partial x_i}(\nabla \cdot \mathbf{v}) + \mu\nabla^2 v_i. \tag{6.11.4}$$

Using notation already familiar this may be written in a number of forms. Equation (6.11.4) is the ith component of

$$\mathbf{a} = \frac{d\mathbf{v}}{dt} = \frac{\partial \mathbf{v}}{\partial t} + (\mathbf{v} \cdot \nabla)\mathbf{v} = \mathbf{f} - \frac{1}{\rho}\nabla p + (\lambda' + \nu)\nabla(\nabla \cdot \mathbf{v}) + \nu\nabla^2\mathbf{v} \tag{6.11.5}$$

where $\nu = \mu/\rho$, $\lambda' = \lambda/\rho$. ν is known as the kinematic viscosity and if Stokes' relation is assumed $\lambda' + \nu = \nu/3$. For an incompressible fluid we have

$$\frac{d\mathbf{v}}{dt} = \mathbf{f} - \frac{1}{\rho}\nabla p + \nu\nabla^2\mathbf{v}, \tag{6.11.6}$$

and for an incompressible inviscid or perfect fluid the equations drop out by setting $\nu = 0$. Using the identity

$$\nabla^2\mathbf{v} = \nabla(\nabla \cdot \mathbf{v}) - \nabla \wedge (\nabla \wedge \mathbf{v}) \tag{6.11.7}$$

(cf. Ex. 3.24.5) the last term is sometimes modified to give

$$\mathbf{a} = \frac{d\mathbf{v}}{dt} = \mathbf{f} - \frac{1}{\rho}\nabla p + (\lambda' + 2\nu)\nabla(\nabla \cdot \mathbf{v}) - \nu\nabla \wedge \mathbf{w}, \tag{6.11.8}$$

or in the incompressible case

$$\mathbf{a} = \mathbf{f} - (\text{grad } p)/\rho - \nu \text{ curl } \mathbf{w}. \tag{6.11.9}$$

This brings out the connection between viscosity and vorticity for we see that for an irrotational flow ($\mathbf{w} = 0$) the viscous term drops out of Eq. (6.11.9) and it reduces to the equation for a perfect fluid.

The equations of hydrostatics are also a very special case obtainable by setting $\mathbf{v} = 0$,

$$\rho\mathbf{f} = \nabla p. \tag{6.11.10}$$

Exercise 6.11.1. Show that for an incompressible fluid

$$\frac{d}{dt}(\rho\mathbf{v}) = \rho\mathbf{f} - \nabla p + \mu\nabla^2\mathbf{v}.$$

Exercise 6.11.2. If $\mathbf{f}$ is irrotational and p is a function of ρ only, show that

$$\mathbf{a} = -\nabla[\Omega + P(p) - (\lambda' + 2\nu)(\nabla \cdot \mathbf{v})] - \nu\nabla \wedge \mathbf{w},$$

where

$$\mathbf{f} = -\nabla\Omega \quad \text{and} \quad P(p) = \int^{p} dp/\rho.$$

$P(p)$ is a strain energy.

Exercise 6.11.3. If the flow is steady (that is, $\partial \mathbf{v}/\partial t = 0$) and $\mathbf{f}$ and p as in the previous question, show that

$$\nabla[\tfrac{1}{2}v^2 + \Omega + P(p) + (\lambda' + 2\nu)(\mathbf{v} \cdot \nabla \ln \rho)] = \mathbf{v} \wedge \mathbf{w} - \nu\nabla \wedge \mathbf{w}$$

where $v^2 = \mathbf{v} \cdot \mathbf{v}$.

Exercise 6.11.4. Show that for the two-dimensional flow of an incompressible Newtonian fluid the vorticity satisfies the diffusion equation

$$\frac{d\mathbf{w}}{dt} = \nu\nabla^2\mathbf{w}.$$

Obtain an analogous equation for three-dimensional flow.

6.12. Boundary conditions

Under ordinary conditions the assumption is made that a viscous fluid sticks to the boundary. Thus if the boundary is moving with known velocity, the fluid velocity is specified and at a stationary boundary $\mathbf{v} = 0$. The existence of solutions of the Navier-Stokes equations is a very deep subject of which comparatively little is known and for which some basic results are only now being obtained.

For an inviscid fluid there may be tangential velocity at the boundary but the normal velocity is specified. The atmosphere at high altitudes exhibits a slip flow though it is not to be regarded as a perfect fluid. The higher terms in the Stokesian constitutive equations may become important and there are other effects present. The name Maxwellian fluid has been coined for a fluid in this condition.

In many problems the stress may be specified on a known or an unknown boundary. In the latter case we have a free boundary problem whose solution requires us to find the form of the boundary.

Exercise 6.12.1. Show that for an incompressible irrotational flow a potential function φ may be found such that $\mathbf{v} = \nabla\varphi$. Deduce that the flow given by specifying the normal velocity on a closed boundary is unique.

6.13. The Reynolds number

From the Navier-Stokes equations there emerges a significant dimensionless number known as the Reynolds number. The technique used to show this is to render the equations dimensionless. This puts the physical principle underlying the equations in its clearest form since they are now free of the arbitrary choice of units. Suppose that a flow is characterized by a certain linear dimension L, a velocity U, and a density ρ_0. For example, if we consider the steady flow round an obstacle, L might be its diameter and U and ρ_0 the velocity and density far from the obstacle. We can make the variables dimensionless by substituting

$$\mathbf{v} = U\mathbf{u}, \qquad \mathbf{x} = L\boldsymbol{\xi}, \qquad t = U\tau/L, \qquad \rho = \sigma\rho_0, \qquad p = \pi\rho_0 U^2,$$
$$\mathbf{f} = U^2\boldsymbol{\varphi}/L. \tag{6.13.1}$$

Then, for example,

$$v_i \frac{\partial v_j}{\partial x_i} \quad \text{becomes} \quad u_i \frac{\partial u_j}{\partial \xi_i} \frac{U^2}{L},$$

$$\frac{1}{\rho}\frac{\partial p}{\partial x_i} \quad \text{becomes} \quad \frac{1}{\sigma}\frac{\partial \pi}{\partial \xi_i}\frac{U^2}{L}, \quad \text{and so on.}$$

Substituting these in Eq. (6.11.4) and for simplicity using Stokes' relation $3\lambda' + 2\nu = 0$, we have

$$\frac{du_i}{d\tau} = \varphi_i - \frac{1}{\sigma}\frac{\partial \pi}{\partial \xi_i} + \frac{\nu}{LU}\left\{\frac{1}{3}\frac{\partial}{\partial \xi_i}\left(\frac{\partial u_j}{\partial \xi_j}\right) + \frac{\partial^2 u_i}{\partial \xi_j\, \partial \xi_j}\right\}. \tag{6.13.2}$$

If ∇' denotes the gradient operator in the dimensionless space variables, we may write this

$$\frac{d\mathbf{u}}{dt} - \boldsymbol{\varphi} + \frac{1}{\sigma}\nabla'\pi = \frac{1}{R}\left\{\frac{1}{3}\nabla'(\nabla' \cdot \mathbf{u}) + \nabla'^2\mathbf{u}\right\}, \tag{6.13.3}$$

where

$$R = UL/\nu. \tag{6.13.4}$$

R is a dimensionless number known as the Reynolds number and it is evidently a measure of the importance of the terms due to viscosity. Thus if R is very large it may be permissible to neglect the terms on the right-hand side of Eqs. (6.13.3) which then reduces to the equation for a perfect fluid. It is important to notice, however, that the nature of the equations for R very large is fundamentally different from the case $(1/R) = 0$. So long as R is finite Eqs. (6.13.3) are second order partial differential equations, but when the right-hand side is set equal to zero we have first order equations.

If we consider a steady incompressible flow with no body forces, the equations of motion can be written

$$(\rho v_i v_j + p\delta_{ij} - 2\mu e_{ij})_{,j} = 0$$

or

$$[R(u_i u_j + \pi\delta_{ij}) - (u_{i,j} + u_{j,i})]_{,j} = 0 \tag{6.13.5}$$

It follows that

$$R(u_i u_j + \pi\delta_{ij}) - (u_{i,j} + u_{j,i})$$

is constant along a streamline. For this reason the Reynolds number is sometimes spoken of as the ratio of the viscous to the inertial stresses. However, particular flow systems must be constructed before this concept has much meaning.

Exercise 6.13.1. An incompressible fluid fills the space between the planes $x_3 = 0$ and $x_3 = L$ and the upper plane is given a velocity $v_1 = U$ so that a steady flow develops. Show that the Reynolds number UL/ν is related to (a) the ratio of the power expended per unit area in moving the upper plane to the mean flow of kinetic energy per unit area across the flow in the fluid, (b) the ratio the shear stress on any plane $x_3 =$ constant to the flow of momentum in that plane. Find these relations.

Exercise 6.13.2. Find similar relations for steady flow through a circular tube.

Exercise 6.13.3. Show that $v_1 v_j = 2\nu e_{1j}$, $j = 2, 3$ for a steady incompressible flow that originates from a region where $v_1 = U$, $v_2 = v_3 = 0$.

6.14. Dissipation of energy by viscous forces

The component of stress on a surface element with normal $\mathbf{n}$ in the direction of the velocity is $(v_i/v)T_{ij}n_j$. The rate at which this stress is doing work is this stress component multiplied by $v\, dS$ and hence the total power for a closed surface S is

$$\iint_S v_i T_{ij} n_j \, dS.$$

However, by the theorem of stress means (Ex. 5.12.2) this is

$$\iiint_V [T_{ij} v_{i,j} + \rho v_i (dv_i/dt) - \rho f_i v_i] \, dV.$$

Rearranging we have

$$\iiint_V \rho \frac{1}{2} \frac{d}{dt} (v^2) \, dV = \frac{d}{dt} \iiint_V \frac{1}{2} \rho v^2 \, dV$$
$$= \iiint_V \rho f_i v_i \, dV - \iiint_V T_{ij} v_{i,j} \, dV + \iint_S v_i t_{(n)i} \, dS. \quad (6.14.1)$$

In words this says that the rate of change of kinetic energy of a material volume is the sum of three parts:

(i) the rate at which the body forces do work,
(ii) the rate at which the internal stresses do work,
(iii) the rate at which the surface stresses do work.

Since T_{ij} is symmetric, the second term may be written

$$-T_{ij} e_{ij} = [p - \lambda\Theta] e_{ii} - 2\mu e_{ij} e_{ij}$$
$$= p\Theta - \lambda\Theta^2 - 2\mu(\Theta^2 - 2\Phi) \quad (6.14.2)$$

where Φ is the second invariant of the deformation tensor. Thus the rate at which kinetic energy per unit volume changes due to the internal stresses to be divided into two parts:

(i) a reversible interchange with strain energy, $p\Theta = -(p/\rho)(d\rho/dt)$,
(ii) a dissipation by viscous forces,

$$-[(\lambda + 2\mu)\Theta^2 - 4\mu\Phi] \tag{6.14.3}$$

Since $\Theta^2 - 2\Phi$ is always positive, this last term is always dissipative. If Stokes' relation is used this term is

$$-\mu[\tfrac{4}{3}\Theta^2 - 4\Phi]; \tag{6.14.4}$$

for incompressible flow it is

$$4\mu\Phi.^* \tag{6.14.5}$$

If $\mathbf{T}$ is not symmetric, we may write

$$\mathbf{T} = \mathbf{T}^s + \mathbf{T}^a$$

where

$$\mathbf{T}^s = \tfrac{1}{2}(\mathbf{T} + \mathbf{T}')$$
$$\mathbf{T}^a = \tfrac{1}{2}(\mathbf{T} - \mathbf{T}').$$

For $v_{i,j}$ we already have a decomposition into symmetric and antisymmetric parts so that

$$T_{ij}v_{i,j} = (T^s_{ij} + T^a_{ij})(e_{ij} + \Omega_{ij}) = T^s_{ij}e_{ij} + T^a_{ij}\Omega_{ij}$$

since the other products vanish identically. Now

$$\Omega_{ij} = \tfrac{1}{2}(v_{i,j} - v_{j,i}) = -\tfrac{1}{2}\epsilon_{ijk}w_k$$

where

$$\mathbf{w} = \nabla \wedge \mathbf{v}, \quad \text{and} \quad T^a_{ij} = \epsilon_{ijk}T_{\times k}$$

where $T_{\times k}$ is the kth component of $\mathbf{T}_\times$ the vector of the antisymmetric part of $\mathbf{T}$. Thus

$$\begin{aligned} T^a_{ij}\Omega_{ij} &= -\tfrac{1}{2}\epsilon_{ijk}\epsilon_{ijp}w_kT_{\times p} \\ &= -\delta_{kp}w_kT_{\times p} \\ &= -\mathbf{w}\cdot\mathbf{T}_\times \end{aligned}$$

Thus the antisymmetric part of the stress tensor does work on the vorticity just as the symmetric part does with the deformation. We shall not pursue these considerations any further here.

* Eq. (6.14.3) is sometimes written $-\mu\Phi$, where Φ is called the dissipation function. We have reserved the symbol Φ for the second invariant, which however is proportional to the dissipation function for incompressible flow. Υ is the symbol used later for the negative of Eq. (6.14.3).

Exercise 6.14.1. Consider the dissipation of energy in the situations of Exs. 6.13.1 and 2.

Exercise 6.14.2. Write down the expression for the dissipation of energy in a plane flow, $v_3 = 0$.

Exercise 6.14.3. Show that the dissipation of energy per unit volume for the flow

$$v_1 = -\frac{\Gamma}{2a}\sinh\frac{2\pi x_2}{a}\Bigg/\left(\cosh\frac{2\pi x_2}{a} - \cos\frac{2\pi x_1}{a}\right)$$

$$v_2 = \frac{\Gamma}{2a}\sin\frac{2\pi x_1}{a}\Bigg/\left(\cosh\frac{2\pi x_2}{a} - \cos\frac{2\pi x_1}{a}\right)$$

is

$$\frac{4\pi^2\Gamma^2}{a^4|\sin 2\pi(x_1 + ix_2)/a|^2}$$

where $i^2 = -1$.

Exercise 6.14.4. For an incompressible Newtonian fluid moving within a fixed stationary boundary with no body forces, show that the total rate of dissipation of kinetic energy is

$$\mu\iiint_V w^2\,dV.$$

6.2. Equations for a Stokesian fluid

For the Stokesian fluid the insertion of the constitutive equations in Cauchy's equation of motion will produce a much more complicated set of equations. From Eq. (5.22.6) we have

$$T_{ij,j} = (-p + \alpha)_{,i} + \beta_{,j}e_{ij} + \beta e_{ij,j} + \gamma_{,j}e_{ik}e_{kj} + \gamma(e_{ik,j}e_{kj} + e_{ik}e_{kj,j}). \tag{6.2.1}$$

The coefficients α, β, and γ are functions of the invariants Θ, Φ, and Ψ and so, for example,

$$\alpha_{,i} = \frac{\partial\alpha}{\partial\Theta}\Theta_{,i} + \frac{\partial\alpha}{\partial\Phi}\Phi_{,i} + \frac{\partial\alpha}{\partial\Psi}\Psi_{,i}.$$

It is not surprising that with equations of this appalling complexity few solutions have been found and that for the special circumstances in which solutions can be obtained, it is best to work more immediately with the stresses themselves.

One point may be worth remarking in connection with the energy dissipation. For the Stokesian fluid we have

$$T_{ij}e_{ij} = (-p + \alpha)e_{ii} + \beta e_{ij}e_{ij} + \gamma e_{ik}e_{kj}e_{ij}$$

of which the first two terms are familiar. To calculate the third we observe that $e_{ik}e_{kj}e_{ji}$ is a scalar (in fact the trace of the cube of the symmetric tensor e_{ij}) and so can be evaluated in any coordinate system. In particular, the principal axes form a suitable one for here

$$e_{ik}e_{kj}e_{ji} = d_1^3 + d_2^3 + d_3^3.$$

However,

$$\begin{aligned} d_1^3 + d_2^3 + d_3^3 &= (d_1 + d_2 + d_3)^3 \\ &\quad - 3(d_1^2 d_2 + d_2^2 d_3 + d_3^2 d_1 + d_1 d_2^2 + d_2 d_3^2 + d_3 d_1^2) - 6d_1 d_2 d_3 \\ &= (d_1 + d_2 + d_3)^3 - 3(d_1 + d_2 + d_3)(d_2 d_3 + d_3 d_1 + d_1 d_2) \\ &\quad + 3d_1 d_2 d_3 \\ &= \Theta^3 - 3\Phi\Theta + 3\Psi, \end{aligned}$$

which gives this term in the dissipation function in terms of the invariants of deformation.

Exercise 6.2.1. A first step away from the Newtonian fluid would a quadratic dependence of the form

$$\alpha = \lambda\Theta + \lambda'\Theta^2 + \lambda''\Phi, \qquad \beta = 2\mu + 2\mu'\Theta, \qquad \gamma = 4\nu.$$

(Notation of Serrin, loc. cit. p. 235). Show that the dissipation term is

$$(4\nu + 2\mu' + \lambda')\Theta^3 - (12\nu + 4\mu' - \lambda'')\Theta\Phi + 12\nu\Psi + (2\mu + \lambda)\Theta^2 - 4\mu\Phi$$

or $4(3\nu\Psi - \mu\Phi)$ in the incompressible case.

Exercise 6.2.2. The name homogeneous N^{th} power fluid might be applied to the case where α, β, and γ are polynomials in Θ, Φ, and Ψ of weight N, $N - 1$, and $N - 2$, respectively. By weight we mean that each term of α is of the form $a_{pqr}\Theta^p\Phi^q\Psi^r$ and $p + 2q + 3r = N$. Show that the dissipation term is a polynomial of weight $N + 1$.

Exercise 6.2.3. Consider the dissipation of energy by viscous forces in the situations of Exs. 5.22.3 and 4.

6.3. The energy equation

We shall attempt no sophisticated formulation of the thermodynamics of flow since it brings out comparatively little of the peculiar virtues of tensor analysis. However, we cannot pass by the formulation of the energy equation entirely since up to this point we have rather more unknowns than equations. In fact we have one continuity equation (involving the density and three velocity components), three equations of motion (involving in addition the pressure and another thermodynamic variable, say the temperature) giving four equations in six unknowns. We have also an *equation of state*, which in incompressible flow asserts that ρ is constant reducing the number of unknowns to five. In the compressible case it is a relation

$$\rho = f(p, T), \tag{6.3.1}$$

which increases the number of equations to five. In either case, there remains a gap of one equation which is filled by the energy equation.

The equations of continuity and motion were derived respectively from principles of conservation of mass and momentum. We now assert the first law of thermodynamics in the form that the increase of total energy (we shall consider only kinetic and internal energies) in a material volume is the sum of the heat transferred and the work done on the volume. Let $\mathbf{q}$ denote the heat flux vector, then, since $\mathbf{n}$ is the outward normal to the surface, $-\mathbf{q} \cdot \mathbf{n}$ is the heat flux into the volume. Let E denote the specific internal energy, then the balance expressed by the first law of thermodynamics is

$$\frac{d}{dt} \iiint_V \rho(\tfrac{1}{2}v^2 + E)\, dV = \iiint_V \rho \mathbf{f} \cdot \mathbf{v}\, dV + \iint_S \mathbf{t}_{(\mathbf{n})} \cdot \mathbf{v}\, dS - \iint_S \mathbf{q} \cdot \mathbf{n}\, dS \tag{6.3.2}$$

This may be simplified by subtracting from it the expression we already have in Eq. (6.14.1) for the rate of change of the kinetic energy. Doing this and using the transport theorem (4.3.4) and Green's theorem

$$\iiint \left[\rho \frac{dE}{dt} + \nabla \cdot \mathbf{q} - \mathbf{T} : (\nabla \mathbf{v}) \right] dV = 0$$

where $\mathbf{T} : (\nabla \mathbf{v})$ is the dyadic notation for $T_{ij}v_{i,j}$ the dissipation of energy by internal stress and the reversible interchange with strain energy. Assuming the continuity of the integrand

$$\rho \frac{dE}{dt} = -\nabla \cdot \mathbf{q} + \mathbf{T} : (\nabla \mathbf{v}). \tag{6.3.3}$$

If we assume Fourier's law for the conduction of heat

$$\mathbf{q} = -k \nabla T \tag{6.3.4}$$

is related to the gradient of temperature T. For a Stokesian fluid we can write

$$\mathbf{T} : (\nabla \mathbf{v}) = -p(\nabla \cdot \mathbf{v}) + \Upsilon \tag{6.3.5}$$

where Υ is the viscous dissipation, which for a Newtonian fluid is

$$\Upsilon = (\lambda + 2\mu)\Theta^2 - 4\mu\Phi. \tag{6.3.6}$$

Substituting back into Eq. (6.3.3) we have

$$\rho \frac{dE}{dt} = \nabla \cdot (k \nabla T) - p(\nabla \cdot \mathbf{v}) + \Upsilon. \tag{6.3.7}$$

Physically we see that the internal energy increases with the influx of heat, the compression and the viscous dissipation.

If we write the equation in the form

$$\rho \frac{dE}{dt} - \frac{p}{\rho}\frac{d\rho}{dt} = \nabla \cdot (k\nabla T) + \Upsilon, \tag{6.3.8}$$

the left-hand side can be transformed by one of the fundamental thermodynamic identities. For if S is the specific entropy,

$$\begin{aligned} T\,dS &= dE + p d(1/\rho) \\ &= dE - (p/\rho^2)\,d\rho, \end{aligned}$$

so that Eq. (6.3.8) is

$$\rho T \frac{dS}{dt} = \nabla \cdot (k\nabla T) + \Upsilon, \tag{6.3.9}$$

giving an equation for the rate of change of entropy. Dividing through Eq. (6.3.9) by T and integrating over a volume gives

$$\begin{aligned} \iiint_V \rho \frac{dS}{dt}\,dV = \frac{d}{dt}\iiint_V \rho S\,dV &= \iiint_V \left[\frac{1}{T}\nabla \cdot (k\nabla T) + \frac{\Upsilon}{T}\right] dV \\ &= \iiint_V \left[\nabla \cdot \left(\frac{k}{T}\nabla T\right) + \frac{k}{T^2}(\nabla T)^2 + \frac{\Upsilon}{T}\right] dV \\ &= \iint_S \frac{k}{T}\frac{\partial T}{\partial n}\,dS + \iiint_V \left[\frac{k}{T^2}(\nabla T)^2 + \frac{\Upsilon}{T}\right] dV. \end{aligned} \tag{6.3.10}$$

Now the first integral is the heat transferred into the material volume divided by the temperature at which it is transferred, that is,

$$\iint_S \frac{-\mathbf{q} \cdot \mathbf{n}}{T}\,dS.$$

The second law of thermodynamics insists that the rate of increase of entropy should not be less than this transfer, and we see that Eq. (6.3.10) is quite consistent with this for the second term on the right-hand side cannot be negative. Indeed it can only be zero if k or ∇T and Υ are zero. For a perfect fluid with no conductivity or viscosity we see that entropy is conserved, for Eq. (6.3.9) becomes

$$\frac{dS}{dt} = 0. \tag{6.3.11}$$

Exercise 6.3.1. If H is the specific enthalpy, show that

$$\rho \frac{dH}{dt} = \nabla \cdot (k\nabla T) + \frac{dp}{dt} + \Upsilon.$$

Exercise 6.3.2. If the specific heats c_p and c_v and the conductivity k of the fluid are constant and it obeys the ideal gas law $p = \rho RT$, show that

$$\rho c_v \frac{dT}{dt} = k\nabla^2 T - p(\nabla \cdot \mathbf{v}) + \Upsilon$$

and

$$\rho c_p \frac{dT}{dt} = k\nabla^2 T + \frac{dp}{dt} + \Upsilon.$$

Exercise 6.3.3. Without assuming Fick's law, show that the second law of thermodynamics will be satisfied if $\Upsilon \geqslant 0$ and $\mathbf{q} \cdot \nabla T \leqslant 0$ and interpret this physically.

Exercise 6.3.4. Modify the energy equation to account for internal sources of heat.

6.41. Résumé of the development of the equations

We have now obtained a sufficient number of equations to match the number of unknown quantities in the flow of a fluid. This does not of course mean that we can solve them nor even that the solutions exist, but it is certainly a necessary beginning. It will be well to review, at this point, the principles that have been used and the assumptions that have been made.

The foundation of the study of fluid motion lies in *kinematics*, the analysis of motion and deformation without reference to the forces that are brought into play. To this is added the concept of mass and the *principle of the conservation of mass*, which leads immediately to the *equation of continuity*,

$$\frac{d\rho}{dt} + \rho(\nabla \cdot \mathbf{v}) = 0. \tag{6.41.1}$$

An analysis of the nature of stress allows us to set up a stress tensor, which together with the *principle of conservation of linear momentum* gives the *equations of motion*

$$\rho \frac{d\mathbf{v}}{dt} = \rho \mathbf{f} + \nabla \cdot \mathbf{T}. \tag{6.41.2}$$

If the conservation of moment of momentum is assumed, it follows that the stress tensor is symmetric, but it is equally permissable to hypothesize the symmetry of the stress tensor and deduce the conservation of moment of momentum. For a certain class of fluids however (here called polar fluids) the stress tensor is not symmetric and there may be an internal angular momentum as well as the external moment of momentum. These may be exchanged subject to the conservation of the total angular momentum.

As yet nothing has been said as to the constitution of the fluid and certain assumptions have to be made as to its behavior. In particular we have noticed the hypotheses of Stokes that lead to the *constitutive equation*

$$T_{ij} = (-p + \alpha)\,\delta_{ij} + \beta e_{ij} + \gamma e_{ik}e_{kj}. \tag{6.41.3}$$

The coefficients in this equation are functions only of the invariants of the deformation tensor and of the thermodynamic state. The latter may be specified by two thermodynamic variables and the nature of the fluid is involved in the *equation of state*, of which one form is

$$p = f(\rho, T). \tag{6.41.4}$$

Finally, the *principle of the conservation of energy* is used to give an *energy equation.* In this, certain assumptions have to be made as to the energy transfer and we have only considered the conduction of heat, giving

$$\rho\frac{dE}{dt} = \nabla\cdot(k\nabla T) - p(\nabla\cdot\mathbf{v}) + \Upsilon. \tag{6.41.5}$$

These equations are both too general and too special. They are too general in the sense that they have to be simplified still further before any large body of results can emerge. They are too special in the sense that we have made some rather restrictive assumptions on the way, excluding for example elastic and electromagnetic effects. We shall broaden our view in a later chapter to consider a reacting mixture, but more than this is beyond our present scope. Rather we will consider some of the specializations of the equations and a few results that exhibit the value of vector and tensor analysis.

6.42. Special cases of the equations

The full equations may be specialized in several ways, of which we shall consider the following:

(i) restrictions on the type of motion,
(ii) specializations of the equations of motion,
(iii) specializations of the constitutive equation or equation of state.

This classification is not the only one and the classes will be seen to overlap. We shall give a selection of examples and of the resulting equations, but the list is by no means exhaustive.

Under the first heading we have any of the specializations of the velocity as a vector field. These are essentially kinematic restrictions. We have already met several of these and the following selection is not exhaustive.

(ia) *Isochoric motion.* The velocity field is solenoidal

$$\Theta = \nabla\cdot\mathbf{v} = 0. \tag{6.42.1}$$

The equation of continuity now gives

$$\frac{d\rho}{dt} = 0, \tag{6.42.2}$$

that is, the density does not change following the motion. This does not mean that it is uniform, though, if the fluid is incompressible so that $\rho =$ constant, the motion is isochoric. The other equations simplify in this case for we have α, β, and γ functions of only Φ and Ψ, in particular for a Newtonian fluid

$$T_{ij} = -p\delta_{ij} + 2\mu e_{ij}, \tag{6.42.3}$$

The energy equation is

$$\rho \frac{dE}{dt} = \nabla \cdot (k\nabla T) + \Upsilon, \tag{6.42.4}$$

and for a Newtonian fluid

$$\Upsilon = -4\mu\Phi. \tag{6.42.5}$$

(ib) *Irrotational motion.* The velocity field is irrotational

$$\mathbf{w} = \nabla \wedge \mathbf{v} = 0. \tag{6.42.6}$$

It follows that there exists a velocity potential $\varphi(\mathbf{x}, t)$ from which the velocity can be derived as

$$\mathbf{v} = \nabla\varphi, \tag{6.42.7}$$

and in place of the three components we seek only one scalar function. The continuity equation becomes

$$\frac{d\rho}{dt} + \rho\nabla^2\varphi = 0 \tag{6.42.8}$$

so that for an isochoric motion or an incompressible fluid, φ is a potential function, satisfying

$$\nabla^2\varphi = 0. \tag{6.42.9}$$

The Navier-Stokes equations become

$$\nabla\left[\frac{\partial\varphi}{\partial t} + \frac{1}{2}(\nabla\varphi)^2\right] = \mathbf{f} - \frac{1}{\rho}\nabla p + (\lambda' + 2\nu)\nabla(\nabla^2\varphi). \tag{6.42.10}$$

In the case of an irrotational body force $\mathbf{f} = -\nabla\Omega$ and when p is a function only of ρ, this has an immediate first integral since every term is a gradient. Thus, if $P(\rho) = \int dp/\rho$,

$$\frac{\partial\varphi}{\partial t} + \tfrac{1}{2}(\nabla\varphi)^2 + \Omega + P(\rho) - (\lambda' + 2\nu)\nabla^2\varphi = g(t) \tag{6.42.11}$$

is a function of time only.

(ic) *Complex lamellar motions, Beltrami motions, etc.* These names can be applied when the velocity field is of this type. Various simplifications are possible by expressing the velocity in terms of scalar fields. We shall not discuss them further here.

(id) *Plane flow.* Here the motion is restricted to two dimensions which may be taken to be the $O12$ plane. Then $v_3 = 0$ and x_3 does not occur in the equations. If the fluid is incompressible,

$$\frac{\partial v_1}{\partial x_1} + \frac{\partial v_2}{\partial x_2} = 0 \tag{6.42.12}$$

so that we may introduce a stream function $\psi(x_1, x_2)$ such that,

$$v_1 = \frac{\partial \psi}{\partial x_2}, \qquad v_2 = -\frac{\partial \psi}{\partial x_1}. \tag{6.42.13}$$

The vorticity has only a single component, that in the $O3$ direction, which we may write without suffix

$$w = \frac{\partial v_2}{\partial x_1} - \frac{\partial v_1}{\partial x_2} = -\nabla^2 \psi. \tag{6.42.14}$$

For an incompressible, irrotational motion we therefore have

$$v_1 = \frac{\partial \varphi}{\partial x_1} = \frac{\partial \psi}{\partial x_2}, \qquad v_2 = \frac{\partial \varphi}{\partial x_2} = -\frac{\partial \psi}{\partial x_1} \tag{6.42.15}$$

which are the Cauchy-Riemann relations and show that $\omega = \varphi + i\psi$ is an analytic function of $z = x_1 + ix_2$. The whole resources of the theory of functions of a complex variable are thus available and many solutions are known.

If the fluid is compressible but the flow is steady (that is, no quantity depends on t) the equation of continuity becomes

$$\frac{\partial}{\partial x_1}(\rho v_1) + \frac{\partial}{\partial x_2}(\rho v_2) = 0. \tag{6.42.16}$$

A stream function can again be introduced, this time in the form

$$v_1 = \frac{1}{\rho}\frac{\partial \psi}{\partial x_2}, \qquad v_2 = -\frac{1}{\rho}\frac{\partial \psi}{\partial x_1}. \tag{6.42.17}$$

The vorticity is now given by

$$\rho w = -\nabla^2 \psi + (\nabla \psi \cdot \nabla \rho)/\rho. \tag{6.42.18}$$

For irrotational, isentropic flow with no body forces this can be developed into an equation for ψ alone. The velocity of sound c is defined by $dp/d\rho$ and in isentropic flow

$$\nabla p = c^2 \nabla \rho.$$

Since the fluid must be inviscid or entropy would not be conserved Eq. (6.42.10) can be written

$$\nabla p = -\rho \nabla(\tfrac{1}{2}v^2). \tag{6.42.19}$$

Also we have identically

$$\begin{aligned} \nabla[\tfrac{1}{2}(\nabla \psi)^2] = \nabla(\tfrac{1}{2}\rho^2 v^2) &= \rho v^2 \nabla \rho + \rho^2 \nabla(\tfrac{1}{2}v^2) \\ &= \rho(v^2 - c^2)\nabla \rho \end{aligned} \tag{6.42.20}$$

by the last two equations. Substituting for $\nabla \rho$ in Eq. (6.42.18) with $w = 0$, and expanding

$$\left[\rho^2 c^2 - \left(\frac{\partial \psi}{\partial x_2}\right)^2\right] \frac{\partial^2 \psi}{\partial x_1^2} + 2 \frac{\partial \psi}{\partial x_1} \frac{\partial \psi}{\partial x_2} \frac{\partial^2 \psi}{\partial x_1 \partial x_2} + \left[\rho^2 c^2 - \left(\frac{\partial \psi}{\partial x_1}\right)^2\right] \frac{\partial^2 \psi}{\partial x_2^2} = 0. \quad (6.42.21)$$

This is a partial differential equation for ψ, more complicated than Laplace's but still amenable to analysis.

(ie) *Axisymmetric flows.* Here the flow is a function only of a coordinate along one axis, say $z = x_3$, and the distance from it $\varpi = (x_1^2 + x_2^2)^{1/2}$. If u and v denote the velocity components in the ϖ and z directions the continuity equation is

$$\varpi \frac{\partial \rho}{\partial t} + \frac{\partial}{\partial \varpi}(\rho \varpi u) + \frac{\partial}{\partial z}(\rho \varpi v) = 0. \quad (6.42.22)$$

For steady flow a stream function can be introduced as in the case of plane flow and similar but more complicated equations follow.

Under the second type of restriction where simplifications are introduced into the equations of motion, we may mention the following:

(iia) *Steady flow.* Examples of this have already been given and indeed it might have been considered as a restriction of the first class. All partial derivatives with respect to time vanish and the material derivative

$$\frac{d}{dt} = \mathbf{v} \cdot \nabla.$$

In particular, the continuity equation is

$$\nabla \cdot (\rho \mathbf{v}) = 0 \quad (6.42.23)$$

so that the mass flux field is solenoidal.

(iib) *Hydrostatics.* This is the ultimate in steady flows when $\mathbf{v}$ itself is set equal to zero, then

$$\rho \mathbf{f} = \nabla p. \quad (6.42.24)$$

(iic) *Creeping flow.* It is sometimes justifiable to assume that the velocity is so small that the square of the velocity is negligible by comparison with the velocity itself. This linearizes the equations and allows them to be solved more readily. For example, the Navier-Stokes equations become

$$\frac{\partial v}{\partial t} = \mathbf{f} - \frac{1}{\rho} \nabla p + (\lambda' + \nu) \nabla(\nabla \cdot \mathbf{v}) + \nu \nabla^2 \mathbf{v}. \quad (6.42.25)$$

In particular, for a steady incompressible flow with no body forces

$$\nabla p = \mu \nabla^2 \mathbf{v}. \quad (6.42.26)$$

However, since the continuity equation is $\nabla \cdot \mathbf{v} = 0$, we have

$$\nabla^2 p = 0 \quad (6.42.27)$$

or p is a potential function. This is the starting point for Stokes' solution of the creeping flow about a sphere and for its various improvements.

Another specialization of this type arises in stability theory when the basic flow is known but is perturbed by a small amount. Here it is the squares and products of the small perturbations that are regarded as negligible.

(iid) *Boundary layer flows.* A circumstance that arises when large gradients of velocity are confined to the neighborhood of a boundary has attracted considerable attention. Here it proves possible to neglect certain terms in the equations of motion by comparison with others. The basic case of steady incompressible flow in two dimensions will be outlined. If a rigid barrier extends along the positive $O1$ axis the velocity components v_1 and v_2 are both zero there. In the region distant from the axis the flow is $v_1 = U(x_1)$, $v_2 = 0$, and the velocity distribution may be expected

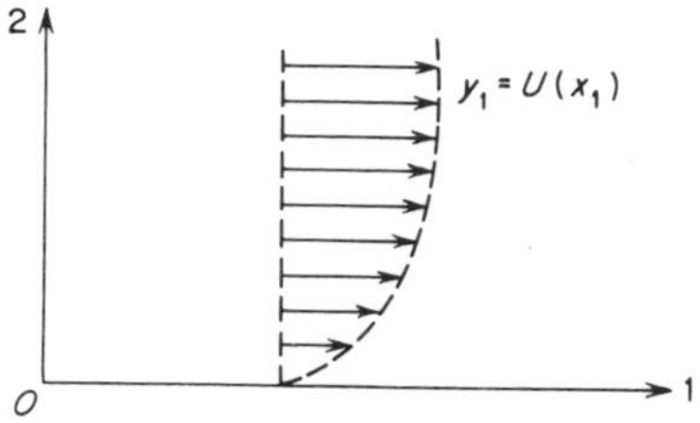

Fig. 6.1

to be of the form shown in Fig. 6.1, in which v_1 differs from U and v_2 from zero only within a comparatively short distance of the plate. To express this we suppose that L is a typical dimension along the plate and δ a typical dimension of this boundary layer and that V_1 and V_2 are typical velocities of the order of magnitude of v_1 and v_2. We then introduce dimensionless variables

$$x = \frac{x_1}{L}, \qquad y = \frac{x_2}{\delta}, \qquad u = \frac{v_1}{V_1}, \qquad v = \frac{v_2}{V_2}, \tag{6.42.28}$$

which will be of the order of unity. This in effect is a stretching upward of the coordinates so that we can compare orders of magnitude, for now all dimensionless quantities will be of order of magnitude one. It is assumed that $\delta \ll L$, but the circumstances under which this is valid will only become apparent later. It is also assumed that the functions are reasonably smooth and no vast variations of gradient occur.

The equation of continuity becomes

$$\frac{V_1}{L}\frac{\partial u}{\partial x} + \frac{V_2}{\delta}\frac{\partial v}{\partial y} = 0 \tag{6.42.29}$$

which would lose its meaning if one of these terms were completely negligible in comparison with the other. It follows that

$$V_2 = O\left(V_1 \frac{\delta}{L}\right), \tag{6.42.30}$$

where the symbol O means "is of the order of." The Navier-Stokes equations become

$$\frac{V_1^2}{L} u \frac{\partial u}{\partial x} + \frac{V_1 V_2}{\delta} v \frac{\partial u}{\partial y} = -\frac{1}{\rho L}\frac{\partial p}{\partial x} + \nu \frac{V_1}{L^2}\left(\frac{\partial^2 u}{\partial x^2} + \frac{L^2}{\delta^2}\frac{\partial^2 u}{\partial y^2}\right)$$

and

$$\frac{V_1 V_2}{L} u \frac{\partial v}{\partial x} + \frac{V_2^2}{\delta} v \frac{\partial v}{\partial y} = -\frac{1}{\rho}\frac{\partial p}{\partial y} + \nu \frac{v_2^2}{\delta^2}\left(\frac{\delta^2}{L^2}\frac{\partial^2 v}{\partial x^2} + \frac{\partial^2 v}{\partial y^2}\right).$$

In the first of these equations the last term on the right-hand side dominates the Laplacian and $\partial^2 u/\partial x^2$ can be neglected. Dividing through by V_1^2/L we see that the other terms will be of the same order of magnitude provided

$$p = O(\rho V_1^2) \quad \text{and} \quad \frac{L\nu}{\delta^2 V_1} = O(1). \tag{6.42.31}$$

Inserting these orders of magnitude in the second equation the dominant term is $\partial p/\partial y$ so that p is a function of x only. Returning to the original variables we have the equations

$$\frac{\partial v_1}{\partial x_1} + \frac{\partial v_2}{\partial x_2} = 0,$$

$$v_1 \frac{\partial v_1}{\partial x_1} + v_2 \frac{\partial v_1}{\partial x_2} = -\frac{1}{\rho}\frac{dp}{dx} + \nu \frac{\partial^2 v_1}{\partial x_2^2}. \tag{6.42.32}$$

The circumstances under which these simplified equations are valid are given by the second term of Eq. (6.42.31), which can be written

$$\frac{\delta}{L} = O\left(\sqrt{\frac{\nu}{V_1 L}}\right). \tag{6.42.33}$$

Since it has been assumed that $\delta \ll L$ this equation shows that this will be the case if $\nu \ll V_1 L$.

Certain of the specializations of the third class have turned up already in the previous cases. We mention here a few important cases.

(iiia) *Incompressible fluid.* The motion of an incompressible fluid is always isochoric and the considerations of (ia) apply. It should be remembered that for an incompressible medium the pressure is not defined thermodynamically, but is an independent variable of the motion.

(iiib) *Perfect fluid.* A perfect fluid has no viscosity so that

$$T_{ij} = -p\delta_{ij}$$

and

$$\rho \frac{d\mathbf{v}}{dt} = \rho \mathbf{f} + -\nabla p \tag{6.42.34}$$

If, in addition, the fluid has zero conductivity the energy equation becomes

$$\frac{dS}{dt} = 0 \qquad (6.42.35)$$

and the flow is *isentropic.*

(iiic) *Ideal gas.* An ideal gas is a perfect fluid with the equation of state

$$p = \rho RT. \qquad (6.42.36)$$

The entropy of an ideal gas is given by

$$S = \int c_v \frac{dT}{T} - R \ln \rho \qquad (6.42.37)$$

which for constant specific heats gives

$$p = e^{S/c_v}\rho^\gamma, \qquad \gamma = c_p/c_v. \qquad (6.42.38)$$

(iiid) *Piezotropic fluid and barotropic flow.* When the pressure and density are directly related, the fluid is said to be piezotropic. If the motion is such that the density and pressure are directly related (for example, for an isentropic flow Eq. (6.42.38) gives $p = k\rho^\gamma$) the motion is called barotropic. Thus all piezotropic fluids (and this includes incompressible fluids as a special case) flow barotropically, but other fluids may also do so. The terms piezotropic and barotropic thus stand in the same relationship as incompressible and isochoric. The simple relation between p and ρ allows us to write

$$\frac{1}{\rho}\nabla p = \nabla P(\rho) = \nabla \int^p \frac{dp}{\rho} \qquad (6.42.39)$$

and such relations as we have noticed above in Eq. (6.42.11) can be obtained from the equations of motion. We shall have occasion to notice these in more detail in the next section.

(iiie) *Newtonian fluids.* Here the assumption of a linear relation between stress and strain leads to the constitutive equation

$$T_{ij} = (-p + \lambda\Theta)\delta_{ij} + 2\mu e_{ij}. \qquad (6.42.40)$$

The equations of motion then become the Navier-Stokes equations (6.11.5) or (6.11.6).

Exercise 6.42.1. Show that for an incompressible, irrotational flow with velocity potential φ, the dissipation function for a Newtonian fluid is

$$\Upsilon = \nabla^2(\nabla\varphi)^2 = \tfrac{1}{2}\nabla^4\varphi^2.$$

Exercise 6.42.2. Obtain an equation similar to Eq. (6.42.21) for the velocity potential of a plane, irrotational, isentropic flow. Show that both can be written

$$(c^2 - v_1^2)\frac{\partial^2\varphi}{\partial x_1^2} - 2v_1v_2\frac{\partial^2\varphi}{\partial x_1\,\partial x_2} + (c^2 - v_2^2)\frac{\partial^2\varphi}{\partial x_2^2} = 0.$$

Exercise 6.42.3. Transform the Navier-Stokes equations in the axisymmetric case (ie).

Exercise 6.42.4. Show that for axisymmetric flow the vorticity

$$w = \frac{1}{\varpi^2}\frac{\partial \psi}{\partial \varpi} - \frac{1}{\varpi}\left[\frac{\partial^2 \psi}{\partial \varpi^2} + \frac{\partial^2 \psi}{\partial z^2}\right]$$

where

$$u = \frac{1}{\rho\varpi}\frac{\partial \psi}{\partial z} \qquad v = -\frac{1}{\rho\varpi}\frac{\partial \psi}{\partial \varpi}$$

Exercise 6.42.5. The plane parallel flow $v_1 = U(x_2)$, $v_2 = 0$, $p = P(x_1, x_2)$ of an incompressible fluid satisfies the Navier-Stokes equations. It suffers a small disturbance so that

$$v_1 = U + v_1', \qquad v_2 = v_2', \qquad p = P + p'$$

Neglecting squares and products of the disturbed quantities show that they satisfy the equations

$$\frac{\partial v_1'}{\partial x_1} + \frac{\partial v_2'}{\partial x_2} = 0,$$

$$\frac{\partial v_1'}{\partial t} + U\frac{\partial v_1'}{\partial x_1} + v_2'\frac{dU}{dx_2} = -\frac{1}{\rho}\frac{\partial p'}{\partial x_1} + \nu\nabla^2 v_1',$$

$$\frac{\partial v_2'}{\partial t} + U\frac{\partial v_2'}{\partial x_1} = -\frac{1}{\rho}\frac{\partial p'}{\partial x_2} + \nu\nabla^2 v_2'.$$

6.51. Bernoulli theorems

Consider the steady barotropic flow of a perfect fluid with irrotational body forces and set

$$\mathbf{f} = -\nabla\Omega, \qquad P(p) = \int^p dp/\rho. \tag{6.51.1}$$

Then the equation of motion is

$$(\mathbf{v}\cdot\nabla)\mathbf{v} = -\nabla(\Omega + P(p))$$

and, in virtue of the identity

$$(\mathbf{v}\cdot\nabla)\mathbf{v} = \nabla(\tfrac{1}{2}v^2) + \mathbf{w}\wedge\mathbf{v},$$

this can be written

$$\nabla(\Omega + P(p) + \tfrac{1}{2}v^2) = \mathbf{v}\wedge\mathbf{w}. \tag{6.51.2}$$

Let H denote the function of which the gradient occurs on the left-hand side of this equation. ∇H is a vector normal to the surfaces of constant H. However, $\mathbf{v}\wedge\mathbf{w}$ is a vector perpendicular to both $\mathbf{v}$ and $\mathbf{w}$ so that these vectors are tangent to the surface. However, $\mathbf{v}$ and $\mathbf{w}$ are tangent to the streamlines and vortex lines respectively so that these must lie in a surface of constant H. It follows that H is constant along streamlines and vortex lines. The surfaces of constant H which are crossed with this network of stream and vortex lines are known as Lamb surfaces and are illustrated in Fig. 6.2.

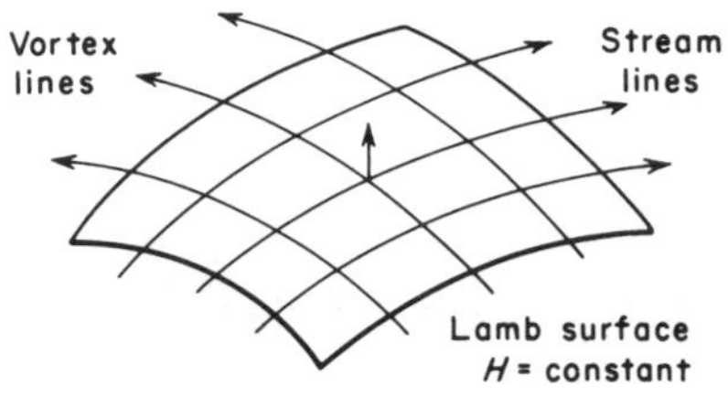

Fig. 6.2

If the flow is irrotational, then $\mathbf{w} = 0$ and hence the energy function

$$H \equiv \Omega + P(p) + \tfrac{1}{2}v^2 \quad (6.51.3)$$

is constant everywhere. This represents a complete integral of the equations of motion and is a result of considerable importance. In general, however, we can only state that for steady barotropic flow H is constant along any streamline, or less importantly, along any vortex line.

Other forms of this theorem, which is known by the name of Bernoulli, can be derived for different flows. Thus for an unsteady irrotational flow a time dependent velocity potential $\varphi(\mathbf{x}, t)$ may be introduced with $\mathbf{v} = \nabla\varphi$. Then

$$\frac{\partial\varphi}{\partial t} + H = f(t) \quad (6.51.4)$$

is a function of time only. Another form has been given for irrotational motion of a Newtonian fluid in Eq. (6.42.11) above. It may be written

$$\frac{\partial\varphi}{\partial t} + H - (\lambda' + 2\nu)\nabla^2\varphi = g(t). \quad (6.51.5)$$

Yet other conditions lead to theorems of this type. Serrin (Handbuch der Physik VIII/1, p. 153) shows that a similar result obtains if the vorticity rather than the flow is steady. Bird (Chem. Engng. Sci. *6*, (1957), 123.) has obtained what he calls the "engineering Bernoulli equation" for the flow through a process in which work is done and friction losses are suffered.

6.52. Some further properties of barotropic flow

We have already seen some of the general properties of vorticity which follow from the fact that the vorticity field is solenoidal. For example, that the strength of a vortex tube is constant and that vortex lines are material lines was noticed above. For steady barotropic flow we also have a theorem due to Kelvin which shows that the circulation around any material closed curve is constant. For

$$\frac{d}{dt}\oint \mathbf{v}\cdot\mathbf{t}\,ds = \frac{d}{dt}\oint \mathbf{v}\cdot d\mathbf{x} = \oint \mathbf{a}\cdot d\mathbf{x} + \oint \mathbf{v}\cdot d\mathbf{v}.$$

The second of these vanishes identically since it is the integral round a closed curve of $\frac{1}{2}v^2$. For barotropic flow of a perfect fluid $\mathbf{a} = \nabla H$ so that $\mathbf{a} \cdot d\mathbf{x} = dH$ and again the integral round a closed circuit must vanish. Hence the circulation remains constant.

If each particle comes from a region of quiet, then a steady barotropic flow is irrotational, since $H = 0$ in a quiescent region and so $\mathbf{w} \wedge \mathbf{v} = 0$ for the whole particle path. However, this means that $\mathbf{w} = k\mathbf{v}$ where k is a scalar variable and since $\nabla \cdot \mathbf{w} = 0$, $\nabla \cdot (k\mathbf{v}) = \rho\mathbf{v} \cdot \nabla(k/\rho) = 0$ and so k/ρ is constant on streamlines. However, if $\mathbf{w}/\rho\mathbf{v}$ is constant on a streamline for which at infinity $\mathbf{v} = 0$, then $\mathbf{w} = 0$ there and since vortex lines are material the flow must be irrotational everywhere.

BIBLIOGRAPHY

6.12. For references on the problem of the existence of solutions of the Navier-Stokes equations see

Serrin, J., Mathematical principles of classical fluid mechanics, in Handbuch der Physik VIII/1, 1959, p. 252.

The Maxwellian fluid is discussed in

Truesdell, C., J. Rat. Mech. and Anal. *1*, (1952), p. 245 et seq.

6.13. The elements of a rigorous development of thermodynamics are to be found in Gibbs' writings. It is only recently, however, that the thermodynamics required for a rational fluid mechanics has been set on a fairly complete and rigorous basis. See

Noll, W. and Coleman, B. D. On the thermostatics of continuous media. Archiv. for Rational Mechanics *4*, (1959), p. 97–128.

Tensors 7

7.11. Coordinate systems and conventions

Let us return for a moment to the style of writing the Cartesian coordinates of a point P as (x, y, z) and consider two other familiar coordinate systems. x, y, and z are the perpendicular distances from the point P to the three coordinate planes through the origin O. They are not the only coordinates that can be used to fix the position of P however, for we might retain z as the height of P above its projection Q on the Oxy plane and take any other pair of coordinates to fix the position of Q in the plane. In particular, Q can be fixed by its distance ρ from the origin and the angle ϕ that OQ makes with a fixed direction, say Ox. This gives a system of coordinates (ρ, ϕ, z) known as cylindrical polars. They are related to the Cartesian coordinates by the equations

$$\begin{aligned} x &= \rho \cos \phi, & y &= \rho \sin \phi, \\ \rho &= (x^2 + y^2)^{1/2}, & \phi &= \tan^{-1}(y/x). \end{aligned} \tag{7.11.1}$$

An alternative description would be to take the distance of P from the origin as one coordinate r. This puts P on a sphere of a certain radius. If as second coordinate we take θ, the angle between OP and Oz, this confines P to a certain latitude on the sphere. On this circle of latitude the position of P can be fixed by taking ϕ again as a coordinate. Thus

$$x = r \sin \theta \cos \phi, \qquad y = r \sin \theta \sin \phi, \qquad z = r \cos \theta,$$

or (7.11.2)

$$r = (x^2 + y^2 + z^2)^{1/2}, \qquad \theta = \tan^{-1}(x^2 + y^2)^{1/2}/z, \qquad \phi = \tan^{-1}(y/x).$$

This system, known as spherical polar coordinates, is another example of curvilinear coordinates, in contrast with the rectilinear Cartesian coordinates. These names arise from the nature of the lines along which only one coordinate varies. In Cartesians these coordinate lines are straight and parallel to the coordinate axes. In cylindrical polars the lines along which ρ, ϕ, and z alone vary are respectively rays through the Oz axis, circles parallel to the Oxy plane and lines parallel to the Oz axis. In spherical polars the coordinate lines are rays through the origin, circles in planes containing Oz and circles parallel to the plane Oxy. In the second and third cases not all the coordinate lines are straight.

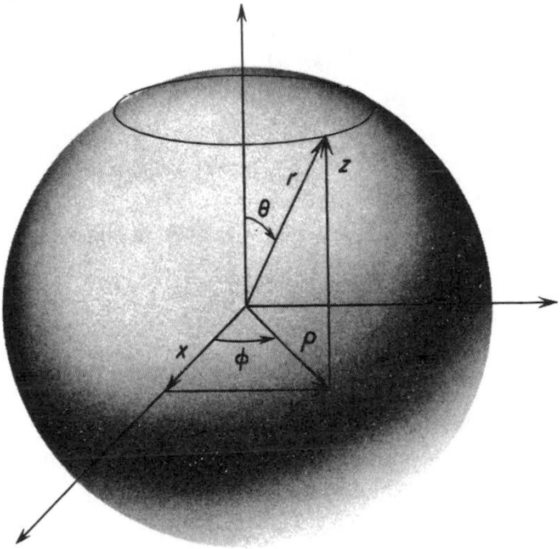

Fig. 7.1

The coordinate surface are those on which only one of the coordinates is constant. For Cartesian coordinates these are all planes but for the other coordinate systems we have cylinders, cones, and spheres appearing as coordinate surfaces. Now such curvilinear coordinate systems have great value and we want to be able to use them and transform from one to another. Moreover, we want to represent physical entities in such a way that when changing the coordinate system our description changes in such a way that we are sure the entity itself has not changed. Up to this point the only transformations we have considered have been rotations of a Cartesian frame of reference, but from now on we need to be able to handle more general transformations.

Before describing the essential features of these transformations we will introduce some new conventions. We shall have occasion to distinguish two types of behavior under transformation and we associate them with affixes in the upper and lower positions. This distinction does not arise when the transformations are only the rotation of Cartesian frames. In the future we shall write the coordinates with an affix in the upper position. Thus x^i, $i = 1, 2, 3$, will stand for the coordinates of P in a certain system in place of the x_i used hitherto for Cartesian systems. Thus we might have $x^1 = \rho$, $x^2 = \phi$, $x^3 = z$ in cylindrical polars. If we transform to another system a bar will again be used to denote the transformed coordinates. For example if the transformed system is that of spherical polars $\bar{x}^1 = r$, $\bar{x}^2 = \theta$, $\bar{x}^3 = \phi$, and the equations defining the transformation are

$$\bar{x}^1 = [(x^1)^2 + (x^3)^2]^{1/2}, \qquad \bar{x}^2 = \tan^{-1}(x^1/x^3), \qquad \bar{x}^3 = x^2. \quad (7.11.3)$$

The summation convention will also be modified as follows:

Summation convention: Any index repeated once in the upper position and once in the lower position in a product of terms is called a dummy index and held to be summed over the range of its values. In three-dimensional space we shall use Latin letters for affixes which will therefore have the range 1, 2, 3. Any index not repeated is called free and may take any value in its range.

In using tensors there is a very convenient check for accidental errors which might be called the conservation of indices. Any free index which appears must appear in the same position in each term of an equation. Thus a formula

$$A^{ij}_k = B_{ip}C^{pj}_q D^q + E^i F^j_{kp} G^p$$

would be obviously incorrect for in the first term on the right-hand side the i is in the wrong position and the k does not appear; the second term however is consistent. The letter used for a dummy index can of course be changed at will, but it cannot be the same as a free index without gross confusion. An upper affix in a denominator counts as a lower index in the numerator and vice versa. Thus, if we were to need a symbol for $\partial/\partial x^i$, we would write $D_i = \partial/\partial x^i$ with its affix in the lower position.

Exercise 7.11.1. Describe the coordinate lines and surfaces of the following coordinate systems (x, y, z are Cartesian coordinates):

(i) $x = x^1x^2$, $y = [\{(x^1)^2 - d^2\}\{1 - (x^2)^2\}]^{1/2}$, $z = x^3$,

(ii) $2x = (x^1)^2 - (x^2)^2$, $y = x^1x^2$, $z = x^3$,

(iii) $x = x^1x^2x^3$, $y = x^1x^2\{1 - (x^3)^2\}^{1/2}$, $2z = (x^1)^2 - (x^2)^2$,

(iv) $x = x^3[\{(x^1)^2 + d^2\}\{1 - (x^2)^2\}]^{1/2}$,
$y = [\{(x^1)^2 + d^2\}\{1 - (x^2)^2\}\{1 - (x^3)^2\}]^{1/2}$, $z = x^1x^2$.

(v) $x(x^1 - x^2) = ax^3\{(x^1)^2 - 1\}^{1/2}$
$y(x^1 - x^2) = a[\{(x^1)^2 - 1\}\{1 - (x^3)^2\}]^{1/2}$
$z(x^1 - x^2) = a\{1 - (x^2)^2\}^{1/2}$

7.12. Proper transformations

We wish to define the class of transformations that we shall regard as proper or admissible. A transformation of coordinates is not of much value if we cannot get back to the original coordinates by inverting the transformation. For example, the transformation from cylindrical to spherical polars is given above by Eq. (7.11.3), but the inverse transformation is

$$x^1 = \bar{x}^1 \sin \bar{x}^2, \qquad x^2 = \bar{x}^3, \qquad x^3 = \bar{x}^1 \cos \bar{x}^2. \tag{7.12.1}$$

Both sets of equations are perfectly definite except at the origin where $x^1 = x^3 = \bar{x}^1 = 0$ and the value of the other coordinates is immaterial. It is shown in Appendix B that the transformation

$$y_i = f_i(x_1, x_2, \ldots, x_n), \qquad i = 1, \ldots, n,$$

can always be inverted in the neighborhood of a point to give

$$x_i = \varphi_i(y_1, y_2, \ldots, y_n), \qquad i = 1, \ldots, n,$$

provided that the Jacobian

$$\frac{\partial(f_1, \ldots, f_n)}{\partial(x_1, \ldots, x_n)}$$

exists and does not vanish. Translating this into our coordinate notation we see that the transformation

$$\bar{x}^i = x^i(\bar{x}^1, x^2, x^3) \tag{7.12.2}$$

can be inverted to give

$$x^j = x^j(\bar{x}^1, \bar{x}^2, \bar{x}^3) \tag{7.12.3}$$

provided that the Jacobian

$$J = \left|\frac{\partial \bar{x}}{\partial x}\right| = \frac{\partial(\bar{x}^1, \bar{x}^2, \bar{x}^3)}{\partial(x^1, x^2, x^3)} \tag{7.12.4}$$

$$= \begin{vmatrix} \dfrac{\partial \bar{x}^1}{\partial x^1} & \dfrac{\partial \bar{x}^2}{\partial x^1} & \dfrac{\partial \bar{x}^3}{\partial x^1} \\ \dfrac{\partial \bar{x}^1}{\partial x^2} & \dfrac{\partial \bar{x}^2}{\partial x^2} & \dfrac{\partial \bar{x}^3}{\partial x^2} \\ \dfrac{\partial \bar{x}^1}{\partial x^3} & \dfrac{\partial \bar{x}^2}{\partial x^3} & \dfrac{\partial \bar{x}^3}{\partial x^3} \end{vmatrix}$$

exists and does not vanish. We have already encountered the geometrical meaning of the Jacobian, namely, the ratio volume elements in the two coordinate systems. Thus, to say that neither J nor its reciprocal is zero, is to say that no infinitesimal region in one coordinate system is collapsed into a single point in the other.

The set of proper transformations forms a group if we define the product of two transformations as the result of applying them successively. A group has the following properties:

(i) the identity transformation I belongs to it,

(ii) if T_1 and T_2 are two transformations of the group their product T_2T_1 also belongs to it,
(iii) if T is a transformation of the group then its inverse T^{-1} also belongs to it, $(TT^{-1} = I)$,
(iv) the product is associative, $T_1(T_2T_3) = (T_1T_2)T_3$.

The identity transformation $\bar{x}^i = x^i$ is obviously proper and our definition of propriety has ensured the existence of the inverse (iii). What we need to show is that the product or successive application of two proper transformations is still proper. Let T_1 be the transformation from x^i to $\bar{x}^i$ and T_2 that from $\bar{x}^i$ to $\hat{x}^i$. Then we can write symbolically $\bar{x} = T_1x$, $\hat{x} = T_2\bar{x}$ and by the product T_2T_1 we mean the transformation from x to $\hat{x}$, that is, $\hat{x} = T_2T_1x$. This product will be proper if the Jacobian $|\partial\hat{x}/\partial x|$ is not infinite or zero. However,

$$\frac{\partial\hat{x}^i}{\partial x^j} = \frac{\partial\hat{x}^i}{\partial\bar{x}^k}\frac{\partial\bar{x}^k}{\partial x^j} \tag{7.12.5}$$

so that the matrix $[J]$ whose typical term is $\partial\hat{x}^i/\partial x^j$ is the product of the matrices $[J_2]$ and $[J_1]$ whose typical terms are $\partial\hat{x}^i/\partial\bar{x}^k$ and $\partial\bar{x}^k/\partial x^j$ respectively. However, the Jacobians of the two transformations and their product are the determinants of these three matrices and the determinant of the product of two matrices is the product of their determinants (see Appendix A9). It follows that

$$\left|\frac{\partial\hat{x}}{\partial x}\right| = \left|\frac{\partial\hat{x}}{\partial\bar{x}}\right|\left|\frac{\partial\bar{x}}{\partial x}\right| = J_2J_1, \tag{7.12.6}$$

and since neither J_1 nor J_2 vanishes or is infinite, their product is neither zero nor infinite and the transformation T_2T_1 is proper.

The idea of a group is a very valuable one. Its importance and the geometry to be associated with it were first brought out by Felix Klein in 1872. A subgroup of the group is formed by a subset of the transformations of a group which themselves enjoy the group property. Thus the identity operation must belong to every subgroup; the inverse of any transformation and the product of any two transformations of the subgroup must also belong to it. For example, the group of transformations given by nonsingular constant matrices A^i_j is a subgroup of the general group of proper transformations,

$$\bar{x}^i = A^i_jx^j. \tag{7.12.7}$$

It is proper since $J = \det A^i_j$ is not zero. Hence the inverse exists and is given by a nonsingular constant matrix. Also, the product of any two constant matrices is a constant matrix and thus belongs to the subgroup. This subgroup induces all forms of Cartesian coordinates including those with oblique axes when operating on the ordinary rectangular Cartesian coordinates. It is itself a subgroup of the more general transformation

$$x^i = A^i_jx^j + b^i \tag{7.12.8}$$

which involves a shift of origin as well as distortion of axes. This last subgroup is called the affine group. It has subgroups such as that given by (7.12.7) with $b^i = 0$ and known as the group of linear homogeneous transformations. This in turn has subgroups of equivoluminar linear homogeneous transformations (for which $J = \pm 1$), the group of orthogonal homogeneous transformations (for which A^i_j is orthogonal and $J = \pm 1$), and the group of rotations (for which A^i_j is orthogonal and $J = +1$). It is this last subgroup which is the only group of transformations considered in constructing Cartesian tensors.

Exercise 7.12.1. If $J = |\partial \bar{x}/\partial x|$, show that

$$2J \frac{\partial x^i}{\partial \bar{x}^j} = \epsilon_{ipq}\epsilon_{jrs} \frac{\partial \bar{x}^p}{\partial x^r} \frac{\partial \bar{x}^q}{\partial x^s} .$$

(Here there is summation on p, q, r and s.)

Exercise 7.12.2. Calculate the Jacobians for the transformation from Cartesian coordinates to those of Ex. 7.11.1, (i)–(v).

7.13. General plan of presentation

It will be useful at this point to sketch the development we shall follow. At first we shall just be paralleling the development of Cartesian vector and tensor analysis and making use of the maturity that the reader should now have attained to minimize some of the more fussy details. We shall have to distinguish at the outset two types of behavior under transformation, which in the subgroup of rotations are identical, but until we reach differentiation the analogy with Cartesians is very close. We shall find, however, that a new kind of differentiation has to be introduced to preserve the tensorial character that partial differentiation has hitherto enjoyed.

A distinction will also emerge between the curvature of the coordinate system and the curvature of the space it is being used to describe. Thus the cylindrical and spherical polar coordinates are curvilinear systems but they describe the Euclidean space in which the ordinary Cartesian coordinates can be constructed. Since the space of our gross experience is Euclidean it is rather difficult to imagine a curved three-dimensional space. The distinction can however be seen in two dimensions by comparing the plane and the surface of a sphere. The latter has an intrinsic curvature which actually precludes the setting up of a Cartesian system of coordinates. Strictly speaking, Cartesian tensors suffice for Euclidean space and if we wanted equations in curvilinear coordinates we could transform the equations in Cartesians to obtain them. However, they do have certain manipulative advantages and this redundancy gives another useful pedagogical bridge from the familiar to the abstract.

When we come to consider flow in surfaces, then a full tensorial treatment is quite essential as the space is not Euclidean. To prepare the way for this, Chapter 9 is devoted to the study of surfaces in space.

7.21. Contravariant vectors

Since the coordinates are in general curved, we cannot expect any linear transformation between the coordinates themselves. However, differentials of the coordinates $d\bar{x}^i$ and dx^j are connected by the laws of partial differentiation and

$$d\bar{x}^i = \frac{\partial \bar{x}^i}{\partial x^j}\, dx^j. \tag{7.21.1}$$

(The affix j in $\partial\bar{x}^i/\partial x^j$ is in the upper position of the denominator and so counts as a lower affix.) The coefficients $\partial\bar{x}^i/\partial x^j$ are generally functions of position, but are calculable from the equations of transformation. Consider, for example, the transformation from Cartesians to cylindrical polars,

$$\bar{x}^1 = \{(x^1)^2 + (x^2)^2\}^{1/2}, \qquad \bar{x}^2 = \tan^{-1}(x^2/x^1), \qquad \bar{x}^3 = x^3.$$

Here we have

$$d\bar{x}^1 = \frac{(x^1\, dx^1 + x^2\, dx^2)}{\{(x^1)^2 + (x^2)^2\}^{1/2}}$$

$$d\bar{x}^2 = \frac{(-x^2\, dx^1 + x^1\, dx^2)}{\{(x^1)^2 + (x^2)^2\}^{1/2}}$$

$$d\bar{x}^3 = dx^3.$$

We call this behavior under transformation the behavior of a contravariant vector and, as before, define a contravariant vector to be anything that behaves in this way.

More precisely, we say that a^i are the components of a contravariant vector at a certain point in the coordinate system $O123$ if under a transformation of coordinates to $O\bar{1}\bar{2}\bar{3}$ the components of the contravariant vector become

$$\bar{a}^i = \frac{\partial \bar{x}^i}{\partial x^j}\, a^j. \tag{7.21.2}$$

Thus a vector is associated with a given point and the coefficients $\partial\bar{x}^i/\partial x^j$ must be evaluated there.

As before, we have to establish this behavior on the part of any entity we wish to show to be a contravariant vector. If $x^i(t)$ is the coordinate of a moving particle with the time t, then

$$v^i = \frac{dx^i}{dt} \tag{7.21.3}$$

is its velocity. In the transformed coordinates the velocity has components $\bar{v}^i = d\bar{x}^i/dt$, but clearly,

$$\bar{v}^i = \frac{d\bar{x}^i}{dt} = \frac{\partial \bar{x}^i}{\partial x^j}\frac{dx^j}{dt} = \frac{\partial \bar{x}^i}{\partial x^j}v^j, \tag{7.21.4}$$

so that velocity is a contravariant vector. Similarly, the acceleration and all higher derivatives are contravariant vectors.

7.22. Covariant vectors

If $f(x^1, x^2, x^3)$ is a scalar function and we transform it to a function of the variables $\bar{x}^1, \bar{x}^2, \bar{x}^3$, its derivatives transform according to the equation

$$\frac{\partial f}{\partial \bar{x}^i} = \frac{\partial x^j}{\partial \bar{x}^i}\frac{\partial f}{\partial x^j}. \tag{7.22.1}$$

We could consistently write

$$a_j = \frac{\partial f}{\partial x^j} \tag{7.22.2}$$

for the upper affix j in the denominator is equivalent to an affix in the lower position. Then, since f is a scalar or invariant function, $\bar{a}_i$ will be the set of quantities $\partial f/\partial \bar{x}^i$. Thus Eq. (7.22.1) is

$$\bar{a}_i = \frac{\partial x^j}{\partial \bar{x}^i}a_j, \tag{7.22.3}$$

and this type of behavior under transformation is the behavior of a covariant vector. More precisely, the quantities a_j are components of a covariant vector in the coordinate system $O123$ if under transformation of $O\bar{1}\bar{2}\bar{3}$ they transform according to Eq. (7.22.3). These two transformation formulae may be recalled readily if the "conservation of indices" is remembered. The index i is associated with the barred component and so must be associated with $\bar{x}$. Since the index on a coordinate is always in the upper position, the $\bar{x}^i$ must be in the denominator and the derivative is $\partial x^j/\partial \bar{x}^i$. Considerations of the summation convention lead to the same reassurance.

For the rotation transformation let us write

$$\bar{x}^j = l_{i\cdot}^{\ j}x^i$$

and its inverse

$$x^i = l^i_{\cdot j}\bar{x}^j.$$

Thus, $\partial \bar{x}^i/\partial x^j = l_{j\cdot}^{\ i}$ and $\partial x^j/\partial \bar{x}^i = l^j_{\cdot i}$. However, $l_{j\cdot}^{\ i}$ and $l^j_{\cdot i}$ are both identical with the l_{ji} of Chapter 2 being the angle between Oj and $O\bar{i}$. Hence Eq. (7.21.2) and (7.22.3) are identical and the distinction between covariance and contravariance does not arise for Cartesian tensors.

Let $y = (y^1, y^2, y^3)$ be the Cartesian coordinates of a point and x^1, x^2, x^3 its coordinates in some other system. For fixed i, $y^i = y^i(x^1, x^2, x^3)$ is a scalar function and so the three quantities

$$\frac{\partial y^i}{\partial x^j} \quad \text{for} \quad j = 1, 2, \text{and } 3$$

are the components of a covariant vector. For fixed j,

$$\mathbf{g}_{(j)} = \frac{\partial \mathbf{y}}{\partial x^j} \tag{7.22.4}$$

is a Cartesian vector. The three Cartesian vectors $\mathbf{g}_{(1)}, \mathbf{g}_{(2)}, \mathbf{g}_{(3)}$ are called base vectors in the coordinate system x^j, and they are not coplanar if the transformation from Cartesian coordinates is a proper one. Any Cartesian vector can be expressed in terms of a system of base vectors. Thus

$$\mathbf{a} = a^1\mathbf{g}_{(1)} + a^2\mathbf{g}_{(2)} + a^3\mathbf{g}_{(3)} = a^i\mathbf{g}_{(i)}, \tag{7.22.5}$$

where the a^i are the components of a with respect to this base system. Now for another system of coordinates $\bar{x}^k$ we would have base vectors

$$\bar{\mathbf{g}}_{(k)} = \frac{\partial \mathbf{y}}{\partial \bar{x}^k} = \frac{\partial x^j}{\partial \bar{x}^k} \mathbf{g}_{(j)}.$$

Hence,

$$\mathbf{a} = \bar{a}^k \bar{\mathbf{g}}_{(k)} = \bar{a}^k \frac{\partial x^j}{\partial \bar{x}^k} \mathbf{g}_{(j)} = a^i \mathbf{g}_{(i)}$$

or

$$a^i = \frac{\partial x^i}{\partial \bar{x}^k} \bar{a}^k. \tag{7.22.6}$$

This shows that the a^i are components of a contravariant vector.

7.23. The metric tensor

The reader will already have guessed that the tensor is to be defined by an extension of these transformation properties which is entirely analogous to the way Cartesian tensors were developed. Before going on to the general definition, it is convenient to consider a special tensor of fundamental importance. We shall only be able to introduce it in an informal style and will have to come back later to establish its nature. In Euclidean space a Cartesian system of coordinates can always be erected. Let us denote these by y^i to distinguish them from the general curvilinear coordinates x^i. The distance between two points P and Q with coordinates y^i and $y^i + dy^i$ is ds, where

$$ds^2 = \sum_{k=1}^{3} dy^k \, dy^k. \tag{7.23.1}$$

Notice that we cannot use the summation convention here since both affixes are at the same level. However,

$$dy^k = \frac{\partial y^k}{\partial x^i}\,dx^i, \tag{7.23.2}$$

hence,

$$\begin{aligned} ds^2 &= \sum_{k=1}^{3}\left(\frac{\partial y^k}{\partial x^i}\,dx^i\right)\left(\frac{\partial y^k}{\partial x^j}\,dx^j\right) \\ &= g_{ij}\,dx^i\,dx^j \end{aligned} \tag{7.23.3}$$

where

$$g_{ij} = \sum_{k=1}^{3}\frac{\partial y^k}{\partial x^i}\frac{\partial y^k}{\partial x^j}. \tag{7.23.4}$$

g_{ij} is called the metric tensor since it relates distance to the infinitesimal coordinate increments.

For example, with the cylindrical polar coordinates

$$x^1 = \{(y^1)^2 + (y^2)^2\}^{1/2}, \qquad x^2 = \tan^{-1}\frac{y^2}{y^1}, \qquad x^3 = y^3$$

we have

$$g_{11} = 1, \qquad g_{22} = (x^1)^2, \qquad g_{33} = 1, \tag{7.23.5}$$

and all $g_{ij} = 0$, $i \neq j$. For spherical polars

$$g_{11} = 1, \qquad g_{22} = (x^1)^2, \qquad g_{33} = (x^1 \sin x^2)^2 \tag{7.23.6}$$

and the off-diagonal terms are again zero. When only the diagonal terms are nonzero, the coordinates are called orthogonal (for reasons which will appear later) and it is convenient to write

$$g_{ii} = h_i^2, \tag{7.23.7}$$

where the h_i are called scale factors. The name arises from the fact that if we make an infinitesimal displacement only in the direction of the x^i coordinate then

$$ds = h_i\,dx^i \qquad \text{(no summation)}$$

so that h_i represents the ratio of distance to coordinate difference. This may be seen in the simplest case with cylindrical polars in Fig. 7.2, where $h_1 = h_3 = 1$, $h_2 = x^1$. Reverting to the notation ρ, ϕ, z we see how this comes about; for $x^1 = \rho$ and $x^3 = z$ are true distances in space, but $x^2 = \phi$ is an angle and the corresponding distance is $\rho\,d\phi$ and varies with ρ.

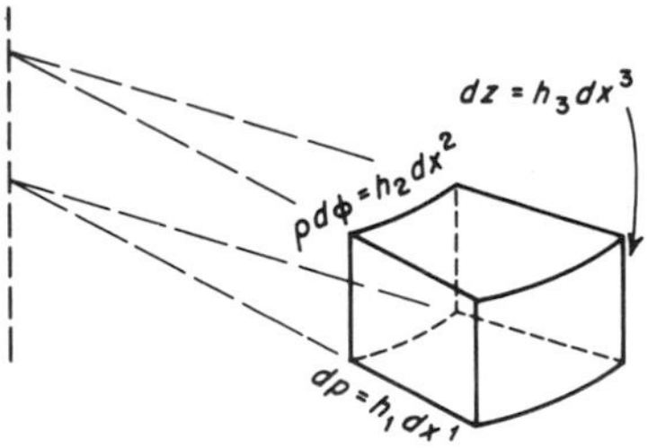

Fig. 7.2

Let g denote the determinant of the matrix whose typical element is g_{ij}. Then g is not zero for it is the square of the Jacobian $|\partial y/\partial x|$ of the transformation to Cartesian coordinates. If g^{ij} denotes the ij^{th} element of the inverse of matrix of g_{ij}, we have $g^{ij}g_{jk} = g_{ij}g^{jk} = 1$ if $i = k$ but zero otherwise. This is the definition of the Kronecker delta which we will now write with one suffix in the upper and one in the lower position,

$$\delta^i_j = \begin{cases} 1 & i = j, \\ 0 & i \neq j. \end{cases} \tag{7.23.8}$$

If the coordinates are orthogonal we observe that $g^{ii} = h_i^{-2}$, $g^{ij} = 0$, $i \neq j$. The notation suggests associating covariance with g_{ij} and contravariance with g^{ij}, and regarding g as a scalar. We must now proceed to formal definitions and establish the tensorial character of these entities.

Exercise 7.23.1. Calculate the metric tensors of the coordinate systems given in Ex. 7.11.1.

Exercise 7.23.2. Show that for the rotation group the metric tensor is always δ_{ij}.

Exercise 7.23.3. Prove that if $J = |\partial y/\partial x|$, then $g = J^2$.

7.24. Absolute and relative tensor fields

An entity specified by 3^{m+n} components $A^{i_1\cdots i_m}_{j_1\cdots j_n}$ is called an absolute tensor of order (or rank) $m + n$ if it transforms, under transformation from coordinate $O123$ to $O\bar{1}\bar{2}\bar{3}$, to a set of quantities given by

$$\overline{A}^{p_1\cdots p_m}_{q_1\cdots q_n} = \frac{\partial \bar{x}^{p_1}}{\partial x^{i_1}} \cdots \frac{\partial \bar{x}^{p_m}}{\partial x^{i_m}} \frac{\partial x^{j_1}}{\partial \bar{x}^{q_1}} \cdots \frac{\partial x^{j_n}}{\partial \bar{x}^{q_n}} A^{i_1\cdots i_m}_{j_1\cdots j_n} \tag{7.24.1}$$

More precisely, we may call it an absolute tensor of contravariant order m and covariant order n, but in many cases it will be sufficient to call it a tensor of order $(m + n)$. When the components are functions of the coordinates we speak of a tensor field though it is often possible to drop the term field and for the most part we shall do this. A tensor of order one is called a vector and one of order zero a scalar. The word absolute will only be used when it is necessary to distinguish it from a relative tensor. A relative tensor of weight w and order $(m + n)$ is defined as above save that on the right-hand side of Eq. (7.24.1) there also appears the factor J^w where $J = |\partial x/\partial \bar{x}|$, is the Jacobian of the transformation.

Let us clothe this bare definition by giving some examples in addition to the absolute covariant and contravariant vectors we have already encountered. The Kronecker delta is a mixed (that is, partly covariant and partly contravariant) tensor of the second order. For we have

$$\bar{\delta}^p_q = \frac{\partial \bar{x}^p}{\partial x^i}\frac{\partial x^j}{\partial \bar{x}^q}\delta^i_j = \frac{\partial \bar{x}^p}{\partial x^i}\frac{\partial x^i}{\partial \bar{x}^q} = \frac{\partial \bar{x}^p}{\partial \bar{x}^q};$$

but the barred coordinates are independent so $\partial \bar{x}^p / \partial \bar{x}^q$ is 1 if $p = q$ but zero otherwise. Thus $\bar{\delta}_q^p = \delta_q^p$ and it is in fact an isotropic tensor. To see that g_{ij} is a covariant tensor of the second order we have only to write down its definition in the new coordinates.

$$\begin{aligned}
\bar{g}_{pq} &= \sum_{k=1}^{3} \frac{\partial y^k}{\partial \bar{x}^p} \frac{\partial y^k}{\partial \bar{x}^q} \\
&= \sum_{k=1}^{3} \frac{\partial y^k}{\partial x^i} \frac{\partial x^i}{\partial \bar{x}^p} \frac{\partial y^k}{\partial x^j} \frac{\partial x^j}{\partial \bar{x}^q} \\
&= \frac{\partial x^i}{\partial \bar{x}^p} \frac{\partial x^j}{\partial \bar{x}^q} \sum_{k=1}^{3} \frac{\partial y^k}{\partial x^i} \frac{\partial y^k}{\partial x^j} \\
&= \frac{\partial x^i}{\partial \bar{x}^p} \frac{\partial x^j}{\partial \bar{x}^q} g_{ij}.
\end{aligned} \tag{7.24.2}$$

Thus g_{ij} is also isotropic in the sense that in the barred coordinates it is again the metric tensor.

The permutation symbol ϵ_{ijk} has already been defined and used. It takes on the values ± 1 according as ijk is an even or odd permutation of 123 and is zero if any of i, j, and k are equal. Let us see how it transforms. By definition of the determinant

$$\frac{\partial x^i}{\partial \bar{x}^p} \frac{\partial x^j}{\partial \bar{x}^q} \frac{\partial x^k}{\partial \bar{x}^r} \epsilon_{ijk} = J \epsilon_{pqr}.$$

Thus

$$\bar{\epsilon}_{pqr} = J^{-1} \frac{\partial x^i}{\partial \bar{x}^p} \frac{\partial x^j}{\partial \bar{x}^q} \frac{\partial x^k}{\partial \bar{x}^r} \epsilon_{ijk} = \epsilon_{pqr} \tag{7.24.3}$$

is again the permutation symbol and is evidently a third order covariant tensor of weight -1. If we define ϵ^{ijk} in the same way, it turns out to be a third order contravariant tensor of weight $+1$. Now $g = J_x^2$ (cf. Ex. 7.23.3) where J_x denotes the Jacobian of the transformation from Cartesians y to the coordinates x. Similarly, $\bar{g} = J_{\bar{x}}^2$ and hence

$$J^2 = \left| \frac{\partial x}{\partial \bar{x}} \right|^2 = \frac{\left| \dfrac{\partial y}{\partial \bar{x}} \right|^2}{\left| \dfrac{\partial y}{\partial x} \right|^2} = \frac{J_{\bar{x}}^2}{J_x^2} = \frac{\bar{g}}{g}. \tag{7.24.4}$$

Substituting for J in Eq. (7.24.3) we have

$$\bar{g}^{1/2} \bar{\epsilon}_{pqr} = \frac{\partial x^i}{\partial \bar{x}^p} \frac{\partial x^j}{\partial \bar{x}^q} \frac{\partial x^k}{\partial \bar{x}^r} g^{1/2} \epsilon_{ijk} \tag{7.24.5}$$

and it appears that

$$\varepsilon_{ijk} = g^{1/2} \epsilon_{ijk} \tag{7.24.6}$$

is an absolute third order covariant tensor. Similarly,

$$\varepsilon^{ijk} = g^{-1/2} \epsilon^{ijk} \tag{7.24.7}$$

is an absolute third order contravariant tensor.

Since $\bar{g} = J^2 g$, g is a relative scalar of weight 2. Now the cofactor of g_{ij} in its matrix is

$$G^{ij} = \frac{1}{2}\epsilon^{imn}\epsilon^{jpq}g_{mp}g_{nq}$$

Hence

$$g^{ij} = \frac{G^{ij}}{g} = \frac{1}{2}\varepsilon^{imn}\varepsilon^{jpq}g_{mp}g_{nq}. \qquad (7.24.8)$$

The right-hand side is a contracted product of absolute tensors and from the result proved below in Section 7.31 it follows that g^{ij} is also an absolute tensor, in fact a second order contravariant one.

Exercise 7.24.1. Show that $\varepsilon^{ijk}\varepsilon_{ipq} = \delta^j_p\delta^k_q - \delta^j_q\delta^k_p$, and $g^{mn}\varepsilon_{mjk}\varepsilon_{npq} = g_{jp}g_{kq} - g_{jq}g_{kp}$.

Exercise 7.24.2. Show that $\delta^i_j\delta^j_i = 3$, $\varepsilon^{ijk}\varepsilon_{ijk} = 3!$

Exercise 7.24.3. Show that $\partial^2 f/\partial x^i\,\partial x^j$ is NOT a covariant second order tensor.

7.25. Isotropic tensors

We have observed that g_{ij} and δ^i_j are isotropic tensors in that their values are defined in the same way in all coordinate systems. The first is defined as the metric tensor in the appropriate coordinates and the second as the Kronecker delta. It follows from its construction that the conjugate metric tensor g^{ij} is also isotropic, and these are the only second order isotropic tensors. Equation (7.24.5) shows that ε_{ijk} is isotropic and so likewise is ε^{ijk}.

The isotropic fourth order tensors may be constructed by arguments analogous to those used in Section 2.7. It turns out that there are again three independent isotropic tensors of each type. The purely contravariant fourth order tensors are

$$g^{ij}g^{kl}, \quad (g^{ik}g^{jl} + g^{il}g^{jk}) \text{ and } (g^{ik}g^{jl} - g^{il}g^{jk}). \qquad (7.25.1)$$

We shall have occasion to use these in the same way as the Cartesian isotropic tensor was used.

Exercise 7.25.1. Find the mixed types of fourth order isotropic tensor.

7.31. Tensor algebra

We may define the operations of tensor algebra with little discussion since they are analogous to the operations already familiar. To avoid ungainly formulas, A^{ij}, B^i_j, C_{ij}, etc. will be used as illustrating typical tensors but it will be evident that all definitions can be applied to tensors of any order. We shall give them only for absolute tensors, their modification for relative tensors being quite easy.

Scalar multiplication. If A^{ij} is a tensor and α a scalar, the scalar product αA^{ij} is the tensor each of whose components is α times the corresponding component of A^{ij}.

Addition. If A_k^{ij} and B_k^{ij} are two tensors of the same order in both covariant and contravariant indices, they can be added. The components of the sum are the sums of the corresponding components of the two tensors. For the difference of two tensors the corresponding elements are subtracted.

Contraction. If a contravariant index is identified with a covariant index and the summation convention invoked, a tensor of order smaller by 2 is obtained. For example, A_j^{ij} is a contravariant vector obtained from the mixed third order tensor A_k^{ij}. It is not the only one for the first and third indices might have been identified to give A_i^{ij}.

Outer product. The outer product of two tensors of orders $m + n$ and $p + q$ is a tensor of contravariant order $m + p$ and covariant order $n + q$ whose elements are the products of one element from each factor. Thus the outer product of A^{ij} and B_m^k is a fourth order tensor $C_m^{ijk} = A^{ij}B_m^k$ contravariant in the first three indices and covariant in the fourth.

Inner product. A contraction of an outer product is an inner product. Thus A^{ij} and B_m^k might have second order products $C^{ik} = A^{ij}B_j^k$ or $C^{jk} = A^{ij}B_i^k$.

Symmetry. If for any tensor the value of a component is unchanged by interchanging two indices, it is said to be symmetric with respect to these. If the magnitude is unchanged but the sign alters, it is said to be anti-symmetric with respect to them. The metric tensor g_{ij} is an example of a symmetric tensor.

Associated tensors. If A_i is a covariant vector, the contravariant vector $g^{ij}A_i = A^j$ is called its associated vector. This operation is known as raising the index and its inverse is the lowering of the index by an inner product with g_{ij}, for example, $B_j = g_{ij}B^i$. A tensor obtained by raising or lowering any index is said to be an associated tensor. Since $g_{ij}g^{jk} = \delta_i^k$ the original tensor is restored when the index is lowered again. g^{ij} has been defined as the conjugate tensor, that is, the element of the inverse of the matrix of g_{ij}. If A_{ij} is a covariant second order tensor, the associated tensor $A^{ij} = g^{ip}g^{jq}A_{pq}$ is not generally the conjugate tensor. The exception to this is when A_{ij} is a scalar multiple of g_{ij} for

$$g^{ip}g^{jq}\,\alpha g_{pq} = \alpha g^{ip}\,\delta_p^j = \alpha g^{ij} \tag{7.31.1}$$

The conjugate of a second order tensor A_{ij} can, however, be calculated in the same way as for g_{ij} and exists provided that the determinant of the A_{ij} does not vanish.

Exercise 7.31.1. Rephrase the definitions for relative tensors.

Exercise 7.31.2. Establish that the tensor character is preserved under addition, contraction or multiplication.

Exercise 7.31.3. Establish the tensorial character of g^{ij} from the previous exercise and Eq. (7.24.8).

Exercise 7.31.4. If the coordinates are orthogonal and $g_{ii} = h_i^2$ show that the associated vector of A^i has components $A_i = h_i^2 A^i$, with no summation on the i.

Exercise 7.31.5. Show that $\varepsilon^{ijk}A_j B_k$ and $\varepsilon_{ijk}A^j B^k$ are associated absolute vectors. They are vector products of **A** and **B**.

Exercise 7.31.6. Show that any tensor can be represented as the sum of outer products of vectors. Take as a typical case A_k^{ij}.

7.32. The quotient rule

The quotient rule stands exactly as for Cartesian tensors and is most easily exhibited in a simple case, the generalization being evident. If $A(ijk)$ is a set of 27 quantities, B^k a contravariant vector independent of A, and it can be shown that the inner product

$$A(ijk)B^k = C^{ij},$$

a contravariant second order tensor, then the $A(ijk)$ are the components of a mixed third order tensor, A_k^{ij}. For

$$\begin{aligned}\bar{C}^{pq} &= \frac{\partial \bar{x}^p}{\partial x^i}\frac{\partial \bar{x}^q}{\partial x^j} C^{ij} = \frac{\partial \bar{x}^p}{\partial x^i}\frac{\partial \bar{x}^q}{\partial x^j} A(ijk)B^k \\ &= \frac{\partial \bar{x}^p}{\partial x^i}\frac{\partial \bar{x}^q}{\partial x^j}\frac{\partial x^k}{\partial \bar{x}^r} A(ijk)\bar{B}^r.\end{aligned}$$

However, $\bar{C}^{pq} = \bar{A}(pqr)\bar{B}^r$ and subtracting

$$\left[\bar{A}(pqr) - \frac{\partial \bar{x}^p}{\partial x^i}\frac{\partial \bar{x}^q}{\partial x^j}\frac{\partial x^k}{\partial \bar{x}^r} A(ijk)\right]\bar{B}^r = 0$$

and since B is independent of A this can only be satisfied by the vanishing of the bracket. This is the transformation law for A_k^{ij}. The generalizations can be written down by invoking the conservation of indices. The essential point is that B should be independent of A. It is not possible to establish the tensorial character of g^{ij} by invoking the quotient rule on the formula $g^{ij}g_{jk} = \delta^i_k$.

7.33. Length of a vector and angle between vectors

For an infinitesimal contravariant vector with components dx^i we already have a measure of length; namely,

$$ds = \{g_{ij}\, dx^i\, dx^j\}^{1/2}. \tag{7.33.1}$$

Any contravariant vector A^i can be thought of as a large multiple of an infinitesimal vector $A^i = M\, dx^i$. The length of A may thus be taken to be $M\, ds = |A|$, and we write

$$|A|^2 = g_{ij}A^iA^j.$$

Since $g_{ij}A^i$ is the associated covariant vector A_j and $A^j = g^{ij}A_i$, we can also write

$$|A|^2 = g_{ij}A^iA^j = A^iA_i = A_jA^j = g^{ij}A_iA_j. \tag{7.33.2}$$

Thus the lengths of a vector and its associate are the same. If $|A| = 1$, we say A is a unit vector.

For example if $x^i(s)$ is the equation of a curve and s is the arc length, then $\tau^i = dx^i/ds$ is a unit tangent vector to the curve. Any vector may be made a unit vector by dividing by its length. The unit contravariant vector tangent to the x^1 coordinate line is obtained by making the vector $A^i = \delta^i_1$ of unit length. Its length is $g_{ij}\delta^i_1\delta^j_1 = g_{11}$, so the unit vector is

$$e^i_{(1)} = \frac{\delta^i_1}{(g_{11})^{1/2}}. \tag{7.33.3}$$

The index 1 is not a tensorial index and so has been enclosed in parentheses. $e^i_{(2)}$ and $e^i_{(3)}$ are similarly defined. If the coordinates are orthogonal

$$e^i_{(j)} = \frac{\delta^i_j}{h_j}, \qquad \text{(no sum on } j). \tag{7.33.4}$$

The associated unit covariant vectors tangent to the coordinate lines are

$$e_{(j)i} = g_{ik}e^k_{(j)} = \frac{g_{ij}}{(g_{jj})^{1/2}} \quad \text{or} \quad \delta^j_i h_j \qquad \text{(no sum on } j). \tag{7.33.5}$$

The angle θ between two unit vectors A^i and B^j is

$$\cos\theta = g_{ij}A^iB^j = A^iB_i = g^{ij}A_iB_j. \tag{7.33.6}$$

To prove this we observe that it is a tensor formula identifying a scalar with the twice contracted product of a tensor and two vectors. It is certainly true in Cartesian coordinates where $g_{ij} = \delta_{ij}$ and hence is true in any coordinate system for we can transform both sides of the equation and the one is a scalar and so stays the same. If A and B are not unit vectors then

$$\cos\theta = \frac{g_{ij}A^iB^j}{|A|\,|B|}. \tag{7.33.7}$$

The angle between the coordinate lines at any point is given by

$$\cos\theta_{pq} = g_{ij}e^i_{(p)}e^j_{(q)} = \frac{g_{pq}}{\{g_{pp}g_{qq}\}^{1/2}} \qquad \text{(no sum on } p \text{ or } q). \tag{7.33.8}$$

Thus, when $g_{pq} = 0$ for $p \neq q$ the angle between the x^p and x^q coordinate lines is a right angle. For this reason, the coordinate system is called orthogonal.

The projection of a vector A^i on the direction tangent to the x^j coordinate line is $|A| \cos\theta$ where θ is the angle between A^i and $e^k_{(j)}$;

$$\begin{aligned} |A|\cos\theta &= \frac{g_{ik}A^i\delta^k_j}{(g_{jj})^{1/2}} \qquad \text{(no sum on } j) \\ &= \frac{g_{ij}A^i}{(g_{jj})^{1/2}}. \end{aligned} \tag{7.33.9}$$

In an orthogonal coordinate system this is

$$|A|\cos\theta = h_jA^j \qquad \text{(no sum on } j). \tag{7.33.10}$$

Exercise 7.33.1. For an orthogonal coordinate system, the angle between two unit vectors is given by

$$\cos\theta = (h_1A^1)(h_1B^1) + (h_2A^2)(h_2B^2) + (h_3A^3)(h_3B^3).$$

Exercise 7.33.2. Show that the length projection of A^i on the direction of B^j is

$$\frac{g_{ij}A^iB^j}{\{g_{pq}B^pB^q\}^{1/2}}.$$

Exercise 7.33.3. Show that if the metric is positive definite (that is, $ds^2 = g_{ij}\,dx^i\,dx^j$ is always positive) then the angle between two vectors is always real.

Exercise 7.33.4. Show that

$$\frac{\partial g^{ij}}{\partial x^k} = -g^{ip}g^{jq}\frac{\partial g_{pq}}{\partial x^k}.$$

7.34. Principal directions of a symmetric second order tensor

The covariant vector $A_{ij}L^j$ derived from the covariant tensor A_{ij} and contravariant vector L^j will have the same direction as the associated vector L_i if

$$A_{ij}L^j = \lambda L_i = \lambda g_{ij}L^j. \tag{7.34.1}$$

This set of three simultaneous equations has nontrivial solutions only if

$$|A_{ij} - \lambda g_{ij}| = 0, \tag{7.34.2}$$

which is a cubic equation for λ. Its roots are the latent or characteristic values of the tensor and the associated L^j are the corresponding characteristic vectors. For a real symmetric tensor, it can be shown that the characteristic values are all real and that characteristic vectors associated with distinct values are orthogonal. It follows that there is a rotation of coordinates that makes the transformed tensor A_{ij} diagonal. These properties are entirely analogous to the Cartesian case.

Exercise 7.34.1. Show that the characteristic values of a tensor are invariant under transformation.

Exercise 7.34.2. Demonstrate the orthogonality of characteristic vectors of a symmetric tensor for distinct latent roots. Construct the transformation to canonical form.

Exercise 7.34.3. Show that if $A_{ij} = g_{ik}A^k_{.j} = g_{jk}A^{k}_{i.} = g_{ik}g_{jl}A^{kl}$, the four tensors A have the same characteristic values and associated characteristic vectors.

7.35. Covariant and contravariant base vectors

In Section 7.22 we defined a set of Cartesian base vectors

$$\mathbf{g}_{(i)} = \frac{\partial \mathbf{y}}{\partial x^i} \tag{7.35.1}$$

where $\mathbf{y}$ is the Cartesian position vector of the point $\mathbf{x}$. (i) is to be regarded as a label on the Cartesian vector but for any given component of the vector, say the first $g^1_{(i)}$, the three quantities $g^1_{(1)}$, $g^1_{(2)}$, $g^1_{(3)}$ are components of a covariant vector. We have also shown that if a Cartesian vector is expressed in the form

$$\mathbf{a} = a^i\mathbf{g}_{(i)} \tag{7.35.2}$$

then the coefficients a^i are the components of a contravariant vector. They may be called the contravariant components of the vector $\mathbf{a}$ in the given coordinate system.

The length of $\mathbf{a}$ is given by

$$\begin{aligned}|\mathbf{a}|^2 &= \mathbf{a} \cdot \mathbf{a} = (a^i\mathbf{g}_{(i)}) \cdot (a^j\mathbf{g}_{(j)}) \\ &= \mathbf{g}_{(i)} \cdot \mathbf{g}_{(j)}a^ia^j.\end{aligned}$$

However, if a^i are the components of a contravariant vector its squared length is $g_{ij}a^ia^j$. So that

$$\mathbf{g}_{(i)} \cdot \mathbf{g}_{(j)} = g_{ij}, \tag{7.35.3}$$

which is another way of writing Eq. 7.23.4. The basis vectors are not usually of unit length but a basis of unit vectors tangent to the coordinate lines is given by $\mathbf{e}_{(i)} = \mathbf{g}_{(i)}/(g_{ii})^{1/2}$. As we have seen the angle between two coordinate lines is

$$\cos \theta_{pq} = \mathbf{e}_{(p)} \cdot \mathbf{e}_{(q)} = \frac{\mathbf{g}_{(p)} \cdot \mathbf{g}_{(q)}}{(g_{pp}g_{qq})^{1/2}} = \frac{g_{pq}}{(g_{pp}g_{qq})^{1/2}}.$$

If Γ denotes the matrix whose i^{th} row is the set of components of $\mathbf{g}_{(i)}$, the matrix of the g_{ij} is $G = \Gamma\Gamma'$. Hence, the determinant γ of Γ is related to the determinant g of the metric tensor by

$$g = \gamma^2. \tag{7.35.4}$$

Since this does not vanish we can construct the reciprocal basis (cf. Section 2.36),

$$\mathbf{g}^{(i)} = \frac{\mathbf{g}_{(j)} \wedge \mathbf{g}_{(k)}}{\gamma} \tag{7.35.5}$$

where i, j, k is an even permutation of 1, 2, 3. The reciprocal basis is such that

$$\mathbf{g}^{(i)} \cdot \mathbf{g}_{(j)} = \delta^i_j. \tag{7.35.6}$$

Let the components of $\mathbf{a}$ with respect to this reciprocal basis be a_k and

$$\mathbf{a} = a_k\mathbf{g}^{(k)}. \tag{7.35.7}$$

Then

$$\mathbf{a} \cdot \mathbf{g}_{(j)} = a_k\mathbf{g}^{(k)} \cdot \mathbf{g}_{(j)} = a_k\delta^k_j = a_j,$$

but by Eq. 7.35.2

$$\mathbf{a} \cdot \mathbf{g}_{(j)} = a^i\mathbf{g}_{(i)} \cdot \mathbf{g}_{(j)} = a^ig_{ij},$$

and comparing these two we see that

$$a_j = a^ig_{ij}. \tag{7.35.8}$$

Thus the covariant components of $\mathbf{a}$ are the components with respect to the reciprocal basis and are also the components of the covariant vector associated with the vector of its contravariant components.

We notice that the matrix whose rows consist of the components of the reciprocal base vectors is $(\Gamma^{-1})'$. Thus the matrix with components

$$g^{ij} = \mathbf{g}^{(i)} \cdot \mathbf{g}^{(j)} \text{ is } (\Gamma^{-1})'(\Gamma^{-1}) = (\Gamma\Gamma')^{-1} = G^{-1}.$$

However, this is just the definition of the conjugate metric tensor, which is evidently formed from the reciprocal basis in just the same way as was the metric from the original basis;

$$g^{ij} = \mathbf{g}^{(i)} \cdot \mathbf{g}^{(j)}. \tag{7.35.9}$$

With curvilinear coördinates the system of basis vectors will depend on the point under consideration. It is this variability, not present in Cartesian coordinates, that makes the construction of a tensor derivative more complicated than a straight forward partial derivative. Light is also thrown on the absence of distinction between covariance and contravariance in Cartesian systems. The basis there is a set of mutually orthogonal unit vectors, whose reciprocal is identical with itself.

7.41. The physical components of a vector in orthogonal coordinate systems

Cartesian coordinates all have the physical dimensions of length but in general we cannot expect this of curvilinear coordinates. For example with cylindrical polars $x^1 = \rho$, $x^2 = \phi$, $x^3 = z$, the first and third have the dimensions of length but the second has no dimensions. Thus the contravariant velocity components $v^i = dx^i/dt$ would not all have the same physical dimensions and can hardly be what we understand by the physical components of velocity. The associated covariant vectors are in no better position for though in this case $v_1 = v^1$ and $v_3 = v^3$ have the dimensions of velocity,

$$v_2 = h_2^2\, v^2 = (x^1)^2(dx^2/dt)$$

has the dimensions of (length)^2/(time). Actually for this system

$$ds^2 = (dx^1)^2 + (x^1\, dx^2)^2 + (dx^3)^2,$$

so that we see that dx^1, $x^1\, dx^2$, and dx^3 all have the physical dimensions of length and dx^1/dt, $x^1(dx^2/dt)$, dx^3/dt would therefore have the dimensions of velocity. For a general orthogonal coordinate system

$$|A|^2 = (h_1 A^1)^2 + (h_2 A^2)^2 + (h_3 A^3)^2 \tag{7.41.1}$$

so that the $h_i A^i$ have the same physical dimensions as the magnitude of the vector A, which is evidently given by Pythagoras' theorem. This is the sort of behavior we expect of a physical component.

The unit contravariant vectors tangent to the three coordinate lines are

$$e^i_{(1)} = \frac{\delta^i_1}{h_1}, \qquad e^i_{(2)} = \frac{\delta^i_2}{h_2}, \qquad e^i_{(3)} = \frac{\delta^i_3}{h_3}.$$

If the contravariant vector A^i is represented as a linear combination of these base vectors

$$A^i = A(1)e^i_{(1)} + A(2)e^i_{(2)} + A(3)e^i_{(3)}, \tag{7.41.2}$$

comparing the three components on each side we see that

$$A(1) = h_1 A^1, \qquad A(2) = h_2 A^2, \qquad A(3) = h_3 A^3. \tag{7.41.3}$$

The $A(i)$ are called the physical components of the contravariant vector A^i

and, as Eq. (7.41.1) shows, they all have the same physical dimensions as the magnitude of A.

The physical components of the covariant vector A_i can be constructed in the same way by the representation

$$A_i = A(1)e_{(1)i} + A(2)e_{(2)i} + A(3)e_{(3)i}. \tag{7.40.3}$$

Since $e_{(j)i} = \delta_i^j h_j$, we have

$$A(1) = \frac{A_1}{h_1}, \qquad A(2) = \frac{A_2}{h_2}, \qquad A(3) = \frac{A_3}{h_3}. \tag{7.41.4}$$

However, $A_1 = h_1^2 A^1$, etc., so that both these sets of formulae define the same physical components.

It is worth remarking that the conservation of indices is a great help in keeping these formulae straight, though there is no summation involved. Thus a parenthetic index has no tensorial significance being merely a label. The lower index to h_1 has covariant significance since h_1^2 is an element of a covariant second order tensor, and the upper index A^1 is contravariant. In the product $h_1 A^1$ these cancel one another out and leave the neutrality of a parenthetic index. The same considerations apply when we write $(g_{11})^{1/2}$ in place of h_1 if we regard the square root as reducing two affixes to one.

The physical components do not of course transform as tensors, but their transformation law can be easily deduced. In a new coordinate system $\bar{A}(i) = \bar{h}_i \bar{A}^i$ (no sum on i) so we have

$$\bar{A}(i) = \bar{h}_i \bar{A}^i = \bar{h}_i \frac{\partial \bar{x}^i}{\partial x^j} A^j = \frac{\bar{h}_i}{h_j} \frac{\partial \bar{x}^i}{\partial x^j} A(j) \tag{7.41.5}$$

where there is no sum on i but a single summation on j. Consider the angle θ_{ij} between $\bar{e}^p_{(i)}$ in the new coordinate system and $e^q_{(j)}$ in the old. We can only calculate this angle by bringing both unit vectors into the same coordinate system. In the new coordinate system $e^q_{(j)}$ becomes

$$\bar{e}^r_{(j)} = \frac{\partial \bar{x}^r}{\partial x^q} e^q_{(j)} = \frac{1}{h_j} \frac{\partial \bar{x}^r}{\partial x^j}.$$

However, this is a unit vector since

$$\bar{g}_{rs} \bar{e}^r_{(j)} \bar{e}^s_{(j)} = \frac{1}{h_j^2} \bar{g}_{rs} \frac{\partial \bar{x}^r}{\partial x^j} \frac{\partial \bar{x}^s}{\partial x^j} = \frac{1}{h_j^2} g_{jj} = 1,$$

and so θ_{ij} is given by

$$\begin{aligned} \cos \theta_{ij} &= \bar{g}_{pr} \bar{e}^p_{(i)} \bar{e}^r_{(j)} = \bar{h}_i^2 \left(\frac{1}{\bar{h}_i}\right) \left(\frac{1}{h_j} \frac{\partial \bar{x}^i}{\partial x^j}\right) \\ &= \frac{\bar{h}_i}{h_j} \frac{\partial \bar{x}^i}{\partial x^j}. \end{aligned} \tag{7.41.6}$$

Hence the transformation law for physical components can be written

$$\bar{A}(i) = \cos \theta_{ij} A(j), \tag{7.41.7}$$

which is just the familiar sum of the projections of the three components on the new direction. In Cartesian systems the physical, covariant and contravariant components are identical.

Exercise 7.41.1. Show that the physical components are the lengths of projections of the vector on the tangents to the coordinate lines in orthogonal systems.

Exercise 7.41.2. Show that the same law of transformation (7.41.5) is obtained by starting with $\bar{A}(i) = \bar{A}_i/\bar{h}_i$.

Exercise 7.41.3. Show that

$$|A|\,|B| \cos\theta = \sum_{i=1}^{3} A(i)B(i).$$

7.42. Physical components of vectors in nonorthogonal coordinate systems

For nonorthogonal coordinates the definition of the physical components is not quite as simple but we start from the same representation as before. The contravariant unit tangents are $e^i_{(j)} = \delta^i_j/(g_{jj})^{1/2}$ and if we set

$$A^i = \sum_{j=1}^{3} A(j)e^i_{(j)} \tag{7.42.1}$$

we have

$$A(j) = A^j(g_{jj})^{1/2} \qquad \text{(no sum on } j) \tag{7.42.2}$$

which reduces to Eq. (7.41.3) for orthogonal coordinates.

For a covariant vector we first construct the associated contravariant vector and apply Eq. (7.42.2). Thus

$$A(j) = (g_{jj})^{1/2} g^{ij} A_i \qquad \text{(no sum on } j). \tag{7.42.3}$$

Only in the orthogonal case where $g^{jj} = 1/g_{jj}$ and $g^{ij} = 0$, $i \neq j$, does this give $A(j) = A_j/(g_{jj})^{1/2}$. However, the relation

$$A_i = \sum_{j=1}^{3} A(j)e_{(j)i} \tag{7.42.4}$$

is preserved since $e_{(j)i} = g_{ij}/(g_{jj})^{1/2}$.

The transformation law of physical components is

$$\begin{aligned}\bar{A}(j) &= (\bar{g}_{jj})^{1/2}\bar{A}^j = (\bar{g}_{jj})^{1/2}\frac{\partial \bar{x}^j}{\partial x^i} A^i \\ &= \left(\frac{\bar{g}_{jj}}{g_{ii}}\right)^{1/2} \frac{\partial \bar{x}^j}{\partial x^i} A(i) \qquad \text{(no sum on } j).\end{aligned} \tag{7.42.5}$$

The scalar product of two vectors can be expressed in terms of the physical components, for it is

$$g_{ij}A^iB^j = \frac{g_{ij}}{(g_{ii}g_{jj})^{1/2}}A(i)B(j)$$
$$= \cos\phi_{ij}A(i)B(j), \tag{7.42.6}$$

where ϕ_{ij} is the angle between the tangents to the x^i and the x^j coordinate lines.

The quantities

$$\varepsilon^{ijk}A_jB_k \quad \text{and} \quad \varepsilon_{ijk}A^jB^k \tag{7.42.7}$$

define a contravariant and a covariant vector respectively, the associated vector products of A and B (cf. Ex. 7.31.5). Denoting these by C^i and C_i we have

$$C(i) = (g_{ii})^{1/2}\varepsilon^{ijk}A_jB_k$$
$$= \left(\frac{g_{ii}}{g}\right)^{1/2}\epsilon^{ijk}g_{jp}\frac{A(p)}{(g_{pp})^{1/2}}g_{kq}\frac{B(q)}{(g_{qq})^{1/2}}.$$

However,

$$\epsilon^{ijk}g_{jp}g_{kq} = g\epsilon_{pqr}g^{ir},$$

so that

$$C(i) = g^{ir}\epsilon_{pqr}\left\{\frac{gg_{ii}}{g_{pp}g_{qq}}\right\}^{1/2}A(p)B(q). \tag{7.42.8}$$

This is a cumbersome formula but reduces in the orthogonal case to

$$C(i) = \epsilon_{ipq}A(p)B(q). \tag{7.42.9}$$

Exercise 7.42.1. Show that the scalar and vector products given by Eqs. (7.42.6 and 8) transform appropriately.

7.43. Physical components of tensors

Higher order tensors usually occur in such formulae as that for stress where $t^i = p^i_j n^j$ and we would like to associate physical components in such a way that $t(i) = p(ij)n(j)$. In orthogonal coordinates we have

$$t(i) = h_it^i = h_ip^i_{\cdot j}n^j = \frac{h_i}{h_j}p^i_{\cdot j}n(j)$$

so that we can write

$$p(ij) = h_ip^i_{\cdot j}/h_j. \tag{7.43.1}$$

Evidently, we may treat each index exactly as the corresponding covariant or contravariant index of the vector is treated. Thus, for example,

$$A(ijk) = \frac{h_ih_jA^{ij}_k}{h_k} \tag{7.43.2}$$

where none of the indices is summed. In orthogonal coordinate systems, we need not distinguish between the physical components, the tensors $p^{i}_{.j}$ and $p_{i.}^{j}$ for

$$p(ij) = \frac{h_i}{h_j}\, p^{i}_{.j} = \frac{h_i}{h_j}\, g^{im} g_{jn} p_{m.}^{n} = \frac{h_j}{h_i}\, p_{i.}^{j}\,.$$

Notice that the diagonal elements of a mixed second order tensor are the same as their physical components

$$p(ii) = p_{i.}^{i} = p^{i}_{.i} \qquad \text{(no sum on } i). \tag{7.43.3}$$

In nonorthogonal systems the relation for a mixed second order tensor leads to

$$p(ij) = \left(\frac{g_{ii}}{g_{jj}}\right)^{1/2} p^{i}_{\ j} \tag{7.43.4}$$

by analogy with Eq. (7.43.1). To find the physical components of a pure covariant or pure contravariant tensor we should lower or raise an index. Thus

$$p(ij) = \left(\frac{g_{ii}}{g_{jj}}\right)^{1/2} g^{im} p_{mj} = \left(\frac{g_{ii}}{g_{jj}}\right)^{1/2} g_{jm} p^{im} \tag{7.43.5}$$

where there is summation on m only. With nonorthogonal coordinate systems a distinction must be made between the tensor formulae $p^{i}_{.j} n^{j}$ and $p_{j.}^{i} n^{j}$. Truesdell writes the physical components of $p_{j.}^{i}$ as $(ji)p$ so that the product would be written $n(j)(ji)p$ in physical form. Thus $(ji)p = (g_{ii}/g_{jj})^{1/2} p_{j.}^{i}$, which for a symmetric tensor is the transpose of $p(ij)$. In orthogonal coordinate systems this distinction does not arise and we will not pursue it farther; the reader is referred to Truesdell's paper in the bibliography. The method which Truesdell elicits in that paper is as follows: Starting with the tensors of lowest order the tensor components are replaced by the physical components as already defined, the final result showing the proper way of defining the physical components of the higher order tensors.

7.44. An example

The discussion of physical components is sufficiently important to warrant an example, which will be made simple to avoid heavy algebraic manipulations. Consider the three coordinate systems:

Cartesian	x^1, x^2, x^3	or	x, y, z
Cylindrical polar	$\bar{x}^1, \bar{x}^2, \bar{x}^3$	or	ρ, ϕ, z
Elliptical cylinder	$\tilde{x}^1, \tilde{x}^2, \tilde{x}^3$	or	σ, χ, z

They are defined and related by

$$x^1 = \bar{x}^1 \cos \bar{x}^2 = \lambda \tilde{x}^1 \cos \tilde{x}^2,$$
$$x^2 = \bar{x}^1 \sin \bar{x}^2 = \mu \tilde{x}^1 \sin \tilde{x}^2,$$
$$x^3 = \bar{x}^3 \qquad = \tilde{x}^3, \qquad (\lambda \text{ and } \mu \text{ constants}).$$

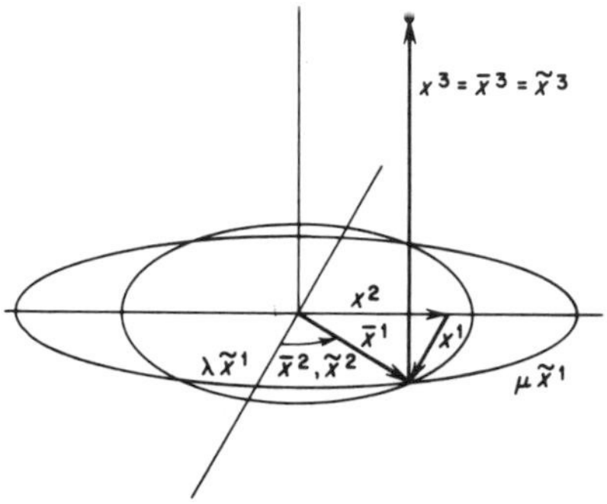

Fig. 7.3

$\bar{x}^1 =$ constant is a circular cylinder and $\tilde{x}^1 =$ constant is an elliptical cylinder,

$$\left(\frac{x^1}{\lambda}\right)^2 + \left(\frac{x^2}{\mu}\right)^2 = (\tilde{x}^1)^2.$$

The coordinate planes $\bar{x}^2 =$ constant and $\tilde{x}^2 =$ constant are both planes through the x^3 axis, as shown in Fig. 7.3. The first two systems are orthogonal but the third is not. Since if $\lambda = \mu = 1$ the second and third coordinates are the same we will do the calculations for the more general elliptical cylinder coordinates.

$$\frac{\partial \tilde{x}^1}{\partial x^1} = \frac{\cos \tilde{x}^2}{\lambda}, \qquad \frac{\partial \tilde{x}^1}{\partial x^2} = \frac{\sin \tilde{x}^2}{\mu}, \qquad \frac{\partial \tilde{x}^1}{\partial x^3} = 0,$$
$$\frac{\partial \tilde{x}^2}{\partial x^1} = -\frac{\sin \tilde{x}^2}{\lambda\mu\tilde{x}^1}, \qquad \frac{\partial \tilde{x}^2}{\partial x^2} = \frac{\cos \tilde{x}^2}{\lambda\mu\tilde{x}^1}, \qquad \frac{\partial \tilde{x}^2}{\partial x^3} = 0,$$
$$\frac{\partial \tilde{x}^3}{\partial x^1} = 0, \qquad \frac{\partial \tilde{x}^3}{\partial x^2} = 0, \qquad \frac{\partial \tilde{x}^3}{\partial x^3} = 1,$$
$$\frac{\partial x^1}{\partial \tilde{x}^1} = \lambda \cos \tilde{x}^2, \qquad \frac{\partial x^1}{\partial \tilde{x}^2} = -\lambda \tilde{x}^1 \sin \tilde{x}^2, \qquad \frac{\partial x^1}{\partial \tilde{x}^3} = 0,$$
$$\frac{\partial x^2}{\partial \tilde{x}^1} = \mu \sin \tilde{x}^2, \qquad \frac{\partial x^2}{\partial \tilde{x}^2} = \mu \tilde{x}^1 \cos \tilde{x}^2, \qquad \frac{\partial x^2}{\partial \tilde{x}^3} = 0,$$
$$\frac{\partial x^3}{\partial \tilde{x}^1} = 0, \qquad \frac{\partial x^3}{\partial \tilde{x}^2} = 0, \qquad \frac{\partial x^3}{\partial \tilde{x}^3} = 1.$$

Thus

$$\tilde{g}_{11} = \lambda^2 \cos^2 \tilde{x}^2 + \mu^2 \sin^2 \tilde{x}^2,$$
$$\tilde{g}_{22} = (\tilde{x}^1)^2\{\lambda^2 \sin^2 \tilde{x}^2 + \mu^2 \cos^2 \tilde{x}^2\},$$
$$\tilde{g}_{33} = 1,$$
$$\tilde{g}_{12} = \tilde{g}_{21} = (\mu^2 - \lambda^2)\tilde{x}^1 \sin \tilde{x}^2 \cos \tilde{x}^2,$$
$$\tilde{g}_{13} = \tilde{g}_{31} = \tilde{g}_{23} = \tilde{g}_{32} = 0.$$

For the orthogonal polars, $\lambda = \mu = 1$,

$$h_1 = 1, \qquad h_2 = \bar{x}^1, \qquad h_3 = 1.$$

Consider a vector **f** with no component in the 3-direction and let it be the force on a particle rotating in a circle of radius r and subject to a retardation proportional to its velocity. If we take its mass and angular velocity to be unity, the physical components of the acceleration in cylindrical polars are

$$f_\rho = r, \qquad f_\phi = -\alpha r, \qquad f_z = 0.$$

In Cartesian coordinates the physical, covariant, and contravariant components are the same and

$$f_x = f^1 = f_1 = x^1 + \alpha x^2, \qquad f_y = f^2 = f_2 = x^2 - \alpha x^1, \qquad f_z = f^3 = f_3 = 0.$$

To get the contravariant components in the other systems we must transform tensorially. Thus

$$\bar{f}^1 = \frac{\partial \bar{x}^1}{\partial x^i} f^i = \frac{\cos \bar{x}^2}{\lambda}\{\lambda \bar{x}^1 \cos \bar{x}^2 + \alpha\mu \bar{x}^1 \sin \bar{x}^2\} + \frac{\sin \bar{x}^2}{\mu}\{\mu \bar{x}^1 \sin \bar{x}^2 - \alpha\lambda \bar{x}^1 \cos \bar{x}^2\}$$

$$= \bar{x}^1 + \alpha\left(\frac{\mu}{\lambda} - \frac{\lambda}{\mu}\right)\bar{x}^1 \cos \bar{x}^2 \sin \bar{x}^2$$

and similarly,

$$\bar{f}^2 = \frac{\partial \bar{x}^2}{\partial x^i} f^i = -\alpha\left(\frac{\cos^2 \bar{x}^2}{\mu} + \frac{\sin^2 \bar{x}^2}{\lambda}\right) + \frac{\mu - \lambda}{\lambda\mu} \sin \bar{x}^2 \cos \bar{x}^2.$$

For the orthogonal case $\lambda = \mu = 1$,

$$\bar{f}^1 = \bar{f}_1 = \bar{x}^1, \qquad \bar{f}^2 = -\alpha, \qquad \bar{f}_2 = -\alpha(x^1)^2.$$

From these we can extract the physical components in the cylindrical polar coordinates

$$\bar{f}(1) = \bar{h}_1 \bar{f}^1 = \frac{\bar{f}_1}{\bar{h}_1} = \bar{x}^1$$

$$\bar{f}(2) = \bar{h}_2 \bar{f}^2 = \frac{\bar{f}_2}{\bar{h}_2} = -\alpha\bar{x}^1$$

and since $\bar{x}^1 = r$ these agree with f_ρ and f_ϕ. The formulae using the oblique coordinates are more cumbersome.

Exercise 7.44.1. Using the apparatus provided in this section, find $\bar{f}(1)$ and $\bar{f}(2)$ and interpret them in the light of the geometry of the situation.

7.45. Anholonomic components of a tensor

If $e^p_{(i)}$, $i = 1, 2, 3$, are three linearly independent contravariant vectors ($\varepsilon_{pqr} e^p_{(1)} e^q_{(2)} e^r_{(3)} \neq 0$) they can be regarded as a basis. The reciprocal basis of covariant vectors can be constructed, a set $e^{(k)}_s$ such that

$$e^p_{(i)} e^{(j)}_p = \delta^j_i, \qquad e^p_{(i)} e^{(i)}_q = \delta^p_q. \tag{7.45.1}$$

The parenthetic index is a label for the member of the triad and the other index is the component index, but we allow summation convention to apply to both. If A^p_q is a mixed second order tensor, the nine quantities $A^p_q e^{(i)}_p e^q_{(j)}$, $i, j = 1, 2, 3$, are scalars which we may denote by $A^{(i)}_{(j)}$. They are called the anholonomic components of the tensor with respect to the given set of base vectors (J. L. Ericksen, Tensor Fields, Section A10, p. 801, in Handbuch der Physik III/1, Berlin, Springer 1960). In the case of a vector its anholonomic components (for example, $A^p e^{(i)}_p$) are clearly the projections on the three base vectors. In general, the anholonomic components of a tensor $A^{p\cdots q}_{r\cdots s}$ are the scalars

$$A^{(i)\cdots(j)}_{(k)\cdots(l)} = A^{p\cdots q}_{r\cdots s} e^{(i)}_p \cdots e^{(j)}_q e^r_{(k)} \cdots e^s_{(l)}. \tag{7.45.2}$$

They are of course as many anholonomic components as tensor components.

If the coordinate system is orthogonal and the base vectors are unit vectors tangent to the coordinate lines, they are

$$\begin{aligned} &\mathbf{e}_{(1)} = (1/h_1, 0, 0), \qquad \mathbf{e}_{(2)} = (0, 1/h_2, 0), \qquad \mathbf{e}_{(3)} = (0, 0, 1/h_3) \\ &\mathbf{e}^{(1)} = (h_1, 0, 0), \qquad \mathbf{e}^{(2)} = (0, h_2, 0), \qquad \mathbf{e}^{(3)} = (0, 0, h_3). \end{aligned}$$

Then the anholonomic components with respect to this system of base vectors of a tensor are

$$T^{(i)\cdots(j)}_{(k)\cdots(l)} = \frac{h_i \cdots h_j}{h_k \cdots h_l} T^{i\cdots j}_{k\cdots l} \tag{7.45.3}$$

where there is no summation on the indices $i, \ldots, l$. We recognize these immediately as the physical components.

7.51. Differentials of tensors

We have noticed that although $\partial f/\partial x^i$ is a covariant vector for any scalar function f the second derivative $\partial^2 f/\partial x^i\, \partial x^j$ does not give a covariant second order tensor (cf. Ex. 7.24.3). In this respect our curvilinear coordinate system is not so convenient as the Cartesian system in which the partial derivatives gave higher order tensors, and it is important to try and define a derivative which preserves tensor character. To be acceptable, such a derivative should tie in with what we already have. That is to say that in Cartesian coordinates and when applied to a scalar, it should reduce to the familiar partial derivative. We should also expect the derivative of a sum to be the sum of the derivatives and the derivative of a product to be given by the usual rule. The name covariant differentiation is applied to this operation and the covariant derivative of a tensor A (the suffixes are suppressed) with respect to x^i is denoted by $A_{,i}$. A further property that we should expect would be that the differential $dA = A(x^i + dx^i) - A(x^i)$ for infinitesimal displacements dx^i should be given by $dA = A_{,i}\, dx^i$.

Our method of approach is as follows. We shall first consider what is meant by the parallel displacement of a vector along a curve and obtain a condition for this. Then follows an interlude in which one of the terms in this condition is expressed in terms of the metric tensor. This defines the Christoffel symbols which we express in various forms. To define the covariant derivative of a contravariant vector field we take an arbitrary covariant parallel vector displaced along an arbitrary curve and form the scalar product of the two. This is a scalar and its derivative with respect to a parameter along the curve is perfectly familiar. However, by using the rule for the derivative of a product and the condition for the second vector to be parallel, we are immediately led to an expression with tensor character and such that $dA^i = A^i_{,j}\,dx^j$. This definition is then extended to an arbitrary tensor.

7.52. Parallel vector fields

In Cartesian coordinates y^i, a vector field B^i with constant components is represented by a field of parallel vectors. If we take any curve $y^i(t)$ in the region of definition of the field, then we can think of the field along the curve as generated by the parallel propagation of the vector along the curve. The constancy of the components B^i is expressed analytically by

$$\frac{\partial B^i}{\partial y^j} = 0 \qquad (7.52.1)$$

and their constancy along the curve by

$$\frac{dB^i}{dt} = \frac{\partial B^i}{\partial y^j}\frac{dy^j}{dt} = 0. \qquad (7.52.2)$$

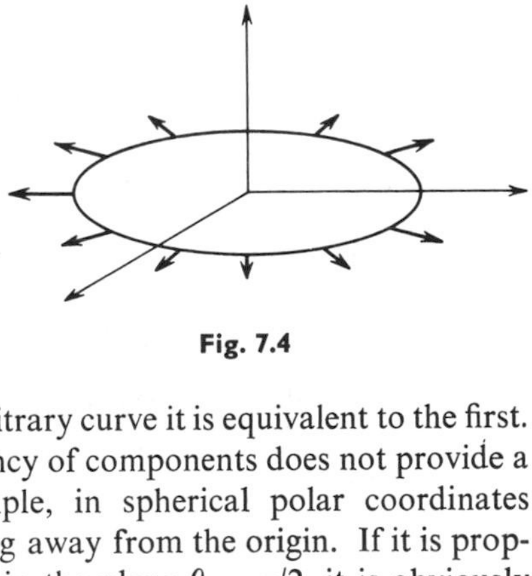

Fig. 7.4

If the second condition holds for an arbitrary curve it is equivalent to the first.

In curvilinear coordinates the constancy of components does not provide a condition for parallelism. For example, in spherical polar coordinates (1, 0, 0) is a unit vector always pointing away from the origin. If it is propagated along a curve, say the unit circle in the plane $\theta = \pi/2$, it is obviously not parallel to itself (see Fig. 7.4). This may be seen by looking at the Cartesian components of the vector, which are y^i/r, $r^2 = (y^1)^2 + (y^2)^2 + (y^3)^2$, and certainly not constant.

Let x^i denote the general coordinate system and A^i be a contravariant vector field. In Cartesian coordinates y^i we will denote the vector by B^i so that

$$B^i = \frac{\partial y^i}{\partial x^j} A^j. \qquad (7.52.3)$$

If $x^i(t)$ and $y^i(t)$ are the parametric equations of an arbitrary curve in the two coordinate systems and the vector field is parallel we must have $dB^i/dt = 0$. However, this condition is

$$\frac{dB^p}{dt} = \frac{d}{dt}\frac{\partial y^p}{\partial x^j}A^j = \frac{\partial y^p}{\partial x^j}\frac{dA^j}{dt} + A^j\frac{d}{dt}\left(\frac{\partial y^p}{\partial x^j}\right)$$

$$= \frac{\partial y^p}{\partial x^j}\frac{dA^j}{dt} + A^j\frac{\partial^2 y^p}{\partial x^j\,\partial x^k}\frac{dx^k}{dt} = 0. \qquad (7.52.4)$$

If we multiply by $\partial y^p/\partial x^i$ and sum on p, we have, by definition of the metric tensor,

$$g_{ij}\frac{dA^j}{dt} + A^j\left(\sum_{i=1}^{3}\frac{\partial y^p}{\partial x^i}\frac{\partial^2 y^p}{\partial x^j\,\partial x^k}\right)\frac{dx^k}{dt} = 0,$$

or multiplying through by g^{ir}

$$\frac{dA^r}{dt} + g^{ir}\left(\sum_{p=1}^{3}\frac{\partial y^p}{\partial x^i}\frac{\partial^2 y^p}{\partial x^j\,\partial x^k}\right)A^j\frac{dx^k}{dt} = 0. \qquad (7.52.5)$$

Exercise 7.52.1. Obtain the condition

$$\frac{dA_q}{dt} = g^{pj}\left(\sum_{i=1}^{3}\frac{\partial y^i}{\partial x^p}\frac{\partial^2 y^i}{\partial x^q\,\partial x^k}\right)A_j\frac{dx^k}{dt}$$

for the parallel propagation of a covariant vector A_i. (*Hint.* Differentiate the scalar A_iC^i for arbitrary parallel C.)

7.53. Christoffel symbols

Since, by definition,

$$g_{ij} = \sum_{p=1}^{3}\frac{\partial y^p}{\partial x^i}\frac{\partial y^p}{\partial x^j}, \qquad (7.53.1)$$

it should be possible to express the sum in Eq. (7.52.5) in terms of the derivatives of the g_{ij}. The summation is always from $p = 1$ to 3 and it is not necessary to write it in full each time. Then we have

$$\frac{\partial g_{ij}}{\partial x^k} = \Sigma\left[\frac{\partial^2 y^p}{\partial x^i\,\partial x^k}\frac{\partial y^p}{\partial x^j} + \frac{\partial y^p}{\partial x^i}\frac{\partial^2 y^p}{\partial x^j\,\partial x^k}\right]$$

$$\frac{\partial g_{ik}}{\partial x^j} = \Sigma\left[\frac{\partial^2 y^p}{\partial x^i\,\partial x^j}\frac{\partial y^p}{\partial x^k} + \frac{\partial y^p}{\partial x^i}\frac{\partial^2 y^p}{\partial x^j\,\partial x^k}\right]$$

$$\frac{\partial g_{jk}}{\partial x^i} = \Sigma\left[\frac{\partial^2 y^p}{\partial x^i\,\partial x^j}\frac{\partial y^p}{\partial x^k} + \frac{\partial y^p}{\partial x^j}\frac{\partial^2 y^p}{\partial x^i\,\partial x^k}\right]$$

from which we have

$$\Sigma\left(\frac{\partial y^p}{\partial x^i}\frac{\partial^2 y^p}{\partial x^j\,\partial x^k}\right) = \frac{1}{2}\left(\frac{\partial g_{ij}}{\partial x^k} + \frac{\partial g_{ik}}{\partial x^j} - \frac{\partial g_{jk}}{\partial x^i}\right). \qquad (7.53.2)$$

This expression is written $[jk, i]$ and is called the Christoffel symbol of the first kind. The factor occurring in Eq. (7.52.5) is $g^{ir}[jk, i]$ which is written as $\begin{Bmatrix} r \\ j \quad k \end{Bmatrix}$ and is the second kind of Christoffel symbol. These symbols are not tensors (cf. Ex. 7.53.1) but are most important functions of the metric. The notation used here is standard and shows the nontensorial character; the notation Γ_{jki} and Γ^r_{jk} is also commonly used. The definition

$$\begin{Bmatrix} i \\ j \quad k \end{Bmatrix} = g^{ip}[jk, p] = \frac{1}{2} g^{ip} \left(\frac{\partial g_{pj}}{\partial x^k} + \frac{\partial g_{pk}}{\partial x^j} - \frac{\partial g_{jk}}{\partial x^p} \right) \qquad (7.53.3)$$

may be best remembered by first writing the last term in $[jk, p]$, which is negative and has the suffixes in this order; namely, $\partial g_{jk}/\partial x^p$. The other terms are positive and are permutations of this order. (Since the metric tensor is symmetric, the parity of the permutation does not matter.) The Christoffel symbols $\begin{Bmatrix} i \\ j \quad k \end{Bmatrix}$ and $[jk, i]$ are clearly symmetric in j and k.

The condition (7.52.5) for a contravariant A^i to be a parallel field may thus be written

$$\frac{dA^i}{dt} + \begin{Bmatrix} i \\ j \quad k \end{Bmatrix} A^j \frac{dx^k}{dt} = 0. \qquad (7.53.4)$$

The analogous condition for a covariant A_i is

$$\frac{dA_i}{dt} - \begin{Bmatrix} j \\ i \quad k \end{Bmatrix} A_j \frac{dx^k}{dt} = 0. \qquad (7.53.5)$$

Since $(d/dt) = (\partial/\partial x^k)(dx^k/dt)$ and the curve was arbitrary, these conditions might be written

$$\frac{\partial A^i}{\partial x^k} + \begin{Bmatrix} i \\ j \quad k \end{Bmatrix} A^j = 0 \qquad (7.53.6)$$

and

$$\frac{\partial A_i}{\partial x^k} - \begin{Bmatrix} j \\ i \quad k \end{Bmatrix} A_j = 0. \qquad (7.53.7)$$

The Christoffel symbols can be interpreted in terms of the variation of the base vectors. The base vector

$$\mathbf{g}_{(i)} = \frac{\partial \mathbf{y}}{\partial x^i}, \qquad (7.53.8)$$

where $\mathbf{y}$ is the Cartesian coordinate of a point (cf. Eq. 7.35.3) and

$$g_{ij} = \mathbf{g}_{(i)} \cdot \mathbf{g}_{(j)}. \qquad (7.53.9)$$

Hence

$$\frac{\partial g_{ij}}{\partial x^k} = \frac{\partial \mathbf{g}_{(i)}}{\partial x^k} \cdot \mathbf{g}_{(j)} + \mathbf{g}_{(i)} \cdot \frac{\partial \mathbf{g}_{(j)}}{\partial x^k}$$

and forming the Christoffel symbol we have

$$[jk, i] = \mathbf{g}_{(i)} \cdot \frac{\partial \mathbf{g}_{(j)}}{\partial x^k}, \tag{7.53.10}$$

since $\partial \mathbf{g}_{(i)}/\partial x^j = \partial \mathbf{g}_{(j)}/\partial x^i$. It follows that

$$\frac{\partial \mathbf{g}_{(j)}}{\partial x^k} - [jk, i]\mathbf{g}^{(i)}$$

and since $\mathbf{g}^{(i)} = g^{im}\mathbf{g}_{(m)}$

$$\frac{\partial \mathbf{g}_{(j)}}{\partial x^k} = \begin{Bmatrix} m \\ j \quad k \end{Bmatrix} \mathbf{g}_{(m)}. \tag{7.53.11}$$

Thus if m is regarded as a contravariant index, the Christoffel symbols are the components of the rate of change of the jth base vector with respect to the kth coordinate.

Exercise 7.53.1. Show that the transformation law of Christoffel symbols is

$$\overline{[lm, n]} = \frac{\partial x^p}{\partial \bar{x}^l} \frac{\partial x^q}{\partial \bar{x}^m} \frac{\partial x^r}{\partial \bar{x}^n} [pq, r] + \frac{\partial^2 x^p}{\partial \bar{x}^l \, \partial \bar{x}^m} \frac{\partial x^q}{\partial \bar{x}^n} g_{pq}$$

and

$$\overline{\begin{Bmatrix} n \\ l \quad m \end{Bmatrix}} = \frac{\partial x^p}{\partial \bar{x}^l} \frac{\partial x^q}{\partial \bar{x}^m} \frac{\partial \bar{x}^n}{\partial x^r} \begin{Bmatrix} r \\ p \quad q \end{Bmatrix} + \frac{\partial^2 x^p}{\partial \bar{x}^l \, \partial \bar{x}^m} \frac{\partial \bar{x}^n}{\partial x^p}.$$

Exercise 7.53.2. Show that if the coordinate system is rectilinear (that is, all coordinate lines are straight), then the Christoffel symbols are all zero.

Exercise 7.53.3. Show that

$$\frac{\partial \mathbf{g}^{(i)}}{\partial x^j} = -\begin{Bmatrix} i \\ j \quad k \end{Bmatrix} \mathbf{g}^{(k)}.$$

7.54. Christoffel symbols in orthogonal coordinates

In an orthogonal system of coordinates $g_{pq} = 0$ if $p \neq q$ and g_{ii} is written as h_i^2. The Christoffel symbols are relatively simple and involve only one term. For example, [12, 3] vanishes identically since it is composed of derivatives of g_{12}, g_{23}, g_{31} which all vanish. Similarly,

$$[12, 1] = \frac{1}{2}\left[\frac{\partial g_{11}}{\partial x^2} + \frac{\partial g_{12}}{\partial x^1} - \frac{\partial g_{12}}{\partial x^1}\right]$$

$$= \frac{1}{2}\frac{\partial}{\partial x^2} h_1^2 = h_1 \frac{\partial h_1}{\partial x^2}$$

and

$$[22, 1] = \frac{1}{2}\left[\frac{\partial g_{12}}{\partial x^2} + \frac{\partial g_{12}}{\partial x^2} - \frac{\partial g_{22}}{\partial x^1}\right] = -h_2 \frac{\partial h_2}{\partial x^1}$$

Thus we can see that

$$[pq, r] = \pm h_i \frac{\partial h_i}{\partial x^j} \tag{7.54.1}$$

when $p = q = r = i = j$, with the positive sign,
or $q = r = i, p = j$, with the positive sign,
or $r = p = i, q = j$, with the positive sign,
or $p = q = i, r = j$, with the negative sign,
and is zero if p, q, and r are all different.

Since $\left\{ {i \atop j\ \ k} \right\} = \frac{1}{h_i^2} [jk, i]$ with no sum on i, we have

$$\left\{ {r \atop p\ \ q} \right\} = \begin{cases} 0 & \text{when } p, q, r \text{ are all different,} \\ \dfrac{1}{h_i} \dfrac{\partial h_i}{\partial x^j} & \begin{array}{l} \text{when } p = q = r = i = j, \\ \text{or } q = r = i,\ p = j, \\ \text{or } r = p = i,\ q = j, \end{array} \\ -\dfrac{h_i}{h_j^2} \dfrac{\partial h_i}{\partial x^j} & \text{when } p = q = i,\ r = j. \end{cases} \tag{7.54.2}$$

In Cartesian coordinates the Christoffel symbols vanish. In cylindrical polars the only nonvanishing symbols are

$$\begin{aligned} &[12, 2] = [21, 2] = x^1, \\ &[22, 1] = \left\{ {1 \atop 2\ \ 2} \right\} = -x^1, \\ &\left\{ {2 \atop 1\ \ 2} \right\} = \left\{ {2 \atop 2\ \ 1} \right\} = \frac{1}{x^1}. \end{aligned} \tag{7.54.3}$$

In spherical polars we have

$$\begin{aligned} &[21, 2] = [12, 2] = -[22, 1] = x^1, \\ &[31, 3] = [13, 3] = -[33, 1] = x^1 \sin^2 x^2, \\ &[32, 3] = [23, 3] = -[33, 2] = (x^1)^2 \sin x^2 \cos x^2, \\ &\left\{ {1 \atop 2\ \ 2} \right\} = -x^1, \qquad \left\{ {1 \atop 3\ \ 3} \right\} = -x^1 \sin^2 x^2, \\ &\left\{ {2 \atop 1\ \ 2} \right\} = \left\{ {2 \atop 2\ \ 1} \right\} = \left\{ {3 \atop 1\ \ 3} \right\} = \left\{ {3 \atop 3\ \ 1} \right\} = \frac{1}{x^1}, \\ &\left\{ {2 \atop 3\ \ 3} \right\} = \sin x^2 \cos x^2, \qquad \left\{ {3 \atop 2\ \ 3} \right\} = \left\{ {3 \atop 3\ \ 2} \right\} = \cot x^2. \end{aligned} \tag{7.54.4}$$

Exercise 7.54.1. Show that

$$\left\{ {i \atop i\ \ j} \right\} = \frac{1}{2} \frac{\partial \log g}{\partial x^j}.$$

Exercise 7.54.2. Calculate the Christoffel symbols for one or more of the coordinate systems given in Ex. 7.11.1.

7.55. Covariant differentiation

A comparison of the conditions for parallelism in Cartesian and general coordinates (see Eqs. 7.52.1 and 7.53.6) suggests that

$$\frac{\partial A^i}{\partial x^k} + \begin{Bmatrix} i \\ j \quad k \end{Bmatrix} A^j$$

may be the generalization of partial differentiation that we are looking for. It certainly satisfies the requirements we have laid down and we could proceed to show that it is a tensor by direct transformation. We shall, however, proceed more slowly and use the quotient rule to establish the tensor character.

Let A^i be any contravariant vector field and $x^i(t)$ a curve within its region of definition. If B_i is an arbitrary parallel covariant vector defined along this curve, then A^iB_i and its derivative dA^iB_i/dt are both scalars. However,

$$\begin{aligned} \frac{d}{dt}(A^iB_i) &= \frac{dA^i}{dt} B_i + A^i \frac{dB_i}{dt} \\ &= \frac{dA^i}{dt} B_i + A^i \begin{Bmatrix} j \\ i \quad k \end{Bmatrix} B_j \frac{dx^k}{dt} \end{aligned}$$

by Eq. (7.53.5) since B_i is a parallel vector. Changing the dummy suffix in the first terms from i to j we can write this

$$\frac{d}{dt}(A^iB_i) = \left[\frac{dA^j}{dt} + \begin{Bmatrix} j \\ i \quad k \end{Bmatrix} A^i \frac{dx^k}{dt}\right] B_j. \tag{7.55.1}$$

Since the left-hand side is a scalar and B_j an arbitrary covariant vector the quotient rule implies that the term in the bracket is a contravariant vector. It is called the intrinsic derivative and written

$$\frac{\delta A^j}{\delta t} = \frac{dA^j}{dt} + \begin{Bmatrix} j \\ i \quad k \end{Bmatrix} A^i \frac{dx^k}{dt}. \tag{7.55.2}$$

Moreover

$$\frac{dA^j}{dt} = \frac{\partial A^j}{\partial x^k} \frac{dx^k}{dt}$$

so that

$$\frac{\delta A^j}{\delta t} = \left[\frac{\partial A^j}{\partial x^k} + \begin{Bmatrix} j \\ i \quad k \end{Bmatrix} A^i\right] \frac{dx^k}{dt}. \tag{7.55.3}$$

The curve was taken to be arbitrary so dx^k/dt is a contravariant vector quite independent of the quantity in brackets. Since the left-hand side is a contravariant vector, the quotient rule asserts that the quantity in the brackets is a mixed second order tensor. We write it

$$A^j_{,k} = \frac{\partial A^j}{\partial x^k} + \left\{\begin{matrix} & j & \\ i & & k\end{matrix}\right\} A^i. \tag{7.55.4}$$

Notice that the same rule applied to a scalar gives the conventional partial derivative since there is no index on which to sum the second term. Also, in Cartesian coordinates the Christoffel symbols vanish identically so that the partial derivative is again recovered.

To find the covariant derivative of a covariant tensor we proceed in the same way as before with an arbitrary curve and parallel contravariant vector. We then arrive at the formula

$$A_{j,k} = \frac{\partial A_j}{\partial x^k} - \left\{\begin{matrix} & i & \\ j & & k\end{matrix}\right\} A_i \tag{7.55.5}$$

and

$$\frac{\delta A_j}{\delta t} = A_{j,k} \frac{dx^k}{dt}. \tag{7.55.6}$$

The general rule may be easily discerned from the case of a mixed second order tensor A^i_j. Let $x^k(t)$ be an arbitrary curve and B_i and C^j arbitrary parallel vectors. Then $A^i_j B_i C^j = E$ is a scalar and

$$\frac{d}{dt}(A^i_j B_i C^j) = \left(\frac{d}{dt} A^i_j\right) B_i C^j + A^i_j \left(\frac{dB_i}{dt} C^j + B_i \frac{dC^j}{dt}\right).$$

Substituting for the derivatives

$$\frac{dB_i}{dt} = B_p \left\{\begin{matrix} & p & \\ i & & k\end{matrix}\right\} \frac{dx^k}{dt},$$

$$\frac{dC^j}{dt} = -C^q \left\{\begin{matrix} & j & \\ q & & k\end{matrix}\right\} \frac{dx^k}{dt},$$

and

$$\frac{dA^i_j}{dt} = \frac{\partial A^i_j}{\partial x^k} \frac{dx^k}{dt},$$

we have

$$\frac{dE}{dt} = \left[\frac{\partial A^i_j}{\partial x^k} B_i C^j + A^i_j B_p C^j \left\{\begin{matrix} & p & \\ i & & k\end{matrix}\right\} - A^i_j B_i C^q \left\{\begin{matrix} & j & \\ q & & k\end{matrix}\right\}\right] \frac{dx^k}{dt}.$$

To get B and C always with the same suffix we now interchange the dummy suffixes i and p in the second term and j and q in the third to give

$$\frac{dE}{dt} = \left[\frac{\partial A^i_j}{\partial x^k} + A^p_j \left\{\begin{matrix} & i & \\ p & & k\end{matrix}\right\} - A^i_q \left\{\begin{matrix} & q & \\ j & & k\end{matrix}\right\}\right] B_i C^j \frac{dx^k}{dt}.$$

Since the left-hand side is a scalar and the vectors B_i, C^j, and dx^k/dt are all independent of the bracket, the quotient rule asserts that

$$A^i_{j,k} = \frac{\partial A^i_j}{\partial x^k} + A^p_j\begin{Bmatrix} i \\ p \quad k\end{Bmatrix} - A^i_q\begin{Bmatrix} q \\ j \quad k\end{Bmatrix} \tag{7.55.7}$$

is a mixed third order tensor. It is evident therefore that

$$A^{i_1\cdots i_m}_{j_1\cdots j_n,k} = \frac{\partial}{\partial x^k} A^{i_1\cdots i_m}_{j_1\cdots j_n} + \sum_{r=1}^{m} A^{i_1\cdots p_r\cdots i_m}_{j_1\cdots j_n}\begin{Bmatrix} i_r \\ p_r \quad k\end{Bmatrix}$$
$$- \sum_{s=1}^{n} A^{i_1\cdots i_m}_{j_1\cdots q_s\cdots j_n}\begin{Bmatrix} q_s \\ j_s \quad k\end{Bmatrix}, \tag{7.55.8}$$

a form that is much more easily perceived than written down.

A delightfully simple and elegant example of the power of tensor methods is the proof that the metric tensor acts as a constant with respect to covariant differentiation,

$$g_{ij,k} = 0. \tag{7.55.9}$$

This is a tensor equation asserting that a certain third order tensor equals the zero tensor. If the equation is true in one coordinate system, it is therefore true in all since we may transform both sides of the equation. However, it is obvious in Cartesian coordinates and so always true. This is known as Ricci's lemma. The same method can also be used to show that the rules of differentiation of sums and products hold for covariant differentiation.

The additional terms that arise in the covariant derivative are due to the variation of the base vectors. To see this let us consider the case of a contravariant vector, which may be written

$$\mathbf{a} = a^i\mathbf{g}_{(i)}.$$

The derivative of $\mathbf{a}$ with respect to x^k is

$$\frac{\partial \mathbf{a}}{\partial x^k} = \frac{\partial a^i}{\partial x^k}\mathbf{g}_{(i)} + a^i\frac{\partial \mathbf{g}_{(i)}}{\partial x^k}.$$

However, by Eq. (7.53.11) this can be written as

$$\frac{\partial a^i}{\partial x^k}\mathbf{g}_{(i)} + a^i\begin{Bmatrix} j \\ i \quad k\end{Bmatrix}\mathbf{g}_{(j)}.$$

Hence

$$\frac{\partial \mathbf{a}}{\partial x^k} = \left[\frac{\partial a^i}{\partial x^k} + \begin{Bmatrix} i \\ j \quad k\end{Bmatrix}a^j\right]\mathbf{g}_{(i)},$$

by interchange of the dummy suffixes i and j. From Ex. 7.53.3 it is evident that a covariant vector would give

$$\frac{\partial \mathbf{a}}{\partial x^k} = \left[\frac{\partial a_j}{\partial x^k} - \begin{Bmatrix} i \\ j \quad k\end{Bmatrix}a_i\right]\mathbf{g}^{(j)}. \tag{7.55.10}$$

These expressions in the brackets are just the appropriate covariant derivative and show that the Christoffel symbol comes in with the variability of the base vectors.

Exercise 7.55.1. Show by direct transformation that

$$A^i_{,j} = \frac{\partial A^i}{\partial x^j} + \begin{Bmatrix} i \\ j \quad k \end{Bmatrix} A^k$$

is a mixed second order tensor.

Exercise 7.55.2. Show from the definition that $g_{ij,k}$ is zero.

Exercise 7.55.3. Show that the covariant derivatives of g^{ij}, g, ε^{ijk}, and ε_{ijk} all vanish.

Exercise 7.55.4. Calculate the components in spherical polar coordinates of the covariant derivatives of the three contravariant tangent vectors to the coordinate lines of that system.

Exercise 7.55.5. Show that

$$\frac{\partial}{\partial x^i}(g^{1/2}g^{ij}) + \begin{Bmatrix} j \\ p \quad q \end{Bmatrix} g^{1/2}g^{pq} = 0.$$

(*Hint.* Use $g^{ij}_{,i} = 0$ and the result of Ex. 7.54.1.)

Exercise 7.55.6. $x^i(t)$ is a curve in space satisfying

$$\frac{d^2x^i}{dt^2} + \begin{Bmatrix} i \\ j \quad k \end{Bmatrix} \frac{dx^j}{dt}\frac{dx^k}{dt} = 0.$$

Show that it is a straight line.

7.56. The Laplacian, divergence, and curl

The covariant derivative of a scalar φ, which reduces to the partial derivative, is a covariant vector $\varphi_{,i}$, the gradient of φ. We can raise the index to give a contravariant form of the gradient $g^{ij}\varphi_{,i}$ and both of these are the vector $\nabla\varphi$ in Cartesian coordinates. If we take the covariant derivative of this vector with respect to x^j and sum on j, this is again a scalar, the *Laplacian* of φ,

$$\nabla^2\varphi = g^{ij}\varphi_{,ij}. \tag{7.56.1}$$

Writing this in full we have

$$\nabla^2\varphi = \frac{\partial}{\partial x^j}\left(g^{ij}\frac{\partial\varphi}{\partial x^i}\right) + \begin{Bmatrix} j \\ j \quad k \end{Bmatrix} g^{ik}\frac{\partial\varphi}{\partial x^i}.$$

However, by Ex. 7.54.1 and switching of dummy indices

$$\begin{Bmatrix} j \\ j \quad k \end{Bmatrix} g^{ik} = \frac{1}{g^{1/2}}\frac{\partial g^{1/2}}{\partial x^k}g^{ik} = \frac{1}{g^{1/2}}\frac{\partial g^{1/2}}{\partial x^j}g^{ij}$$

so that

$$\nabla^2\varphi = \frac{1}{g^{1/2}}\frac{\partial}{\partial x^j}\left\{g^{1/2}g^{ij}\frac{\partial\varphi}{\partial x^i}\right\}. \tag{7.56.2}$$

In orthogonal coordinates this takes the form

$$\nabla^2\varphi = \frac{1}{h_1h_2h_3}\left[\frac{\partial}{\partial x^1}\left(\frac{h_2h_3}{h_1}\frac{\partial\varphi}{\partial x^1}\right) + \frac{\partial}{\partial x^2}\left(\frac{h_3h_1}{h_2}\frac{\partial\varphi}{\partial x^2}\right) + \frac{\partial}{\partial x^3}\left(\frac{h_1h_2}{h_3}\frac{\partial\varphi}{\partial x^3}\right)\right]. \tag{7.56.3}$$

The covariant derivative with respect to x^i of a contravariant vector A^i summed on i is called the *divergence* of the vector A

$$\operatorname{div} A = A^i_{,i}. \tag{7.56.4}$$

If A_i is a covariant vector, the suffix must first be raised to give

$$\operatorname{div} A = g^{ij}A_{i,j}. \tag{7.56.5}$$

By the same methods as before we can write these

$$\frac{1}{g^{1/2}}\frac{\partial}{\partial x^i}(g^{1/2}A^i) \quad \text{or} \quad \frac{1}{g^{1/2}}\frac{\partial}{\partial x^j}(g^{1/2}g^{ij}A_i). \tag{7.56.6}$$

These are scalars and so are identical with their physical components but we can express them in terms of the physical component of A by using the relation $A(i) = (g_{ii})^{1/2}A^i = (g_{ii})^{1/2}g^{ij}A_j$ both become

$$\operatorname{div} A = \frac{1}{g^{1/2}}\frac{\partial}{\partial x^i}\left[\left(\frac{g}{g_{ii}}\right)^{1/2}A(i)\right] \tag{7.56.7}$$

or

$$= \frac{1}{h_1h_2h_3}\left[\frac{\partial}{\partial x^1}\{A(1)h_2h_3\} + \frac{\partial}{\partial x^2}\{h_1A(2)h_3\} + \frac{\partial}{\partial x^3}\{h_1h_2A(3)\}\right] \tag{7.56.8}$$

in orthogonal coordinates.

The expression $\varepsilon^{ijk}A_{k,j}$ is an absolute contravariant vector which reduces in Cartesian coordinates to the familiar expression for the curl of the vector. We therefore define

$$\operatorname{curl} A = \varepsilon^{ijk}A_{k,j} \quad \text{or} \quad \varepsilon^{ijk}g_{kp}A^p_{,j}. \tag{7.56.9}$$

To obtain the physical components of the curl we need first the physical components of $A_{k,j}$. In an orthogonal system

$$A(k,j) = \frac{1}{h_jh_k}A_{k,j} = \frac{1}{h_jh_k}\left[\frac{\partial}{\partial x^j}\{h_kA(k)\} - h_pA(p)\left\{{p \atop j\ k}\right\}\right]$$

We only need this for $k \neq j$ so that the two Christoffel symbols that appear are with $p = j$ or $p = k$. However, these are

$$\left\{{j \atop j\ k}\right\} = \frac{1}{h_j}\frac{\partial h_j}{\partial x^k} \quad \text{and} \quad \left\{{k \atop j\ k}\right\} = \frac{1}{h_k}\frac{\partial h_k}{\partial x^j}.$$

Then

$$A(k,j) = \frac{1}{h_j}\frac{\partial A(k)}{\partial x^j} + A(k)\frac{1}{h_jh_k}\frac{\partial h_k}{\partial x^j} - A(j)\frac{1}{h_jh_k}\frac{\partial h_j}{\partial x^k} - A(k)\frac{1}{h_jh_k}\frac{\partial h_k}{\partial x^j}$$

$$= \frac{1}{h_j}\frac{\partial A(k)}{\partial x^j} - A(j)\frac{1}{h_jh_k}\frac{\partial h_j}{\partial x^k}.$$

Now $\varepsilon^{ijk} = 1/h_1h_2h_3$ if i, j, k is an even permutation of 123 and the negative of this if an odd permutation. Let ijk be fixed as an even permutation of 123 then the ith physical component of curl A is

$$\begin{aligned} h_i\varepsilon^{ijk}A_{k,j} &= \frac{h_i}{h_1h_2h_3}\, h_jh_k\{A(k, j) - A(j, k)\} \\ &= A(k, j) - A(j, k) \qquad (7.56.10) \\ &= \left\{\frac{1}{h_j}\frac{\partial A(k)}{\partial x^j} - \frac{1}{h_k}\frac{\partial A(j)}{\partial x^k}\right\} + \frac{1}{h_jh_k}\left\{A(k)\frac{\partial h_k}{\partial x^j} - A(j)\frac{\partial h_j}{\partial x^k}\right\}. \end{aligned}$$

The second line shows that the definition of the curl in physical components is exactly its Cartesian definition.

Exercise 7.56.1. Obtain the Laplacian in cylindrical and spherical polar coordinates.

Exercise 7.56.2. If X^{ij} is an antisymmetric contravariant tensor, show that

$$X^{ij}_{,j} = \frac{1}{g^{1/2}}\frac{\partial}{\partial x^j}(g^{1/2}X^{ij}).$$

Exercise 7.56.3. Find the physical components of the curl in a nonorthogonal coordinate system.

Exercise 7.56.4. The physical Christoffel symbol may be written

$$(ijk) = \frac{h_i}{h_jh_k}\begin{Bmatrix} i \\ j \quad k \end{Bmatrix}.$$

Show that the physical components of $A^i_{j,k}$ are

$$\begin{aligned} A(ij, k) &= \frac{1}{h_k}\frac{\partial}{\partial x^k}A(ij) + \frac{A(ij)}{h_k}\frac{\partial}{\partial x^k}\left(\frac{h_j}{h_i}\right) \\ &\quad + (ikp)A(pj) - A(iq)(qjk) \end{aligned}$$

where summation is only on p and q.

7.57. Green's and Stokes' theorems

In Cartesian coordinates we can write the theorems of Green and Stokes replacing the partial derivatives by covariant derivatives since the two are identical. Thus, for any contravariant vector,

$$\iiint_V A^i_{,i}\, dV = \iint_S A^in_i\, dS \qquad (7.57.1)$$

where n_i is the covariant outward normal. Similarly,

$$\iiint_V g^{ij}A_{i,j}\, dV = \iint_S g^{ij}A_in_j\, dS = \iint_S A_in^i\, dS. \qquad (7.57.2)$$

Since this formula is constructed in full tensorial form, it is true not only in Cartesian coordinates (for which we have proved it above) but also in any coordinate system.

The special forms of Green's theorem are also valid;

$$\iiint_V g^{ij}\varphi_{,ij}\,dV = \iint_S \varphi_{,i}n^i\,dS = \iint_S \frac{\delta\varphi}{\delta n}\,dS \tag{7.57.3}$$

where $\delta\varphi/\delta n$ is the intrinsic derivative normal to the surface.

Stokes' theorem may be written

$$\iint_S \varepsilon^{ijk}A_{k,j}n_i\,dS = \oint_C A_k t^k\,dS. \tag{7.57.4}$$

Exercise 7.57.1. Show that

$$\iiint_V \psi g^{ij}\varphi_{,ij}\,dV + \iint_V g^{ij}\varphi_{,i}\psi_{,j}\,dV = \iint_S \psi\frac{\delta\varphi}{\delta n}\,dS.$$

7.6. Euclidean and other spaces

We have been dealing all through this chapter with a Euclidean space of three dimensions. By definition this is one in which a Cartesian coordinate system can be set up. We therefore defined the metric tensor from the Cartesian definition of the length element and transformed it into general coordinates. It is possible to proceed more abstractly and say that a Riemannian space is one in which there exists a metric

$$ds^2 = g_{ij}\,dx^i\,dx^j$$

where the g_{ij} is a covariant second order tensor field. Euclidean space is then certainly Riemannian for we know how to construct its metric tensor, but Riemannian space is more general for it includes spaces that are not Euclidean. An example already mentioned is the surface of a sphere, which is a two-dimensional, non-Euclidean space. We may therefore ask what conditions the metric has to satisfy for it to be the metric of a Euclidean space.

Now in Cartesian coordinates a suitably continuous vector field A_i certainly enjoys the property that

$$\frac{\partial A_i}{\partial x^j\,\partial x^k} = \frac{\partial A_i}{\partial x^k\,\partial x^j}$$

or

$$A_{i,jk} = A_{i,kj}. \tag{7.6.1}$$

However, this last formula is a valid tensor formula and so must hold in any coordinate system we care to transform to. Evidently Eq. (7.6.1) is a necessary condition for the space to be Euclidean. Let us consider the left-hand side in detail. It is

$$\begin{aligned}
(A_{i,j})_{,k} &= \frac{\partial A_{i,j}}{\partial x^k} - \begin{Bmatrix} p \\ i \quad k \end{Bmatrix} A_{p,j} - \begin{Bmatrix} q \\ j \quad k \end{Bmatrix} A_{i,q} \\
&= \frac{\partial}{\partial x^k}\left[\frac{\partial A_i}{\partial x^j} - \begin{Bmatrix} r \\ i \quad j \end{Bmatrix} A_r\right] \\
&\quad - \begin{Bmatrix} p \\ i \quad k \end{Bmatrix}\left[\frac{\partial A_p}{\partial x^j} - \begin{Bmatrix} r \\ p \quad j \end{Bmatrix} A_r\right] \\
&\quad - \begin{Bmatrix} q \\ j \quad k \end{Bmatrix}\left[\frac{\partial A_i}{\partial x^q} - \begin{Bmatrix} r \\ i \quad q \end{Bmatrix} A_r\right] \\
&= \frac{\partial^2 A_i}{\partial x^k\,\partial x^j} - \left[\begin{Bmatrix} r \\ i \quad j \end{Bmatrix}\frac{\partial A_r}{\partial x^k} + \begin{Bmatrix} p \\ k \quad i \end{Bmatrix}\frac{\partial A_p}{\partial x^j} + \begin{Bmatrix} q \\ j \quad k \end{Bmatrix}\frac{\partial A_q}{\partial x^i}\right] \\
&\quad + \left[\begin{Bmatrix} p \\ i \quad k \end{Bmatrix}\begin{Bmatrix} r \\ p \quad j \end{Bmatrix} + \begin{Bmatrix} q \\ j \quad k \end{Bmatrix}\begin{Bmatrix} r \\ i \quad q \end{Bmatrix} - \frac{\partial}{\partial x^k}\begin{Bmatrix} r \\ i \quad j \end{Bmatrix}\right] A_r.
\end{aligned}$$

Now the first two terms are symmetric in j and k and so is the second expression within the third bracket; thus

$$\begin{aligned}
A_{i,jk} - A_{i,kj} &= \left[\frac{\partial}{\partial x^j}\begin{Bmatrix} r \\ k \quad i \end{Bmatrix} - \frac{\partial}{\partial x^k}\begin{Bmatrix} r \\ i \quad j \end{Bmatrix} + \begin{Bmatrix} p \\ i \quad k \end{Bmatrix}\begin{Bmatrix} r \\ p \quad j \end{Bmatrix} - \begin{Bmatrix} p \\ i \quad j \end{Bmatrix}\begin{Bmatrix} r \\ p \quad k \end{Bmatrix}\right] A_r \\
&= R^r_{.ijk} A_r \qquad (7.6.2)
\end{aligned}$$

since by the quotient rule the bracket on the right-hand side must be a tensor. Now by Eq. (7.6.1) the left-hand side must be zero for any A if the space is to be Euclidean. Evidently this is equivalent to the condition

$$R^r_{.ijk} = 0. \qquad (7.6.3)$$

This tensor is known as the Riemann-Christoffel tensor of the space and its vanishing is a condition for the space to be Euclidean. Since the Christoffel symbols are combinations of derivatives of the metric tensor, the Riemann-Christoffel tensor is really a condition on the metric. It can be shown that $R^r_{.ijk}$ really has only six independent components so that there are six conditions on the metric of a Euclidean space. The associated tensor $R_{pijk} = g_{pr}R^r_{.ijk}$ is called the curvature tensor since it embodies information about the non-Euclidean character or curvature of the space. We shall have no occasion to use this in three dimensions, but we shall meet it for a surface where it will appear that it has only one independent component. Some of its properties are given below as exercises.

In deriving the covariant derivatives we again proceeded from the idea of parallelism in Euclidean space. We might, however, have written down the definition of Eq. (7.55.8) and proved it a tensor from the transformation properties of the Christoffel symbols. This would have founded everything on the metric tensor and would show that covariant differentiation holds in any Riemannian space. However, it is possible to be even more abstract and define covariant differentiation by Eq. (7.55.8) asserting only that the Christoffel symbols are given functions of the coordinates and that they transform like Christoffel symbols. The space so defined is called a space of linear connection (or of affine connection if the symbols are symmetric). Just as there are Riemannian spaces for which no Cartesian coordinates can be found, so there are spaces of linear connection for which no metric can be found. Such spaces are more general than Riemannian spaces, but still retain a certain structure. In three dimensions we can be content with Euclidean space, but for the two-dimensional surface flows we shall need Riemannian space. Although we shall not advert to it in this book, the space-time of relativity is a four-dimensional Riemannian space in which the equations of continuum mechanics have an extraordinarily compact expression.

Exercise 7.6.1. Show that $R^i_{.ijk} = 0$.

Exercise 7.6.2. Show that

$$R_{pijk} = \frac{\partial}{\partial x^j}[ik, p] - \frac{\partial}{\partial x^k}[ij, p] + \left\{ {m \atop i\ \ j} \right\}[kp, m] - \left\{ {n \atop i\ \ k} \right\}[jp, n].$$

Exercise 7.6.3. Show that

$$R_{pijk} = -R_{pikj} = -R_{ipjk} = R_{ipkj} = R_{jkpi}.$$

Exercise 7.6.4. Show that

$$R_{pijk} + R_{pjki} + R_{pkij} = 0.$$

Exercise 7.6.5. Show that in two dimensions there is only one independent nonvanishing component which can be taken as R_{1212}. (The general result in N dimensions is $N^2(N^2 - 1)/12$.)

Exercise 7.6.6. Show that the symmetry relations of R_{pijk} are satisfied by $\varepsilon_{mpi}\varepsilon_{njk}S^{mn}$, where S^{mn} is a symmetric second order tensor. Hence, conclude that there are six independent components of the curvature tensor in three dimensions.

Exercise 7.6.7. Show that for an orthogonal metric

$$S^{ii} = -\frac{1}{h_i^2 h_j h_k}\left[\frac{\partial}{\partial x^j}\left(\frac{1}{h_j}\frac{\partial h_k}{\partial x^j}\right) + \frac{\partial}{\partial x^k}\left(\frac{1}{h_k}\frac{\partial h_j}{\partial x^k}\right) + \frac{1}{h_i^2}\frac{\partial h_j}{\partial x^i}\frac{\partial h_k}{\partial x^i}\right]$$

$$S^{ij} = \frac{1}{h_i^2 h_j^2 h_k}\left[\frac{\partial^2 h_k}{\partial x^i\,\partial x^j} - \frac{1}{h_i}\frac{\partial h_k}{\partial x^i}\frac{\partial h_i}{\partial x^j} - \frac{1}{h_j}\frac{\partial h_k}{\partial x^j}\frac{\partial h_j}{\partial x^i}\right]$$

where i, j, k are all different and no summation occurs.

BIBLIOGRAPHY

The bulk of this chapter is standard matter and considered in almost all texts. The following specific points may be noted.

7.4. The clearest discussion of physical components is to be found in a paper by

Truesdell, C., "The physical components of vectors and tensors," Z.a.M.M. **33,** 345–356, 1953.

7.6. For an introduction to more general spaces see

Synge, J. L. and A. Schild, Tensor calculus, University of Toronto Press: Toronto, 1949, Chapter 8.

The Equations of Fluid Flow in Euclidean Space 8

We are now in a position to apply our understanding of general coordinate frames and transformations to the equations of fluid motion. This can be done quite rapidly when we have agreed on certain conventions of covariance and contravariance. For, since we are dealing with Euclidean space, we already have the basic equations in Cartesian tensors and have only to express these in a form which is invariant under more general transformation.

From the outset the convention of contravariant coordinate differentials has been used. It follows that the derivatives of position with respect to time are also contravariant vectors in particular the velocity v^i and acceleration a^i. Since the normal to a surface $\varphi(\mathbf{x}) =$ constant is in the direction $\varphi_{,i} = \partial\varphi/\partial x^i$, we take the normal to an element to be a covariant vector n_i. The force vector we take to be contravariant, and, since the contracted product of the stress tensor and the normal to an element is the stress or force per unit area, the stress tensor itself must be purely contravariant. We can of course associate vectors and tensors of different variance with these by raising and lowering an index.

8.11. Intrinsic derivatives

As before we may take the material coordinate of a particle to be its position $\boldsymbol{\xi}$ at some arbitrary origin of time. The path of a particle $\mathbf{x} = \mathbf{x}(\boldsymbol{\xi}, t)$

is a curve in space and $\mathbf{v} = d\mathbf{x}/dt$ (d/dt denoting differentiation with $\boldsymbol{\xi}$ constant) is its velocity. The intrinsic derivative along this curve of any function A of position x has already been defined as

$$\frac{\delta A}{\delta t} = A_{,i}\frac{dx^i}{dt} = A_{,i}v^i. \tag{8.11.1}$$

A can be any scalar or tensor and this derivative has been shown to be a tensor of the same order (see Section 7.55). If A is a function of t as well as of position, we must write

$$\frac{\delta A}{\delta t} = \frac{\partial A}{\partial t} + A_{,i}v^i. \tag{8.11.2}$$

If A is a scalar we have the same material derivative as before, but if A is a tensor of order greater than zero, there will be extra terms arising. In particular, the acceleration is given by

$$\begin{aligned} a^i &= \frac{\partial v^i}{\partial t} + v^j v^i_{,j} \\ &= \frac{\partial v^i}{\partial t} + \frac{\partial v^i}{\partial x^j}v^j + \begin{Bmatrix} & i & \\ j & & k \end{Bmatrix} v^j v^k. \end{aligned} \tag{8.11.3}$$

In an orthogonal coordinate system the physical components of velocity and acceleration are

$$v(i) = h_i v^i \quad \text{and} \quad a(i) = h_i a^i. \tag{8.11.4}$$

Thus

$$a(i) = \frac{\partial v(i)}{\partial t} + \frac{h_i}{h_j}v(j)\frac{\partial}{\partial x^j}\frac{v(i)}{h_i} + \frac{h_i}{h_j h_k}\begin{Bmatrix} & i & \\ j & & k \end{Bmatrix} v(j)v(k). \quad (\text{no sum on } i) \tag{8.11.5}$$

Exercise 8.11.1. Using the relations (7.54.2), show that

$$a(i) = \frac{\partial v(i)}{\partial t} + \sum_{j=1}^{3}\frac{v(j)}{h_j}\left[\frac{\partial v(i)}{\partial x^j} + \frac{v(i)}{h_i}\frac{\partial h_i}{\partial x^j} - \frac{v(j)}{h_i}\frac{\partial h_j}{\partial x^i}\right].$$

Exercise 8.11.2. Obtain the acceleration in cylindrical and spherical polar coordinates.

8.12. The transport theorem and equation of continuity

The divergence of the velocity field is the scalar $v^i_{,i}$. Since J, the Jacobian of the transformation from material to spatial coordinates, is a scalar, the same proof as in Section 4.21 can be used to show that

$$\frac{\delta J}{\delta t} = \frac{dJ}{dt} = v^i_{,i}J. \tag{8.12.1}$$

We again have Reynolds' transport theorem

$$\begin{aligned}\frac{\delta}{\delta t}\iiint F\,dV &= \iiint\left[\frac{\delta F}{\delta t} + Fv^i_{,i}\right] dV \\ &= \iiint\left[\frac{\partial F}{\partial t} + F_{,i}v^i + Fv^i_{,i}\right] dV \\ &= \iiint \frac{\partial F}{\partial t}\,dV + \iint Fv^i n_i\,dS\end{aligned} \tag{8.12.2}$$

for any scalar or tensor F.

In particular, putting $F = \rho$, the density, we have, by the conservation of mass,

$$\frac{d\rho}{dt} + \rho v^i_{,i} = \frac{\partial \rho}{\partial t} + (\rho v^i)_{,i} = 0. \tag{8.12.3}$$

This is the equation of continuity and gives the corollary to the transport theorem,

$$\frac{\delta}{\delta t}\iiint \rho F\,dV = \iiint \rho\frac{\delta F}{\delta t}\,dV. \tag{8.12.4}$$

Exercise 8.12.1. Obtain the continuity equation in terms of the physical components of the velocity for an orthogonal coordinate system.

Exercise 8.12.2. Show that in cylindrical polar coordinates with velocity components v_r, v_ϕ, v_z the continuity equation is

$$\frac{\partial \rho}{\partial t} + \frac{1}{r}\frac{\partial}{\partial r}(r\rho v_r) + \frac{1}{r}\frac{\partial}{\partial \phi}(\rho v_\phi) + \frac{\partial}{\partial z}(\rho v_z) = 0.$$

Exercise 8.12.3. Show that in spherical polar coordinates with velocity components v_r, v_θ, v_ϕ the continuity equation is

$$\frac{\partial \rho}{\partial t} + \frac{1}{r^2}\frac{\partial}{\partial r}(r^2\rho v_r) + \frac{1}{r\sin\theta}\frac{\partial}{\partial \theta}(\sin\theta\,\rho v_\theta) + \frac{1}{r\sin\theta}\frac{\partial}{\partial \phi}(\rho v_\phi) = 0.$$

8.13. The equations of motion

To form the equations of motion we take an arbitrary constant parallel field of unit vectors l_i and equate the rate of change of momentum to the net force in this direction. This is necessary since the direction in space of a given coordinate line is changing—a contrast to the Cartesian case. Since the field is parallel, $l_{i,j} = 0$.

The contravariant stress tensor T^{ij} can be constructed by identifying its physical components with the corresponding components of the Cartesian stress tensor. Then the force on an element of area dS with normal n_j is $T^{ij}n_j\,dS$. In the absence of stress couples and intrinsic angular momentum the stress tensor is symmetric. The external body forces per unit mass are represented by a contravariant vector f^i. The total net force on an arbitrary volume V with surface S resolved in the direction l_i is

$$\iiint_V \rho f^i l_i\,dV + \iint_S T^{ij} n_j l_i\,dS = \iiint_V [\rho f^i + T^{ij}_{,j}]l_i\,dV, \qquad (8.13.1)$$

where the last integral has been transformed by Green's theorem.

Equating this to the rate of change of linear momentum in the direction l_i we get

$$\frac{\delta}{\delta t}\iiint_V \rho v^i l_i\,dV = \iiint_V \rho a^i l_i\,dV = \iiint_V [\rho f^i + T^{ij}_{,j}]l_i\,dV. \qquad (8.13.2)$$

Since both the volume V and the parallel vector field l_i are arbitrary, this equation can only be satisfied if

$$a^i = \frac{\partial v^i}{\partial t} + v^j v^i_{,j} = f^i + \frac{T^{ij}_{,j}}{\rho}. \qquad (8.13.3)$$

A slight extension of Ex. 7.56.2 shows that

$$T^{ij}_{,j} = \frac{1}{g^{1/2}}\frac{\partial}{\partial x^j}(g^{1/2}T^{ij}) + T^{jk}\left\{ {i \atop j\ k} \right\}. \qquad (8.13.4)$$

In orthogonal coordinates the physical components of this may be written

$$\rho\{a(i) - f(i)\} = T(ij, j)$$
$$= \frac{h_i}{h_1h_2h_3}\frac{\partial}{\partial x^j}\left[\frac{h_1h_2h_3}{h_ih_j}T(ij)\right] + \frac{h_i}{h_jh_k}\left\{ {i \atop j\ k} \right\}T(jk) \qquad (8.13.5)$$

(no sum on i).

A covariant form of the equations of motion can be obtained by lowering the index to give

$$a_i = f_i + \frac{T^j_{i,j}}{\rho}$$

where (8.13.6)

$$T^j_{i\cdot} = g_{ik}T^{kj}.$$

Exercise 8.13.1. Obtain the equations of motion in terms of physical components in cylindrical polar coordinates.

Exercise 8.13.2. Obtain the equations of motion in terms of physical components in spherical polar coordinates.

Exercise 8.13.3. Show that the assumption of conservation of angular momentum is equivalent to the symmetry of the stress tensor.

8.21. The Newtonian fluid

We will first carry through a discussion of the Newtonian fluid and arrive at the Navier-Stokes equations before returning to a more fundamental consideration of the constitutive equations.

If the velocities of two nearby points x^i and $x^i + dx^i$ are v^j and $v^j + dv^j$, their relative velocity is

$$dv^j = v^j_{,i}\, dx^i. \tag{8.21.1}$$

This can be written as the sum of symmetric and antisymmetric parts

$$v^k_{,i} = g^{jk}(e_{ij} + \Omega_{ij}) \tag{8.21.2}$$

where

$$e_{ij} = \tfrac{1}{2}(v_{j,i} + v_{i,j}), \tag{8.21.3}$$

$$\Omega_{ij} = \tfrac{1}{2}(v_{j,i} - v_{i,j}), \tag{8.21.4}$$

and

$$v_i = g_{ij}v^j. \tag{8.21.5}$$

As before, the interpretation of $\Omega_{ij}\, dx^j$ is that it represents a rigid body rotation of the element. The symmetric part of the rate of strain tensor is the deformation tensor e_{ij}, and this may be interpreted as before (see, for example, Ex. 8.21.1, 2).

The stress tensor corresponding to a hydrostatic pressure p is $-pg^{ij}$. For the stress on an element with normal n_j is $t^i = -pg^{ij}n_j = -pn^i$ which is in the direction of the normal and of magnitude $-p$. Again we write

$$T^{ij} = -pg^{ij} + P^{ij} \tag{8.21.6}$$

where P^{ij}, the viscous stress tensor, is to be related to the deformation. For an isotropic Newtonian fluid this relation must be linear and isotropic so that

$$P^{ij} = G^{ijmn}e_{mn} \tag{8.21.7}$$

where G^{ijmn} is an isotropic tensor of the fourth order. It must be symmetric both in i, j and m, n and so according to Section 7.25 is a linear combination of $g^{ij}g^{mn}$ and $(g^{im}g^{jn} + g^{in}g^{jm})$. We therefore write this relation as follows:

$$P^{ij} = \lambda g^{ij}g^{mn}e_{mn} + \mu(g^{mi}g^{jn} + g^{in}g^{jm})e_{mn}$$

or

$$T^{ij} = (-p + \lambda e^m_m)g^{ij} + 2\mu e^{ij}. \tag{8.21.8}$$

If Stokes' hypothesis is invoked and p identified with the mean stress $-\frac{1}{3}T^i_i$, then we again have

$$3\lambda + 2\mu = 0$$

and

$$T^{ij} = -(p + \tfrac{2}{3}\mu e^m_m)g^{ij} + 2\mu e^{ij}. \tag{8.21.9}$$

$e^m_m = v^i_{,i}$ is the dilatation so that for an incompressible fluid

$$T^{ij} = -pg^{ij} + 2\mu e^{ij}. \tag{8.21.10}$$

For a perfect fluid with no viscous stress

$$T^{ij} = -pg^{ij}, \tag{8.21.11}$$

Exercise 8.21.1. Show that $(\delta g_{ij}/\delta t) = 0$, and deduce that

$$\frac{\delta}{\delta t}\log ds = \frac{e_{ij}\,dx^i\,dx^j}{g_{ij}\,dx^i\,dx^j}.$$

Exercise 8.21.2. Show that there are generally three mutually orthogonal directions in which the rate of strain is stationary.

Exercise 8.21.3. Show that in cylindrical polar coordinates $x^1 = r$, $x^2 = \phi$, $x^3 = z$ the physical components of stress for an incompressible fluid are

$$T_{rr} = -p + 2\mu\frac{\partial v_r}{\partial r}, \qquad T_{\phi z} = \mu\left(\frac{\partial v_\phi}{\partial z} + \frac{1}{r}\frac{\partial v_z}{\partial \phi}\right),$$

$$T_{\phi\phi} = -p + 2\mu\left(\frac{1}{r}\frac{\partial v_\phi}{\partial \phi} + \frac{v_r}{r}\right), \qquad T_{zr} = \mu\left(\frac{\partial v_z}{\partial r} + \frac{\partial v_r}{\partial z}\right),$$

$$T_{zz} = -p + 2\mu\frac{\partial v_z}{\partial z}, \qquad T_{r\phi} = \mu\left(\frac{1}{r}\frac{\partial v_r}{\partial \phi} + \frac{\partial v_\phi}{\partial r} - \frac{v_\phi}{r}\right).$$

Exercise 8.21.4. Show that in spherical polar coordinates $x^1 = r$, $x^2 = \theta$, $x^3 = \phi$ the physical components of stress for an incompressible fluid are

$$T_{rr} = -p + 2\mu\frac{\partial v_r}{\partial r}, \qquad T_{\theta\phi} = \mu\left(\frac{1}{r\sin\theta}\frac{\partial v_\theta}{\partial \phi} + \frac{1}{r}\frac{\partial v_\phi}{\partial \theta} - \frac{v_\phi}{r}\cot\theta\right)$$

$$T_{\theta\theta} = -p + 2\mu\left(\frac{1}{r}\frac{\partial v_\theta}{\partial \theta} + \frac{v_r}{r}\right), \qquad T_{\phi r} = \mu\left(\frac{\partial v_\phi}{\partial r} + \frac{1}{r\sin\theta}\frac{\partial v_r}{\partial \phi} - \frac{v_\phi}{r}\right)$$

$$T_{\phi\phi} = -p + 2\mu\left(\frac{1}{r\sin\theta}\frac{\partial v_\phi}{\partial \phi} + \frac{v_r}{r} + \frac{v_\theta}{r}\cot\theta\right), \quad T_{r\theta} = \left(\frac{1}{r}\frac{\partial v_r}{\partial \theta} + \frac{\partial v_\theta}{\partial r} - \frac{v_\theta}{r}\right)$$

Exercise 8.21.5. Obtain the additional terms in the stress components for a compressible fluid in these two coordinate systems.

8.22. The Navier-Stokes equations

To make use of the relation between stress and strain in the equations of motion we must first calculate $T^{ij}_{,j}$. From Eq. (8.21.8) we have, for constant λ and μ,

$$\begin{aligned} T^{ij}_{,j} &= (-p_{,j} + \lambda e^m_{m,j})g^{ij} + 2\mu e^{ij}_{,j} \\ &= (-p_{,j} + \lambda e^m_{m,j})g^{ij} + 2\mu g^{ik}e_{km,j} \\ &= (-p_{,j} + \lambda v^k_{,kj})g^{ij} + \mu g^{ik}g^{jm}(v_{k,mj} + v_{m,kj}) \\ &= (-p_{,j} + (\lambda + \mu)v^k_{,kj})g^{ij} + \mu g^{jm}v^i_{,jm}. \end{aligned} \qquad (8.22.1)$$

Thus the Navier-Stokes equations can be written

$$\rho a^i = \rho f^i - g^{ij}p_{,j} + (\lambda + \mu)v^k_{,kj}g^{ij} + \mu g^{jk}v^i_{,jk} \tag{8.22.2}$$

or in covariant form

$$\rho a_i = \rho f_i - p_{,i} + (\lambda + \mu)v^k_{,ki} + \mu g^{jk}v_{i,jk}. \tag{8.22.3}$$

The term $v^k_{,ki}$ can be written grad div **v** and the last term is the Laplacian of **v**. Thus we have

$$\rho \mathbf{a} = \rho \mathbf{f} - \text{grad}\, p + (\lambda + \mu)\, \text{grad div}\, \mathbf{v} + \mu \nabla^2 \mathbf{v} \tag{8.22.4}$$

which is exactly as before.

In physical components we have

$$\rho a(i) = \rho f(i) - g_{ii}^{1/2} g^{ij} p_{,j} + (\lambda + \mu) g_{ii}^{1/2} g^{ij} (\text{div}\, v)_{,j} + \mu g_{ii}^{1/2} \nabla^2 [g_{ii}^{1/2} v(i)] \tag{8.22.5}$$

and we can use the divergence expression in Eq. (7.56.7) for the penultimate term of this last equation. In orthogonal coordinates this term becomes

$$(\lambda + \mu) \frac{1}{h_i} \frac{\partial}{\partial x^i} \left\{ \frac{1}{h_1 h_2 h_3} \frac{\partial}{\partial x^k} \left[\frac{h_1 h_2 h_3}{h_k} v(k) \right] \right\},$$

where there is summation on k but not on i. The expression in Eq. (7.56.2) for the Laplacian of a scalar cannot be used for the last term of Eq. (8.22.5), since this is the Laplacian of a vector. Even in orthogonal coordinates the form of this is too complicated to be given in generality; some simple examples occur below.

Exercise 8.22.1. Obtain the following Navier-Stokes equations for incompressible flow in cylindrical polar coordinates. ($\nu = \mu/\rho$),

$$\frac{\partial v_r}{\partial t} + v_r \frac{\partial v_r}{\partial r} + \frac{v_\phi}{r} \frac{\partial v_r}{\partial \phi} + v_z \frac{\partial v_r}{\partial z} - \frac{v_\phi^2}{r} = -\frac{1}{\rho} \frac{\partial p}{\partial r}$$

$$+ \nu \left\{ \frac{\partial^2 v_r}{\partial r^2} + \frac{1}{r^2} \frac{\partial^2 v_r}{\partial \phi^2} + \frac{\partial^2 v_r}{\partial z^2} + \frac{1}{r} \frac{\partial v_r}{\partial r} - \frac{2}{r^2} \frac{\partial v_\phi}{\partial \phi} - \frac{v_r}{r^2} \right\},$$

$$\frac{\partial v_\phi}{\partial t} + v_r \frac{\partial v_\phi}{\partial r} + \frac{v_\phi}{r} \frac{\partial v_\phi}{\partial \phi} + v_z \frac{\partial v_\phi}{\partial z} + \frac{v_r v_\phi}{r} = -\frac{1}{\rho r} \frac{\partial p}{\partial \phi}$$

$$+ \nu \left\{ \frac{\partial^2 v_\phi}{\partial r^2} + \frac{1}{r^2} \frac{\partial^2 v_\phi}{\partial \phi^2} + \frac{\partial^2 v_\phi}{\partial z^2} + \frac{1}{r} \frac{\partial v_\phi}{\partial r} + \frac{2}{r^2} \frac{\partial v_r}{\partial \phi} - \frac{v_\phi}{r^2} \right\},$$

$$\frac{\partial v_z}{\partial t} + v_r \frac{\partial v_z}{\partial r} + \frac{v_\phi}{r} \frac{\partial v_z}{\partial \phi} + v_z \frac{\partial v_z}{\partial z} = -\frac{1}{\rho} \frac{\partial p}{\partial z}$$

$$+ \nu \left\{ \frac{\partial^2 v_z}{\partial r^2} + \frac{1}{r^2} \frac{\partial^2 v_z}{\partial \phi^2} + \frac{\partial^2 v_z}{\partial z^2} + \frac{1}{r} \frac{\partial v_z}{\partial r} \right\}.$$

Exercise 8.22.2. Obtain the following equations in spherical polar coordinates r, θ, ϕ.

$$\frac{\partial v_r}{\partial t} + v_r \frac{\partial v_r}{\partial r} + \frac{v_\theta}{r}\frac{\partial v_r}{\partial \theta} + \frac{v_\phi}{r \sin\theta}\frac{\partial v_r}{\partial \phi} - \frac{v_\theta^2 + v_\phi^2}{r} = -\frac{1}{\rho}\frac{\partial p}{\partial r}$$
$$+ \nu\left\{\frac{1}{r}\frac{\partial^2(rv_r)}{\partial r^2} + \frac{1}{r^2}\frac{\partial^2 v_r}{\partial \theta^2} + \frac{1}{r^2 \sin^2\theta}\frac{\partial^2 v_r}{\partial \phi^2} + \frac{\cot\theta}{r^2}\frac{\partial v_r}{\partial \theta}\right.$$
$$\left. - \frac{2}{r^2}\frac{\partial v_\theta}{\partial \theta} - \frac{2}{r^2 \sin\theta}\frac{\partial v_\phi}{\partial \phi} - \frac{2v_r}{r^2} - \frac{2\cot\theta}{r^2}v_\theta\right\},$$

$$\frac{\partial v_\theta}{\partial t} + v_r \frac{\partial v_\theta}{\partial r} + \frac{v_\theta}{r}\frac{\partial v_\theta}{\partial \theta} + \frac{v_\phi}{r \sin\theta}\frac{\partial v_\theta}{\partial \phi} + \frac{v_r v_\theta}{r} - \frac{v_\phi^2 \cot\theta}{r} = -\frac{1}{\rho r}\frac{\partial p}{\partial \theta}$$
$$+ \nu\left\{\frac{1}{r}\frac{\partial^2(rv_\theta)}{\partial r^2} + \frac{1}{r^2}\frac{\partial^2 v_\theta}{\partial \theta^2} + \frac{1}{r^2 \sin^2\theta}\frac{\partial^2 v_\theta}{\partial \phi^2} + \frac{\cot\theta}{r^2}\frac{\partial v_\theta}{\partial \theta}\right.$$
$$\left. - \frac{2\cot\theta}{r^2 \sin\theta}\frac{\partial v_\phi}{\partial \phi} + \frac{2}{r^2}\frac{\partial v_r}{\partial \theta} - \frac{v_\theta}{r^2 \sin^2\theta}\right\},$$

$$\frac{\partial v_\phi}{\partial t} + v_r \frac{\partial v_\phi}{\partial r} + \frac{v_\theta}{r}\frac{\partial v_\phi}{\partial \theta} + \frac{v_\phi}{r \sin\theta}\frac{\partial v_\phi}{\partial \phi} + \frac{v_r v_\phi}{r} + \frac{v_\theta v_\phi \cot\theta}{r} = \frac{-1}{\rho r \sin\theta}\frac{\partial p}{\partial \phi}$$
$$+ \nu\left\{\frac{1}{r}\frac{\partial^2(rv_\phi)}{\partial r^2} + \frac{1}{r^2}\frac{\partial^2 v_\phi}{\partial \theta^2} + \frac{1}{r^2 \sin^2\theta}\frac{\partial^2 v_\phi}{\partial \phi^2} + \frac{\cot\theta}{r^2}\frac{\partial v_\phi}{\partial \theta}\right.$$
$$\left. + \frac{2}{r^2 \sin\theta}\frac{\partial v_r}{\partial \phi} + \frac{2\cot\theta}{r^2 \sin\theta}\frac{\partial v_\theta}{\partial \phi} - \frac{v_\phi}{r^2 \sin^2\theta}\right\}.$$

Exercise 8.22.3. Obtain the coefficients of $(\lambda + \mu)$ in the Navier-Stokes equations in these two cases.

Exercise 8.22.4. Specialize the equations in cylindrical polar coordinates to the cases where v_z and v_ϕ are the only nonvanishing components of velocity.

8.31. Convected coordinates

Before going on to discuss more general constitutive equations let us look again at the material or convected coordinate system. When this was introduced before we did not have the full panoply of tensor analysis with which to handle the curvilinear coordinates.

Consider, for example, the flow of a Newtonian fluid in a circular tube for which $v_r = v_\phi = 0$, $v_z = 2U\{1 - r^2/a^2\}$. The surface $z = z_0$ at time $t = t_0$ becomes at some later time the paraboloid

$$z = z_0 + 2U\{1 - r^2/a^2\}(t - t_0). \tag{8.31.1}$$

The surfaces r = constant and ϕ = constant are not distorted by the flow as is shown in Fig. 8.1. Let us take as spatial coordinates the cylindrical polars already used

$$x^1 = r, \qquad x^2 = \phi, \qquad x^3 = z.$$

Then the material coordinates will be

$$\xi^1 = r, \qquad \xi^2 = \phi, \qquad \xi^3 = z_0.$$

The two are related by

$$\xi^1 = x^1, \qquad \xi^2 = x^2, \qquad \xi^3 = x^3 - 2U\{1 - (x^1/a)^2\}t, \tag{8.31.2}$$

where for convenience we set $t_0 = 0$. This is a time dependent transformation of coordinates.

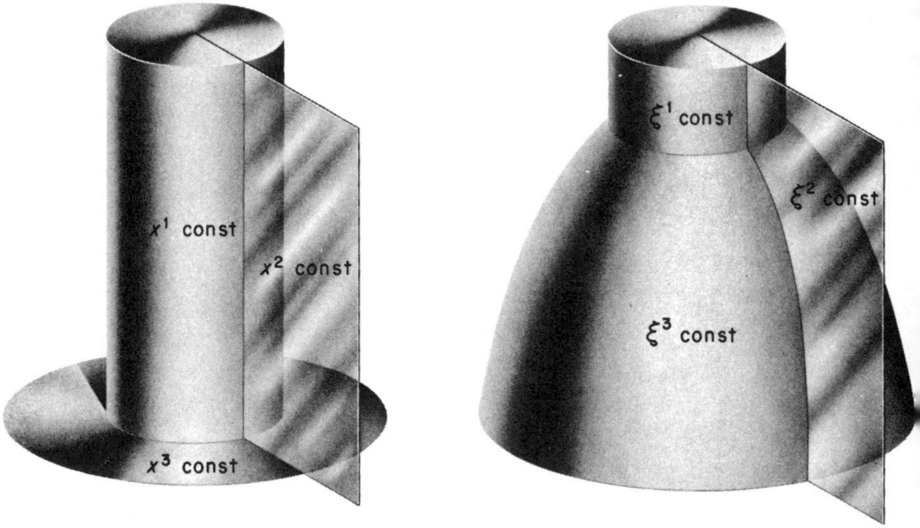

Fig. 8.1

In the fixed coordinate system the metric is given by

$$ds^2 = (dx^1)^2 + (x^1\,dx^2)^2 + (dx^3)^2 = g_{ij}\,dx^i\,dx^j. \tag{8.31.3}$$

In the convected coordinate system

$$ds^2 = \gamma_{ij} d\xi^i d\xi^j \tag{8.31.4}$$

where

$$\gamma_{ij} = \frac{\partial x^p}{\partial \xi^i}\frac{\partial x^q}{\partial \xi^j} g_{pq}.$$

Thus

$$\begin{aligned}\gamma_{11} &= 1 + (4Ut\xi^1/a^2)^2, \\ \gamma_{22} &= (\xi^1)^2, \\ \gamma_{33} &= 1, \\ \gamma_{13} &= \gamma_{31} = -(4Ut\xi^1/a^2), \\ \gamma_{12} &= \gamma_{21} = \gamma_{23} = \gamma_{32} = 0.\end{aligned} \tag{8.31.5}$$

The metric in the convected coordinate system is time dependent. However, we notice that the determinant $\gamma = (\xi^1)^2$ is independent of time. Moreover, the Jacobian of the transformation from convected to fixed coordinates, $J = 1$, which accords with the fact that the motion is isochoric.

The convected coordinate system will in general be a curvilinear coordinate system with a metric tensor that is a function of time, $\gamma_{ij}(\boldsymbol{\xi}, t)$. The transformation from fixed coordinates $\mathbf{x}$ to convected coordinates $\boldsymbol{\xi}$ is not quite the same as the transformation between two fixed coordinate systems $\mathbf{x}$ and $\bar{\mathbf{x}}$, for time does not enter the latter. The convected coordinate system need not necessarily be the material coordinate system such as has been constructed, for any transformation to coordinates $\bar{\boldsymbol{\xi}}$ that does not depend on time could be made and would give an equally valid system of convected coordinates.

Exercise 8.31.1. Show that no proper transformation of convected coordinates can make all the elements of γ_{ij} be independent of time.

8.32. Convective differentiation

Suppose we have a tensor $\alpha^i_j(\boldsymbol{\xi}, t)$ in convected coordinates. $d\alpha^i_j/dt$ denotes the rate of change of α^i_j keeping the convected coordinates constant. The same tensor in a fixed coordinate system we denote by $A^p_q(\mathbf{x}, t)$ and we wish to calculate the components in the fixed coordinate system of $d\alpha^i_j/dt$. This derivative known as the convective derivative of A^p_q was first described by Oldroyd (Proc. Roy. Soc. A200 (1950), p. 523), who used the notation $\mathfrak{d}A/\mathfrak{d}t$. We shall use the notation d_cA/dt for this derivative.

To calculate $d_cA^p_q/dt$ we set down the transformation in the form

$$\frac{\partial x^p}{\partial \xi^i}\alpha^i_j = \frac{\partial x^q}{\partial \xi^j}A^p_q \tag{8.32.1}$$

and differentiate both sides with respect to t holding $\boldsymbol{\xi}$ constant. Then recalling that

$$\frac{d}{dt}\frac{\partial x^p}{\partial \xi^i} = \frac{\partial v^p}{\partial \xi^i} = \frac{\partial v^p}{\partial x^n}\frac{\partial x^n}{\partial \xi^i},$$

we have

$$\frac{\partial x^p}{\partial \xi^i}\frac{d\alpha^i_j}{dt} + \alpha^i_j\frac{\partial v^p}{\partial x^n}\frac{\partial x^n}{\partial \xi^i} = \frac{\partial x^q}{\partial \xi^j}\left[\frac{\partial A^p_q}{\partial t} + \frac{\partial A^p_q}{\partial x^m}v^m\right] + \frac{\partial v^q}{\partial x^s}\frac{\partial x^s}{\partial \xi^j}A^p_q.$$

Multiplying by $\partial\xi^j/\partial x^r$ and rearranging we have

$$\frac{\partial\xi^j}{\partial x^r}\frac{\partial x^p}{\partial\xi^i}\frac{d\alpha_j^i}{dt}=\frac{d_cA_r^p}{dt}=\left(\frac{\partial\xi^j}{\partial x^r}\frac{\partial x^q}{\partial\xi^j}\right)\left[\frac{\partial A_q^p}{\partial t}+\frac{\partial A_q^p}{\partial x^m}v^m\right]$$
$$+\frac{\partial v^q}{\partial x^s}\left(\frac{\partial x^s}{\partial\xi^j}\frac{\partial\xi^j}{\partial x^r}\right)A_q^p-\frac{\partial v^p}{\partial x^n}\left(\frac{\partial x^n}{\partial\xi^i}\frac{\partial\xi^j}{\partial x^r}\alpha_j^i\right)$$

or

$$\frac{d_cA_r^p}{dt}=\frac{\partial A_r^p}{\partial t}+v^m\frac{\partial A_r^p}{\partial x^m}+\frac{\partial v^q}{\partial x^r}A_q^p-\frac{\partial v^p}{\partial x^n}A_r^n. \tag{8.32.2}$$

In this form it is not clear that we have a tensor on the right-hand side. However,

$$\frac{\partial A_r^p}{\partial x^m}=A_{r,m}^p-\begin{Bmatrix}p\\m\ \ n\end{Bmatrix}A_r^n+\begin{Bmatrix}q\\m\ \ r\end{Bmatrix}A_q^p$$

$$\frac{\partial v^q}{\partial x^r}=v_{,r}^q-\begin{Bmatrix}q\\m\ \ r\end{Bmatrix}v^m$$

$$\frac{\partial v^p}{\partial x^n}=v_{,n}^p-\begin{Bmatrix}p\\m\ \ n\end{Bmatrix}v^m$$

so that the last three terms on the right-hand side can be written as covariant derivatives and we have

$$\frac{d_cA_r^p}{dt}=\frac{\partial A_r^p}{\partial t}+A_{r,m}^pv^m+A_q^pv_{,r}^q-A_r^nv_{,n}^p. \tag{8.32.3}$$

In this form the tensor character is evident.

A little thought will show that for a tensor of higher order this formula must be modified by writing the last two terms as sums. The first is a sum of products $A_{\cdot q\cdot\cdot}^{\cdot p\cdot\cdot}v_{,r}^q$ for each covariant index and the second a sum of products $-A_{\cdot q\cdot\cdot}^{\cdot n\cdot\cdot}v_{,n}^p$ for each contravariant index. We shall not torture the typesetter by trying to write this down in glaring generality.

The convective derivative of the velocity v^i is

$$\frac{d_cv^i}{dt}=\frac{\partial v^i}{\partial t}+v^mv_{,m}^i-v^nv_{,n}^i=\frac{\partial v^i}{\partial t}. \tag{8.32.4}$$

This makes sense, for an observer moving with the fluid sees the velocity to be zero but an acceleration which in spatial coordinates is $\partial v^i/\partial t$.

Exercise 8.32.1. Show that if A_q^p is a relative tensor of weight W, then

$$\frac{d_cA_q^p}{dt}=\frac{\partial A_q^p}{\partial t}+v^mA_{q,m}^p+v_{,q}^mA_m^p-v_{,m}^pA_q^m+Wv_{,m}^mA_q^p.$$

8.33. Strain and rate of strain in convected coordinates

In the convected coordinate system

$$ds^2 = \gamma_{ij}(\boldsymbol{\xi}, t)\, d\xi^i\, d\xi^j. \tag{8.33.1}$$

If ds_0 denote the length of the element at time $t = t_0$ we may write

$$\begin{aligned} ds^2 - ds_0^2 &= [\gamma_{ij}(\boldsymbol{\xi}, t) - \gamma_{ij}(\boldsymbol{\xi}, t_0)]\, d\xi^i\, d\xi^j \\ &= 2\eta_{ij}(\boldsymbol{\xi}, t, t_0)\, d\xi^i\, d\xi^j, \end{aligned} \tag{8.33.2}$$

where

$$\eta_{ij}(\boldsymbol{\xi}, t, t_0) = \tfrac{1}{2}[\gamma_{ij}(\boldsymbol{\xi}, t) - \gamma_{ij}(\boldsymbol{\xi}, t_0)]. \tag{8.33.3}$$

This tensor may be called the strain tensor since it expresses the state of strain at time t relative to that at time t_0.

Its derivative is the rate of strain or deformation tensor

$$\epsilon_{ij}(\boldsymbol{\xi}, t) = \frac{d}{dt}\eta_{ij}(\boldsymbol{\xi}, t) = \frac{1}{2}\frac{d}{dt}\gamma_{ij}(\xi, t). \tag{8.33.4}$$

which is still expressed in convected coordinates. Differentiating Eq. (8.33.2)

$$(ds)\frac{d}{dt}(ds) = \epsilon_{ij}(\boldsymbol{\xi}, t)\, d\xi^i\, d\xi^j$$

or

$$\frac{d}{dt}\log(ds) = \frac{\epsilon_{ij}\, d\xi^i\, d\xi^j}{\gamma_{ij}\, d\xi^i\, d\xi^j}. \tag{8.33.5}$$

If $e_{ij}(x, t)$ is the rate of strain tensor in fixed coordinates, then Eq. (8.33.4) transforms to

$$\begin{aligned} e_{ij} &= \frac{1}{2}\frac{d_c}{dt} g_{ij} \\ &= \frac{1}{2}\left[\frac{\partial g_{ij}}{\partial t} + v^m g_{ij,m} + g_{ip}v^p_{,j} + g_{pj}v^p_{,i}\right] \end{aligned}$$

or

$$e_{ij} = \tfrac{1}{2}(v_{i,j} + v_{j,i}), \tag{8.33.6}$$

since g_{ij} is independent of time and $g_{ij,m}$ is always zero. We have obtained the deformation tensor by considering the rate of strain in the convected coordinates and it has therefore emerged without its antisymmetric associate.

The volume element is proportional to $\gamma^{1/2}$, the square root of the determinant of the metric tensor. Now

$$\begin{aligned} \frac{d\gamma}{dt} = \frac{d}{dt}\epsilon^{pqr}\gamma_{1p}\gamma_{2q}\gamma_{3r} &= 2\epsilon^{pqr}(\epsilon_{1p}\gamma_{2q}\gamma_{3r} + \gamma_{1p}\epsilon_{2q}\gamma_{3r} + \gamma_{1p}\gamma_{2q}\epsilon_{3r}) \\ &= 2\gamma\epsilon^{i}_{\cdot i}. \end{aligned} \tag{8.33.7}$$

The quantity

$$\Delta = \frac{1}{2} \log \frac{\gamma(\xi, t)}{\gamma(\xi, t_0)}$$

is known as the relative dilatation. Thus

$$\frac{d\Delta}{dt} = \epsilon^{i}_{\cdot i},$$

or, in fixed coordinates,

$$\frac{d_c\Delta}{dt} = e^{i}_{\cdot i} = v^{i}_{,i}. \tag{8.33.8}$$

Exercise 8.33.1. Show that

$$\frac{d_c e_{ij}}{dt} = \frac{1}{2}\left\{\frac{\partial v_{i,j}}{\partial t} + \frac{\partial v_{j,i}}{\partial t} + v^m v_{i,jm} + v^m_{,i} v_{j,m} + 2v^m_{,i} v_{m,j} + v_{i,m} v^m_{,j}\right\}.$$

Exercise 8.33.2. Show that

$$\frac{d_c g^{ij}}{dt} = -2e^{ij}.$$

Exercise 8.33.3. Show that $\dfrac{d_c}{dt} A_{pq} - g_{iq} \dfrac{d_c}{dt} A^i_p = 2e_{iq} A^i_p$.

Exercise 8.33.4. Show that $\dfrac{d_c}{dt} A^{pq} - g^{pj} \dfrac{d_c}{dt} A^q_j = -2e^{pj} A^q_j$.

Exercise 8.33.5. Show that $\dfrac{d_c}{dt} (g^M A^p_q) - g^M \dfrac{d_c}{dt} A^p_q = 2Mg^M e^r_r A^p_q$.

8.34. Constitutive equations

The constitutive equations should be formulated in convected coordinates for they represent the intrinsic behavior of the material or its inherent response in terms of stress and deformation. This point was first clearly brought out by Oldroyd (loc. cit.) who stated that "only those tensor quantities need be considered which have a significance for the material element independent of its motion as a whole in space." This idea has now been enlarged to the principle of material indifference and its consequences have been more fully developed as we shall indicate in Section 8.4. For the moment we will recover some results we already have and indicate some new ones.

The stress tensor $T^{ij} = -pg^{ij} + P^{ij}$ becomes, in the convected coordinates, $\tau^{ij} = -p\gamma^{ij} + \pi^{ij}$. p still represents the hydrostatic pressure present in this form for all fluids at rest. It is the relation of π^{ij} to ϵ_{ij} that distinguishes different fluids. A constitutive equation in general must be set up as the relation between π^{ij}, ϵ_{ij} the thermodynamic variables and material constants. In its broadest case this would be a set of integro-differential equations which might represent the stress as a functional of the whole past history of the material, but in the first cases we assume that the fluid has no elastic properties and so no "memory" of the past.

For the Newtonian fluid we assume a linear isotropic relation between π^{ij} and ϵ_{ij}. The most general form of this is

$$\begin{aligned}\pi^{ij} - \lambda\gamma^{ij}\gamma^{pq}\epsilon_{pq} + \mu(\gamma^{ip}\gamma^{jq} + \gamma^{iq}\gamma^{jp})\epsilon_{pq} \\ = \lambda\epsilon^{m}_{m}\gamma^{ij} + 2\mu\epsilon^{ij}.\end{aligned} \tag{8.34.1}$$

When this is translated into fixed coordinates it becomes

$$p^{ij} = \lambda e^{m}_{m}g^{ij} + 2\mu e^{ij}, \tag{8.34.2}$$

which we have already used in Eq. (8.21.8).

For the Stokesian fluid just such an argument as we have used earlier in Section 5.22 shows that the most general isotropic relation between stress and rate of strain is

$$\pi^{ij} = \alpha\gamma^{ij} + \beta\epsilon^{ij} + \gamma'\epsilon^{ik}\epsilon^{j}_{k}. \tag{8.34.3}$$

Here α, β, and γ' are functions only of the invariants of the rate of strain tensor. This gives

$$p^{ij} = \alpha g^{ij} + \beta e^{ij} + \gamma' e^{ik}e^{j}_{k} \tag{8.34.4}$$

which is the constitutive relation Eq. (5.22.5), we have obtained before in Cartesian coordinates.

Up to this point there seems to have been little advantage to be had from going into convected coordinates for the correct equations should have been obtained in the fixed coordinate system. The importance of what Oldroyd did comes to light when we consider a fluid with elastic as well as viscous properties. A model of the colloid suspension suggested by Fröhlich and Sack (Proc. Roy. Soc. A185, (1946) p. 415) serves well to illustrate this. They considered a suspension of elastic spheres in a viscous fluid and on this basis suggested the equation

$$\left(1 + \lambda\frac{d}{dt}\right)\pi_{ij} = 2\mu\left(1 + \tau\frac{d}{dt}\right)\epsilon_{ij} \tag{8.34.5}$$

where λ and τ are two additional constants. In the fixed coordinate system this would be

$$\left(1 + \lambda\frac{d_c}{dt}\right)P_{ij} = 2\mu\left(1 + \tau\frac{d_c}{dt}\right)e_{ij}$$

or

$$\begin{aligned}P_{ij} + \lambda\left\{\frac{\partial P_{ij}}{\partial t} + v^m P_{ij,m} + v^m_{,i}P_{mj} + v^m_{,j}P_{im}\right\} \\ = 2\mu e_{ij} + 2\mu\tau\left\{\frac{\partial e_{ij}}{\partial t} + v^m e_{ij,m} + v^m_{,i}e_{mj} + v^m_{,j}e_{im}\right\},\end{aligned} \tag{8.34.6}$$

which destroys the apparent simplicity of the constitutive relation. However, a major difficulty in making an invariant generalization such as Eq. (8.34.5) from Cartesian coordinates is that its consequences may be rather sensitive to the exact form of equation. It might be thought that the equation

$$\left(1 + \lambda \frac{d_c}{dt}\right) P^{ij} = 2\mu\left(1 + \tau \frac{d_c}{dt}\right) e^{ij} \tag{8.34.7}$$

would be an equally valid form of the equations. However, Oldroyd shows that the behavior of the two materials represented by Eqs. (8.34.6) and (8.34.7) in a rotating cylinder viscometer would be totally different.*

In other forms of equations that have been suggested, integral expressions such as

$$\epsilon_{ij}(\boldsymbol{\xi}, t) - \int_{-\infty}^{t} \psi(t - s)\epsilon_{ij}(\boldsymbol{\xi}, s)\, ds$$

have been used. Oldroyd has also shown how this can be consistently converted into fixed coordinates. We shall not discuss this here but the result is worth quoting, namely,

$$e_{ij}(\mathbf{x}, t) - \int_{-\infty}^{t} \psi(t - s)e_{pq}(\mathbf{y}, s)\frac{\partial y^p}{\partial x^i}\frac{\partial y^q}{\partial x^j}\, ds,$$

where $\mathbf{y}$ is the position of the particle with convected coordinates $\boldsymbol{\xi}$ at time s.

8.4. The general theory of constitutive equations

It should be evident that we have run into quite sufficient complication in our treatment of the Newtonian fluid. Rather than continuing to plough on into more and more obscure convolutions let us back away and consider rather broadly the field that has to be cultivated. We should emphasize again the distinction that has been made between the equations of motion or field equations and the constitutive equations. Thus for the former we have Cauchy's equation of motion (8.13.3)

$$\rho a^i = \rho f^i + T^{ij}_{,j}.$$

This relates the motion to the body forces and stress tensor but cannot say anything about the material itself. It is a universal law valid for all continuous media. Moreover, any stress distribution is compatible with these equations and any motion, for we may look at them as defining the body force vector f^i that brings it to the motion about. Thus no help can be obtained from this equation in defining the internal constitution of the fluid. If we are dealing with the situation where there is no internal angular momentum, then the

* A good discussion of the different concepts of the rates of change of tensors is given in a paper by L. I. Sedov, Prik. Mat. i Mekh., 24 (1960) pp. 393–398.

conservation of angular momentum requires the stress tensor to be symmetric

$$T^{ij} = T^{ji}.$$

This is a restriction which must be satisfied by any constitutive equations valid in this situation.

The great distinction between the field and constitutive equations is that the latter represent the intrinsic response of the material. This response must be seen to be the same by all observers for otherwise it would not be intrinsic to the material. Actually, the hypothesis of Stokes that the constitutive relation should be independent of the rigid body motion of an element is expressing this for we have seen that the rate of strain can be represented by pure stretching along three mutually perpendicular axes plus the rotation of these three axes. Now if we choose to observe the fluid from a frame of reference which is moving and rotating with these axes, all that is left is the deformation or stretching along the three axes. Thus the stress tensor cannot depend on the antisymmetric part of the rate of strain or $\mathbf{T} = f(\mathbf{e})$. This is the hypothesis I of Section 5.21. It is possible therefore to assert a much more general principle and prove that Stokes' hypotheses are consequences of this. This is the principle of material indifference: *The response of the material is the same for all observers*. It has been used implicitly in all correct work on the constitutive equations but was first isolated and explicitly stated by Noll (J. Rat. Mech. and Anal. 4 (1955), pp. 3–81).

Truesdell (Handbuch der Physik Bd. III/1, Sect. 293) lists six other principles that must be observed in the setting up of constitutive equations. They are:

(i) *Consistency* with the general principles of balance of mass, momentum and energy.

(ii) *Coordinate invariance*. The constitutive equation must be framed in tensor language to ensure that the relation is the same in all coordinate frames.

(iii) *Isotropy*. A material is isotropic if any rotation of the *material* coordinates leaves the constitutive equation invariant. It is aeolotropic if invariant only under certain subgroups of the full rotation group.

(iv) *Just setting*. When the constitutive equations are combined with the equations of motion, energy, and continuity, they should be a unique solution for physically sensible initial and boundary conditions. This requirement has only been met in the very simplest of cases. A necessary, but far from sufficient, condition is that the number of unknowns should be equal to the number of equations.

(v) *Dimensional invariance*. The material constants (such as viscosity, etc.) on which the behavior of the fluid depends must be specified in a way which is consistent with the classical π-theorem of dimensional analysis.

(vi) *Material indifference.* This is the principle we have discussed above which, as we have seen, is something more than the coordinate invariance (ii).

(vii) *Equipresence.* In the simpler approaches to continuum mechanics certain coupling effects are ruled out. For example, we associate stress with rate of strain and heat flux with temperature gradient but do not cross link them. From a general viewpoint such separations are quite arbitrary and an independent variable present in one constitutive equation should be present in all, In practice this Olympian approach has seldom been possible and more commonly one or two coupling effects have been brought in to explain phenomena actually occurring. Its value and importance, however, have been brought out by Truesdell's analysis of the so-called Maxwellian fluid. This is a fluid in which the stress and energy flux depend on the derivatives of the thermodynamic variables and velocity and for which a viscosity, thermal conductivity, and temperature, but no other material coefficients, are defined. From very general considerations of invariance it can be shown that only certain combinations of terms can occur in the stress and energy flux tensors. This discrimination of effects is no longer an arbitrary one but arises naturally from the requirements of invariance.

A full understanding and exploitation of these results is currently growing. While the mathematician chiefly rejoices in the clarity of thought that is being attained, the engineer must always remember that the applications and results that he needs will come from such rationality and from such alone.

The Geometry of Surfaces in Space 9

9.11. Surface coordinates

We have had occasion several times to express the equation of a two-dimensional surface in three-dimensional space in the form

$$x^3 = f(x^1, x^2) \quad \text{or} \quad F(x^1, x^2, x^3) = 0. \tag{9.11.1}$$

For example, the sphere is given either by the equation

$$x^3 = \pm\{a^2 - (x^1)^2 - (x^2)^2\}^{1/2}$$

or by

$$(x^1)^2 + (x^2)^2 + (x^3)^2 - a^2 = 0.$$

The second representation is perhaps to be preferred as the first one is suitable only for surfaces we have called 3-elementary or 3-composite. An alternative and more interesting way of representing the surface is by constructing a coordinate grid on the surface itself and expressing the position of a point in space in terms of these surface coordinates. Thus, if in Fig. 9.1 u^1 is the latitude of a point on the sphere and u^2 its longitude from some fixed meridian, say $x^2 = 0$, then

$$x^1 = a \sin u^1 \cos u^2,$$

$$x^2 = a \sin u^1 \sin u^2,$$

$$x^3 = a \cos u^1.$$

We therefore have the equation of a surface in the form

$$x^i = x^i(u^1, u^2). \tag{9.11.2}$$

u^1 and u^2 are called *surface coordinates* and the lines u^1 = constant or u^2 = constant are called *coordinate lines.*

The surface coordinates u^1 and u^2 are intrinsic to the surface. That is to say, they have been set up as if by creatures living entirely within the surface

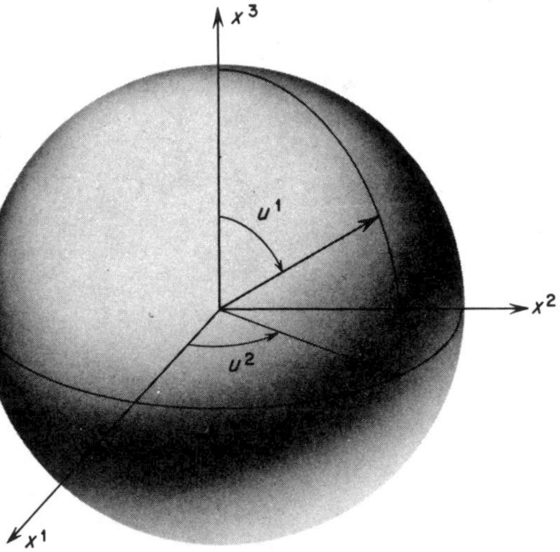

Fig. 9.1

itself and without any reference to the three-dimensional space in which the surface is embedded. It will be important therefore to find out how much information about the surface can be gained from this intrinsic viewpoint and how this is supplemented when the connection between the surface and the surrounding space is brought in. We shall find, for example, that a certain impression of curvature can be gained from the intrinsic geometry but that this is enlarged when the surface is viewed from without.

Greek letters will be used for the affixes of surface quantities and the summation will apply. Thus

$$u^\alpha u_\alpha = u^1 u_1 + u^2 u_2$$

and a free Greek index may take on either of the values 1 or 2.

9.12. Transformations of surface coordinates—surface tensors

There is nothing sacrosanct about a particular system of surface coordinates and we wish to express physical quantities in such a way that their relations are seen to be independent of the coordinate system. Suppose that $\bar{u}^\alpha = \bar{u}^\alpha(u^1, u^2)$ is another coordinate system; then we shall call this transformation a proper transformation provided that it is continuous and can be inverted to give $u^\alpha = u^\alpha(\bar{u}^1, \bar{u}^2)$. This we know is equivalent to the requirement that the Jacobian

$$j = \frac{\partial(\bar{u}^1, \bar{u}^2)}{\partial(u^1, u^2)} \tag{9.12.1}$$

should neither vanish nor be infinite.

The differentials of the coordinates du^α become

$$d\bar{u}^\beta = \frac{\partial \bar{u}^\beta}{\partial u^\alpha} du^\alpha, \tag{9.12.2}$$

which is the prototype of the contravariant surface vector. More generally we say that the entity A^α with two components A^1 and A^2 in the u^1, u^2 coordinate system is a contravariant surface vector if in the coordinate system $\bar{u}^1$, $\bar{u}^2$ its components are

$$\bar{A}^\beta = \frac{\partial \bar{u}^\beta}{\partial u^\alpha} A^\alpha. \tag{9.12.3}$$

Similarly, if $f(u^1, u^2)$ is any function defined on the surface, its partial derivatives $\partial f/\partial u^1$ and $\partial f/\partial u^2$ transform as

$$\frac{\partial f}{\partial \bar{u}^\beta} = \frac{\partial u^\alpha}{\partial \bar{u}^\beta} \frac{\partial f}{\partial u^\alpha}. \tag{9.12.4}$$

Again we define a covariant surface vector as an entity with components A_1 and A_2 that transform according to the rule

$$\bar{A}_\beta = \frac{\partial u^\alpha}{\partial \bar{u}^\beta} A_\alpha. \tag{9.12.5}$$

A surface tensor of any order is defined in a precisely similar fashion. For example, a mixed third order tensor $A^\alpha_{\beta\gamma}$ would obey the transformation law

$$\bar{A}^\lambda_{\mu\nu} = \frac{\partial \bar{u}^\lambda}{\partial u^\alpha} \frac{\partial u^\beta}{\partial \bar{u}^\mu} \frac{\partial u^\gamma}{\partial \bar{u}^\nu} A^\alpha_{\beta\gamma}. \tag{9.12.6}$$

The algebra of tensors carries over immediately to surface tensors with the trivial adjustment that the summations only contain two terms instead of three. In particular, the quotient rule holds for surface tensors.

Exercise 9.12.1. If a particle moves in a surface, its surface coordinates being functions of time $u^\alpha(t)$, show that its velocity, du^α/dt and higher rates of change of position are all contravariant vectors.

Exercise 9.12.2. Show that the surface flux of any covariant vector property can be expressed as a mixed second order tensor.

Exercise 9.12.3. Prove the quotient rule for surface tensors.

Exercise 9.12.4. Show that $\epsilon^{11} = \epsilon^{22} = 0,\ \epsilon^{12} = -\epsilon^{21} = 1$ are the components of a relative contravariant second order tensor of weight 1.

9.13. The metric tensor

The same denizens of the surface who originally set up the coordinate grid could also obtain a metric tensor if they had an infinitesimal yardstick of length ds. Being infinitesimal it would always lie flat on the surface and by turning it around at a point they could measure the coordinate increments du^1 and du^2 for which the diagonal distance was ds. They would be gratified to find that they did not require a complete table of all their measurements at each point, for their findings would be expressed by the formula

$$\begin{aligned} ds^2 &= a_{11}(du^1)^2 + 2a_{12}\, du^1\, du^2 + a_{22}(du^2)^2 \\ &= a_{\alpha\beta}(u^1, u^2)\, du^\alpha\, du^\beta. \end{aligned} \tag{9.13.1}$$

Being educated they would conclude that they lived in a Riemannian space with metric tensor $a_{\alpha\beta}$.

This is the intrinsic construction of the metric tensor and of course the mathematician would simply say that the space is Riemannian if the metric tensor exists. Since we are free to leave the surface and look at it from the space in which it is embedded we can see why its inhabitants would discover it to be Riemannian. Suppose y^i are Cartesian coordinates in the Euclidean space in which the surface lies and that it is given by the equations

$$y^i = y^i(u^1, u^2). \tag{9.13.2}$$

Then, since $ds^2 = \Sigma\,(dy^i)^2$ in Euclidean space,

$$\begin{aligned} ds^2 = \sum_{i=1}^{3} (dy^i)^2 &= \sum_{i=1}^{3} \frac{\partial y^i}{\partial u^\alpha}\frac{\partial y^i}{\partial u^\beta}\, du^\alpha\, du^\beta \\ &= a_{\alpha\beta}\, du^\alpha\, du^\beta, \end{aligned}$$

where

$$a_{\alpha\beta} = \sum_{i=1}^{3} \frac{\partial y^i}{\partial u^\alpha}\frac{\partial y^i}{\partial u^\beta}\,. \tag{9.13.3}$$

Evidently it is because the space is Euclidean (or more generally Riemannian, Ex. 9.13.1) and the derivatives $\partial y^i/\partial u^\alpha$ exist that the surface is a Riemannian space. This, however, is not an intrinsic viewpoint. From within the surface it has to be established that Eq. (9.13.1) correlates the observations, just as our conviction that the space of our gross experience is Euclidean rests on the observation that $ds^2 = \Sigma\,(dy^i)^2$.

Since ds^2 is a scalar the quotient rule assures us that $a_{\alpha\beta}$ is a symmetric covariant second order tensor. Its determinant is

$$a = a_{11}a_{22} - (a_{12})^2 \tag{9.13.4}$$

and its conjugate is

$$a^{11} = \frac{a_{22}}{a}, \qquad a^{12} = a^{21} = -\frac{a_{12}}{a}, \qquad a^{22} = \frac{a_{11}}{a}. \tag{9.13.5}$$

The Kronecker delta δ^α_β which is 1 if $\alpha = \beta$ but zero otherwise is related to the metric tensor by

$$a^{\alpha\gamma}a_{\gamma\beta} = a^{\gamma\alpha}a_{\gamma\beta} = a^{\alpha\gamma}a_{\beta\gamma} = \delta^\alpha_\beta . \tag{9.13.6}$$

The metric tensor and its conjugate can be used to raise and lower suffixes as before. Thus

$$A^\alpha = a^{\alpha\beta}A_\beta$$

is the contravariant vector associated with the covariant A_β.

Since $\epsilon^{\alpha\beta}$, where $\epsilon^{11} = \epsilon^{22} = 0$, $\epsilon^{12} = -\epsilon^{21} = 1$, is a relative tensor of weight 1 (Ex. 9.12.4) and

$$j^2 = \frac{a}{\bar{a}}, \tag{9.13.7}$$

it follows that

$$\varepsilon^{\alpha\beta} = \frac{\epsilon^{\alpha\beta}}{a^{1/2}} \tag{9.13.8}$$

is an absolute contravariant tensor. If $\epsilon_{\alpha\beta} = \epsilon^{\alpha\beta}$ then similarly

$$\varepsilon_{\alpha\beta} = \epsilon_{\alpha\beta}a^{1/2} \tag{9.13.9}$$

is an absolute covariant tensor.

Exercise 9.13.1. g_{ij} is the metric tensor of the coordinate system x^1, x^2, x^3 in a Riemannian space in which the surface $x^i = x^i(u^1, u^2)$ is embedded. Show that the metric tensor of the surface is related to the metric tensor of the space by

$$a_{\alpha\beta} = g_{ij}\frac{\partial x^i}{\partial u^\alpha}\frac{\partial x^j}{\partial u^\beta} .$$

Exercise 9.13.2. Show that the conjugate metric tensor $a^{\alpha\beta}$ is indeed a contravariant second order tensor.

Exercise 9.13.3. Show that $j^2 = a/\bar{a}$.

Exercise 9.13.4. Show that $\varepsilon^{\alpha\beta}\varepsilon_{\gamma\beta} = \delta^\alpha_\gamma$.

Exercise 9.13.5. Show that $\varepsilon^{\alpha\lambda}\varepsilon^{\beta\mu}a_{\lambda\mu} = a^{\alpha\beta}$.

Exercise 9.13.6. Obtain the metric and ε tensors for the sphere

$$y^1 = a \sin u^1 \cos u^2, \qquad y^2 = a \sin u^1 \sin u^2, \qquad y^3 = a \cos u^1.$$

Exercise 9.13.7. Show that

$$\varepsilon^{\alpha\lambda}\varepsilon^{\beta\mu} + \varepsilon^{\alpha\mu}\varepsilon^{\beta\lambda} = 2a^{\alpha\beta}a^{\lambda\mu} - a^{\alpha\lambda}a^{\beta\mu} - a^{\alpha\mu}a^{\beta\lambda}$$

Exercise 9.13.8. Show that

$$\varepsilon_{\lambda\alpha}a^{\alpha\beta}\varepsilon_{\beta\gamma}a^{\gamma\mu} = -\delta^{\mu}_{\lambda}.$$

9.14. Length and direction of surface vectors

The metric tensor associates a length ds with an infinitesimal vector whose components are du^1 and du^2. A vector with finite components A^1 and A^2 can be regarded as a very large multiple of an infinitesimal vector; $A^\alpha = M\,du^\alpha$, where M is very large. We would naturally think of the length A of the finite vector as the same multiple of the infinitesimal length ds as its components are of the du^α. Thus

$$|A|^2 = a_{\alpha\beta}A^\alpha A^\beta = A^\alpha A_\alpha. \tag{9.14.1}$$

Any vector can be made a unit vector in the same direction by dividing by its length; thus $A^\alpha/\,|A|$ are components of a unit vector.

If du^α and dv^β are two infinitesimal vectors in the surface $y^i = y^i(u^1, u^2)$, they correspond to two infinitesimal space vectors dy^i and dz^i in Cartesians. Thus

$$dy^i = \frac{\partial y^i}{\partial u^\alpha}\,du^\alpha, \qquad dz^i = \frac{\partial y^i}{\partial u^\beta}\,dv^\beta,$$

the expressions in the parentheses both being partial derivatives of the equations of the surface evaluated at the point. Now the angle between the two infinitesimal vectors in space is θ where

$$\begin{aligned}|dy|\,|dz|\cos\theta &= \sum_{i=1}^{3} dy^i\,dz^i = \sum_{i=1}^{3}\left(\frac{\partial y^i}{\partial u^\alpha}\frac{\partial y^i}{\partial u^\beta}\right)du^\alpha\,dv^\beta\\ &= a_{\alpha\beta}\,du^\alpha\,dv^\beta.\end{aligned}$$

Thus

$$\cos\theta = \frac{a_{\alpha\beta}\,du^\alpha\,dv^\beta}{\{a_{\gamma\delta}\,du^\gamma\,du^\delta a_{\lambda\mu}\,dv^\lambda\,dv^\mu\}^{1/2}}. \tag{9.14.2}$$

For finite vectors we have, similarly,

$$\cos\theta = \frac{a_{\alpha\beta}A^\alpha B^\beta}{|A|\,|B|} = \frac{A^\alpha B_\alpha}{|A|\,|B|}. \tag{9.14.3}$$

If

$$a_{\alpha\beta}A^\alpha B^\beta = 0, \tag{9.14.4}$$

the two vectors are said to be orthogonal, and if $a_{12} = 0$, the coordinate system is called orthogonal. The angle between the coordinate lines is θ_{12}, where $\cos\theta_{12} = a_{12}\{a_{11}a_{22}\}^{-1/2}$. To give a sense to rotation we say that the rotation from a vector A^α to a vector B^β is positive if $\varepsilon_{\alpha\beta}A^\alpha B^\beta$ is positive (cf. Ex. 9.14.2).

Exercise 9.14.1. If b_α is a unit vector, show that $\varepsilon^{\alpha\beta}b_\beta = c^\alpha$ is a unit vector orthogonal to it.

Exercise 9.14.2. Show that the angle between any two unit vectors is given by

$$\sin\theta = \varepsilon_{\alpha\beta}b^\alpha c^\beta.$$

Exercise 9.14.3. In an orthogonal coordinate system $a_{\alpha\alpha} = b_\alpha^2$, $a_{\alpha\beta} = 0$, $\alpha \neq \beta$, show that unit vectors tangent to the coordinate lines are $(b_1, 0)$, $(0, b_2)$, $(b_1^{-1}, 0)$, $(0, b_2^{-1})$.

Exercise 9.14.4. Find the unit vectors bisecting the angles between the coordinate lines of a general coordinate system.

Exercise 9.14.5. The surface $y^3 = f(y^1, y^2)$ is given by a single valued continuous function f. Show that an orthogonal surface coordinate grid with a projection on the $O12$ plane which is also orthogonal can only be found if the surface is generated by lines parallel to a fixed line in $O12$ plane.

9.21. Christoffel symbols

The Christoffel symbols in three dimensions were defined as certain combinations of the derivatives of the metric tensor. We take over these definitions into two dimensions and write

$$[\alpha\beta, \gamma] = \frac{1}{2}\left[\frac{\partial a_{\gamma\alpha}}{\partial u^\beta} + \frac{\partial a_{\beta\gamma}}{\partial u^\alpha} - \frac{\partial a_{\alpha\beta}}{\partial u^\gamma}\right], \tag{9.21.1}$$

and

$$\begin{Bmatrix} \delta \\ \alpha \quad \beta \end{Bmatrix} = a^{\gamma\delta}[\alpha\beta, \gamma]. \tag{9.21.2}$$

They are symmetric with respect to the indices α and β. They will appear as before in connection with parallelism and covariant differentiation, but it is useful first to see the form they take in certain cases.

If the surface is given by

$$y^3 = f(y^1, y^2),$$

where f is a single valued continuous function, y^1 and y^2 can be used as the surface coordinates. Then

$$y^1 = u^1, \qquad y^2 = u^2, \qquad y^3 = f(u^1, u^2). \tag{9.21.3}$$

If f_α and $f_{\alpha\beta}$ denote $\partial f/\partial u^\alpha$ and $\partial^2 f/\partial u^\alpha\,\partial u^\beta$ respectively, the metric may be written

$$ds^2 = (1 + f_1^2)(du^1)^2 + 2f_1 f_2\, du^1\, du^2 + (1 + f_2^2)(du^2)^2. \tag{9.21.4}$$

Thus

$$a = 1 + f_1^2 + f_2^2 \tag{9.21.5}$$

and

$$a^{11} = \frac{1 + f_2^2}{1 + f_1^2 + f_2^2},$$

$$a^{12} = a^{21} = \frac{-f_1 f_2}{1 + f_1^2 + f_2^2}, \tag{9.21.6}$$

$$a^{22} = \frac{1 + f_1^2}{1 + f_1^2 + f_2^2}.$$

To calculate the Christoffel symbols we notice

$$\frac{\partial a_{11}}{\partial u^1} = 2f_1 f_{11}, \qquad \frac{\partial a_{11}}{\partial u^2} = 2f_1 f_{12}, \qquad \frac{\partial a_{12}}{\partial u^1} = f_{11} f_2 + f_1 f_{21},$$

$$\frac{\partial a_{21}}{\partial u^2} = f_{12} f_2 + f_1 f_{22}, \qquad \frac{\partial a_{22}}{\partial u^1} = 2f_2 f_{21}, \qquad \frac{\partial a_{22}}{\partial u^2} = 2f_2 f_{22},$$

all of which can be subsumed under the formula

$$\frac{\partial a_{\alpha\beta}}{\partial u^\gamma} = f_{\alpha\gamma} f_\beta + f_\alpha f_{\beta\gamma}. \tag{9.21.7}$$

whence,

$$[\alpha\beta, \gamma] = f_{\alpha\beta} f_\gamma. \tag{9.21.8}$$

Also, since $a^{\gamma\delta} f_\gamma = f_\delta / a$, we have

$$\begin{Bmatrix} \delta \\ \alpha \quad \beta \end{Bmatrix} = \frac{f_{\alpha\beta} f_\delta}{a}. \tag{9.21.9}$$

For the orthogonal surface coordinates

$$ds^2 = (b_1\, du^1)^2 + (b_2\, du^2)^2, \tag{9.21.10}$$

we have

$$\begin{aligned} \frac{\partial a_{\alpha\beta}}{\partial u^\gamma} &= 2b_\alpha \frac{\partial b_\alpha}{\partial u^\gamma}, \qquad \alpha = \beta, \\ &= 0, \qquad \alpha \neq \beta. \end{aligned}$$

Accordingly,

$$[11, 1] = b_1 \frac{\partial b_1}{\partial u^1}, \qquad [22, 2] = b_2 \frac{\partial b_2}{\partial u^2}$$

$$[12, 1] = [21, 1] = -[11, 2] = b_1 \frac{\partial b_1}{\partial u^2} \qquad (9.21.11)$$

$$[12, 2] = [21, 2] = -[22, 1] = b_2 \frac{\partial b_2}{\partial u^1}$$

and

$$\begin{Bmatrix} 1 \\ 1 \quad 1 \end{Bmatrix} = \frac{1}{b_1} \frac{\partial b_1}{\partial u^1}, \qquad \begin{Bmatrix} 1 \\ 1 \quad 2 \end{Bmatrix} = \begin{Bmatrix} 1 \\ 2 \quad 1 \end{Bmatrix} = \frac{1}{b_1} \frac{\partial b_1}{\partial u^2}, \qquad \begin{Bmatrix} 1 \\ 2 \quad 2 \end{Bmatrix} = -\frac{b_2}{b_1^2} \frac{\partial b_2}{\partial u^1},$$

$$(9.21.12)$$

$$\begin{Bmatrix} 2 \\ 1 \quad 1 \end{Bmatrix} = -\frac{b_1}{b_2^2} \frac{\partial b_1}{\partial u^2}, \qquad \begin{Bmatrix} 2 \\ 1 \quad 2 \end{Bmatrix} = \begin{Bmatrix} 2 \\ 2 \quad 1 \end{Bmatrix} = \frac{1}{b_2} \frac{\partial b_2}{\partial u^1}, \qquad \begin{Bmatrix} 2 \\ 2 \quad 2 \end{Bmatrix} = \frac{1}{b_2} \frac{\partial b_2}{\partial u^2}.$$

Exercise 9.21.1. Show that the Christoffel symbols all vanish if and only if the surface $y^3 = f(y^1, y^2)$ is a plane.

Exercise 9.21.2. Obtain the Christoffel symbols for the cylinder

$$y^1 = a \cos u^1, \qquad y^2 = a \sin u^1, \qquad y^3 = u^2.$$

Exercise 9.21.3. Obtain the Christoffel symbols for the sphere

$$y^1 = a \sin u^1 \cos u^2, \qquad y^2 = a \sin u^1 \sin u^2, \qquad y^3 = a \cos u^1.$$

Exercise 9.21.4. Show that

$$\frac{\partial a_{\alpha\beta}}{\partial u^\gamma} = [\alpha\gamma, \beta] + [\beta\gamma, \alpha]$$

and

$$\frac{\partial a^{\alpha\beta}}{\partial u^\gamma} = -a^{\alpha\epsilon} \begin{Bmatrix} \beta \\ \gamma \quad \epsilon \end{Bmatrix} - a^{\epsilon\beta} \begin{Bmatrix} \alpha \\ \gamma \quad \epsilon \end{Bmatrix}.$$

9.22. Geodesics

A curve on the surface can be specified by giving the surface coordinates as functions of a single parameter t along it, $u^\alpha = u^\alpha(t)$. The arc length of the curve between two points $t = a$ and $t = b$ is

$$s = \int_a^b \{a_{\alpha\beta} \dot{u}^\alpha(t) \dot{u}^\beta(t)\}^{1/2} \, dt$$

where $\dot{u}^\alpha(t) = du^\alpha/dt$. Let us use the abbreviation

$$U^2(u, \dot{u}) = a_{\alpha\beta}(u^1, u^2) \dot{u}^\alpha \dot{u}^\beta \qquad (9.22.1)$$

then

$$s = \int_a^b U(u, \dot{u}) \, dt. \qquad (9.22.2)$$

We now wish to find the curve between the two fixed points $u^\alpha(a)$ and $u^\alpha(b)$ which has the least arc length.

To determine this curve we suppose that the curve $u^\alpha(t)$ does achieve the minimum length. If $v^\alpha(t)$ is any continuous function which vanishes for $t = a$ and $t = b$, then $u^\alpha = u^\alpha(t) + \epsilon v^\alpha(t)$ is a distortion of the minimizing curve as shown in Fig. 9.2. We can make this distortion as small as we please by

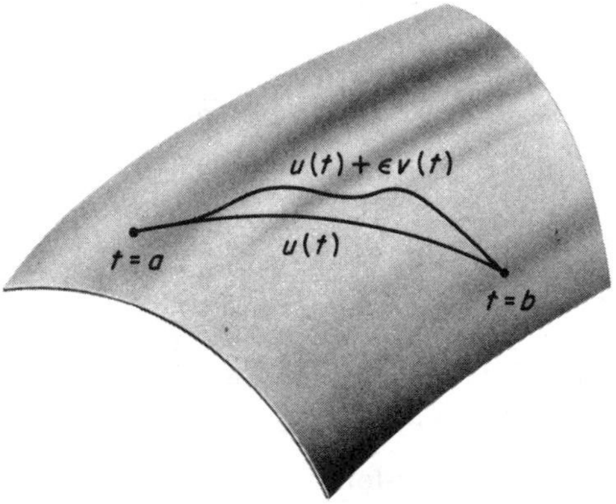

Fig. 9.2

making ϵ small and setting $\epsilon = 0$ gives the minimum. The length of the distorted curve is a function of ϵ for any fixed distorting function $v(t)$,

$$s(\epsilon) = \int_a^b U(u + \epsilon v, \dot{u} + \epsilon \dot{v})\, dt.$$

Now if s is a minimum for $\epsilon = 0$, $ds/d\epsilon = 0$ when $\epsilon = 0$. However,

$$\left(\frac{ds}{d\epsilon}\right)_{\epsilon=0} = \int_a^b \left\{\frac{\partial U}{\partial u^\alpha} v^\alpha + \frac{\partial U}{\partial \dot{u}^\alpha} \dot{v}^\alpha\right\} dt = 0$$

and this can be transformed by partially integrating the second term into

$$\left[\frac{\partial U}{\partial \dot{u}} v^\alpha\right]_a^b - \int_a^b v^\alpha(t) \left\{\frac{d}{dt}\left(\frac{\partial U}{\partial \dot{u}^\alpha}\right) - \frac{\partial U}{\partial u^\alpha}\right\} dt = 0. \qquad (9.22.3)$$

Of this, the first term disappears since $v = 0$ at both end points and the second has to be zero for any distortion $v(t)$. It follows that the integrand must vanish, and U must satisfy

$$\frac{d}{dt}\left(\frac{\partial U}{\partial \dot{u}^\alpha}\right) - \frac{\partial U}{\partial u^\alpha} = 0. \qquad (9.22.4)$$

To express this as an equation in $u^\alpha(t)$ we return to the definition of U and recall that it is a function of the u^α only through the $a_{\beta\gamma}$.

$$U^2 = a_{\alpha\beta}\dot{u}^\alpha \dot{u}^\beta.$$

Then

$$2U\frac{\partial U}{\partial u^\alpha} = \frac{\partial a_{\beta\gamma}}{\partial u^\alpha}\dot{u}^\beta \dot{u}^\gamma \tag{9.22.5}$$

and

$$U\frac{\partial U}{\partial \dot{u}^\alpha} = a_{\alpha\beta}\dot{u}^\beta. \tag{9.22.6}$$

Thus

$$\frac{d}{dt}\left(\frac{\partial U}{\partial \dot{u}^\alpha}\right) = \frac{d}{dt}\left(\frac{a_{\alpha\beta}\dot{u}^\beta}{U}\right) = \frac{1}{U}\left[a_{\alpha\beta}\ddot{u}^\beta + \frac{\partial a_{\alpha\beta}}{\partial u^\gamma}\dot{u}^\beta \dot{u}^\gamma\right] - \frac{a_{\alpha\beta}\dot{u}^\beta}{U^2}\frac{dU}{dt}.$$

Now take the parameter t to be the arc length so that $U = 1$ and $(dU/dt) = 0$; then Eq. (9.22.4) simplifies to

$$a_{\alpha\beta}\ddot{u}^\beta + \left\{\frac{\partial a_{\alpha\beta}}{\partial u^\gamma} - \frac{1}{2}\frac{\partial a_{\beta\gamma}}{\partial u^\alpha}\right\}\dot{u}^\beta \dot{u}^\gamma = 0.$$

Recalling the definition of the Chrisotffel symbol this becomes

$$a_{\alpha\beta}\ddot{u}^\beta + [\beta\gamma, \alpha]\dot{u}^\beta \dot{u}^\gamma = 0$$

or raising the index

$$\ddot{u}^\alpha + \left\{\begin{matrix} & \alpha & \\ \beta & & \gamma \end{matrix}\right\}\dot{u}^\beta \dot{u}^\gamma = 0. \tag{9.22.7}$$

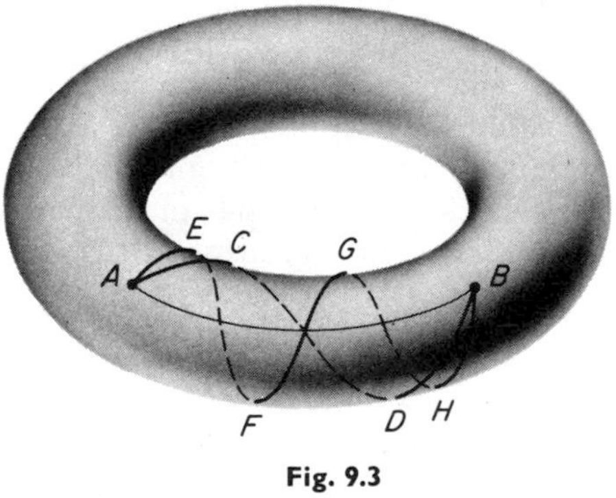

Fig. 9.3

This then is the equation satisfied by a geodesic when the parameter along it is the arc length.

The geodesic as we have constructed it is only a local extremum and we have not even checked to see that it is a minimum. Between two points on a sphere, for example, there is the shortest arc along the great circle connecting them, but the remainder of this great circle constitutes the longest arc connecting them and also satisfies the geodesic equation. On a torus there is not only the shortest path connecting A and B (see Fig. 9.3) but also the shortest path that goes once ($ACDB$) or that goes twice ($AEFGHB$) around the torus and hence an infinite sequence of local minima. We have obtained a pair of second order nonlinear differential equations which have to be integrated between two end points. As an elementary example consider the cylinder of radius a, given by

$$y^1 = a \cos u^1, \qquad y^2 = a \sin u^1, \qquad y^3 = u^2.$$

Here $a_{11} = a^2$, $a_{12} = 0$, $a_{22} = 1$ so that all the Christoffel symbols vanish and the geodesic equation becomes $\ddot{u}^\alpha = 0$ or $u^\alpha = a^\alpha + c^\alpha s$, where a^α and c^α are constants. However, if s is indeed the arc length

$$ds^2 = a^2(c^1\, ds)^2 + (c^2\, ds)^2$$

so that $(ac^1)^2 + (c^2)^2 = 1$ and we can write $u^1 = a^1 + \sin\theta(s/a)$, $u^2 = a^2 + \cos\theta\, s$. a^1 and a^2 can be fixed by one end point, say $s = 0$, and if the other end point is b^1, b^2 we have to choose θ so that

$$\tan\theta = a\,\frac{b^1 - a^1}{b^2 - a^2}.$$

This gives an infinite number of geodesics corresponding to paths that wrap themselves around the cylinder any given number of times. All these paths are helices however for we notice that θ is the angle that they make with the generators, $u^1 =$ constant.

Another way of getting a definite curve out of the differential equation for the geodesic would be to specify one end point and the direction through it that the geodesic should take; that is, we specify u^α and $\dot{u}^\alpha$ at one point instead of u^α at two. A geodesic may be realized approximately as the track of a very small two-wheeled buggy. If the wheels are fixed rigidly to a single shaft, they must turn through the same angle and so run on paths of equal length. Two such paths will lie on either side of a geodesic and the center point of the shaft will approximate the geodesic if the shaft is short enough.

Exercise 9.22.1. Show that the geodesics in a plane are straight lines.

Exercise 9.22.2. A cylinder is a surface generated by the family of parallel straight lines which pass through a closed curve. Show that the geodesics on the cylinder are always helices as defined in Ex. 3.12.2.

Exercise 9.22.3. Show that the shortest distance between two points on a sphere is a great circular arc.

9.23. Geodesic coordinates

It is possible to introduce a special system of coordinates in which all the Christoffel symbols vanish at a given point. We cannot hope for more than this, for only in a Euclidean space can a Cartesian coordinate system, in which the Christoffel symbols vanish everywhere, be set up. The Euclidean two-dimensional space is a plane, so that, except for the plane, the most we can do is to secure the vanishing of the Christoffel symbols at a point. This system is called the geodesic coordinate system and offers locally some of the virtues of the Cartesian system. It may be possible to establish a certain relation more readily in geodesic coordinates and, if it is a valid tensor relation, it consequently holds in all coordinate systems.

Let v^α be the point at which the new Christoffel symbols are to vanish and introduce the coordinates

$$\bar u^\alpha = (u^\alpha - v^\alpha) + \frac{1}{2}\left\{\begin{matrix} \alpha \\ \beta \quad \gamma \end{matrix}\right\}_0 (u^\beta - v^\beta)(u^\gamma - v^\gamma), \tag{9.23.1}$$

where the Christoffel symbol is evaluated at $u^\alpha = v^\alpha$. Evidently $\bar u^\alpha = 0$ at the point v^α, which we may call the polc. Differentiating we have

$$\frac{\partial \bar u^\alpha}{\partial u^\beta} = \delta^\alpha_\beta + \left\{\begin{matrix} \alpha \\ \beta \quad \gamma \end{matrix}\right\}_0 (u^\gamma - v^\gamma)$$

and multiplying this by $\partial u^\lambda / \partial \bar u^\alpha$ we have

$$\frac{\partial u^\lambda}{\partial \bar u^\alpha}\frac{\partial \bar u^\alpha}{\partial u^\beta} = \delta^\lambda_\beta = \frac{\partial u^\lambda}{\partial \bar u^\beta} + \left\{\begin{matrix} \alpha \\ \beta \quad \gamma \end{matrix}\right\}_0 \frac{\partial u^\lambda}{\partial \bar u^\alpha}(u^\gamma - v^\gamma). \tag{9.23.2}$$

At the pole $u^\alpha = v^\alpha$ we therefore have

$$\left(\frac{\partial u^\lambda}{\partial \bar u^\beta}\right)_0 = \delta^\lambda_\beta . \tag{9.23.3}$$

Now differentiate Eq. 9.23.2 again with respect to $\bar u^{\mu}$:

$$0 = \frac{\partial^2 u^\lambda}{\partial \bar u^\beta\, \partial \bar u^\mu} + \left\{\begin{matrix} \alpha \\ \beta \quad \gamma \end{matrix}\right\}_0 \left[\frac{\partial u^\lambda}{\partial \bar u^\alpha}\frac{\partial u^\gamma}{\partial \bar u^\mu} + (u^\gamma - v^\gamma)\frac{\partial^2 u^\lambda}{\partial \bar u^\alpha\, \partial \bar u^\mu}\right]$$

and at the pole this gives

$$\left(\frac{\partial^2 u^\lambda}{\partial \bar u^\beta\, \partial \bar u^\mu}\right)_0 = -\left\{\begin{matrix} \alpha \\ \beta \quad \gamma \end{matrix}\right\}_0 \delta^\lambda_\alpha\, \delta^\gamma_\mu = -\left\{\begin{matrix} \lambda \\ \beta \quad \mu \end{matrix}\right\}_0 . \tag{9.23.4}$$

The transformation law of the Christoffel symbols is exactly as in the three-dimensional case (cf. Ex. 7.53.1),

$$\overline{\left\{\begin{matrix} \lambda \\ \mu \quad \nu \end{matrix}\right\}} = \frac{\partial \bar u^\lambda}{\partial u^\alpha}\frac{\partial u^\beta}{\partial \bar u^\mu}\frac{\partial u^\gamma}{\partial \bar u^\nu}\left\{\begin{matrix} \alpha \\ \beta \quad \gamma \end{matrix}\right\} + \frac{\partial^2 u^\alpha}{\partial \bar u^\mu \bar u^\nu}\frac{\partial \bar u^\lambda}{\partial u^\alpha} . \tag{9.23.5}$$

However, at the pole we have

$$\overline{\left\{\begin{matrix} \lambda \\ \mu \quad \nu \end{matrix}\right\}}_0 = \delta^\lambda_\alpha \delta^\beta_\mu \delta^\gamma_\nu \left\{\begin{matrix} \alpha \\ \beta \quad \gamma \end{matrix}\right\}_0 - \left\{\begin{matrix} \alpha \\ \mu \quad \nu \end{matrix}\right\}_0 \delta^\lambda_\alpha = 0 \tag{9.23.6}$$

so that the Christoffel symbols all vanish. Along the coordinate line $u^2 = 0$ the value of u^1 in the neighborhood of the pole is simply s, the arc length. It follows that $u^1 = 1$, $u^1 = 0$ and, since the Christoffel symbols vanish, the coordinate lines satisfy the geodesic equation in the neighborhood of the pole; hence the name geodesic coordinates.

Exercise 9.23.1. Construct the geodesic coordinate system for an arbitrary point on the sphere.

Exercise 9.23.2. Geodesic polar coordinates are constructed by taking all the geodesics through a point as the u^1 coordinate lines and their orthogonal trajectories as the u^2 coordinate lines. Show that the metric is

$$ds^2 = (du^1)^2 + (b_2\, du^2)^2$$

where b_2/u_1 tends to a constant at the pole.

9.24. Parallel vectors in a surface

We, now, have to establish the proper rules for covariant differentiation within a surface and it will be recalled that to arrive at this we considered parallel vectors along a curve in three dimensions. In Euclidean space it was possible to use our intuitive notions of parallelism in Cartesian coordinates to obtain a condition for the parallel propagation of a vector along a curve. In a nonplanar surface, however, we have no Euclidean space to make our appeal to and must follow a different route. The condition for a vector A^i in three dimensions to remain parallel when propagated along the curve $x^i(t)$ was that its intrinsic derivative

$$\frac{\delta A^i}{\delta t} = \frac{dA^i}{dt} + \begin{Bmatrix} i \\ j \quad k \end{Bmatrix} A^j \frac{dx^k}{dt}$$

should vanish. This is a piece of experience on which we can build by first asking whether the intrinsic derivative can be analogously defined in two dimensions and then seeing whether the vanishing of the intrinsic derivative gives any of the properties we expect of parallelism.

The important thing about the intrinsic derivative of a vector is that it is again a vector, so we should first see if the analogue has the correct transformation property. Let

$$\frac{\delta A^\alpha}{\delta t} = \frac{dA^\alpha}{dt} + \begin{Bmatrix} \alpha \\ \beta \quad \gamma \end{Bmatrix} A^\beta \frac{du^\gamma}{dt} \tag{9.24.1}$$

for some curve $u^\gamma = u^\gamma(t)$ in the surface. Then

$$\begin{aligned} \frac{\delta \bar{A}^\lambda}{\delta t} &= \frac{d\bar{A}^\lambda}{dt} + \overline{\begin{Bmatrix} \lambda \\ \mu \quad \nu \end{Bmatrix}} \bar{A}^\mu \frac{d\bar{u}^\nu}{dt} \\ &= \frac{d}{dt}\left(\frac{\partial \bar{u}^\lambda}{\partial u^\alpha} A^\alpha\right) + \overline{\begin{Bmatrix} \lambda \\ \mu \quad \nu \end{Bmatrix}} \frac{\partial \bar{u}^\mu}{\partial u^\beta} A^\beta \frac{\partial \bar{u}^\nu}{\partial u^\gamma} \frac{du^\gamma}{dt} \\ &= \frac{\partial \bar{u}^\lambda}{\partial u^\alpha} \frac{dA^\alpha}{dt} + \overline{\begin{Bmatrix} \lambda \\ \mu \quad \nu \end{Bmatrix}} \frac{\partial \bar{u}^\mu}{\partial u^\beta} \frac{\partial \bar{u}^\nu}{\partial u^\gamma} A^\beta \frac{du^\gamma}{dt} + \frac{\partial^2 \bar{u}^\lambda}{\partial u^\alpha \partial u^\beta} \frac{du^\beta}{dt} A^\alpha. \end{aligned}$$

Using the transformation of Christoffel symbols in the form

$$\frac{\partial \bar{u}^\lambda}{\partial u^\alpha}\begin{Bmatrix} \alpha \\ \beta \;\; \gamma \end{Bmatrix} = \frac{\partial \bar{u}^\mu}{\partial u^\beta}\frac{\partial \bar{u}^\nu}{\partial u^\gamma}\overline{\begin{Bmatrix} \lambda \\ \mu \;\; \nu \end{Bmatrix}} + \frac{\partial^2 \bar{u}^\lambda}{\partial u^\beta\,\partial u^\gamma}$$

and changing the dummy indices, this can be written

$$\begin{aligned}\frac{\delta \bar{A}^\lambda}{\delta t} &= \frac{\partial \bar{u}^\lambda}{\partial u^\alpha}\left[\frac{dA^\alpha}{dt} + \begin{Bmatrix} \alpha \\ \beta \;\; \gamma \end{Bmatrix} A^\beta \frac{du^\gamma}{dt}\right] \\ &= \frac{\partial \bar{u}^\lambda}{\partial u^\alpha}\frac{\delta A^\alpha}{\delta t}. \end{aligned} \tag{9.24.2}$$

Hence the intrinsic derivative has the correct transformation properties.

If the surface is a plane so that Cartesian coordinates may be used, the intrinsic derivative becomes the ordinary derivative, the vanishing of which is the condition for parallel propagation. The equation

$$\frac{\delta A^\alpha}{\delta t} = \frac{dA^\alpha}{dt} + \begin{Bmatrix} \alpha \\ \beta \;\; \gamma \end{Bmatrix} A^\beta \frac{du^\gamma}{dt} = 0 \tag{9.24.3}$$

is a first order differential equation, so that if A^α were specified at one point of a given curve $u^\gamma = u^\gamma(t)$, the solution of this equation would give a vector field $A^\alpha(t)$ along the curve. The question is whether this retains any properties we associate with parallel fields.

Consider, first, the square of the magnitude of the vector. It is a scalar whose rate of change is

$$\begin{aligned}\frac{d\,|A|^2}{dt} &= \frac{d}{dt}(a_{\alpha\beta}A^\alpha A^\beta) \\ &= \frac{da_{\alpha\beta}}{dt}A^\alpha A^\beta + a_{\alpha\beta}\left[\frac{dA^\alpha}{dt}A^\beta + A^\alpha\frac{dA^\beta}{dt}\right] \\ &= A^\alpha A^\beta\left\{\frac{da_{\alpha\beta}}{dt} - \left[a_{\delta\beta}\begin{Bmatrix} \delta \\ \gamma \;\; \alpha \end{Bmatrix} + a_{\alpha\delta}\begin{Bmatrix} \delta \\ \beta \;\; \gamma \end{Bmatrix}\right]\frac{du^\gamma}{dt}\right\} \\ &= 0\end{aligned}$$

since

$$\frac{\partial a_{\alpha\beta}}{\partial u^\gamma} = [\alpha\gamma, \beta] + [\beta\gamma, \alpha].$$

Hence the magnitude of a vector does not change under parallel propagation. It may be similarly shown that the angle between two vectors does not change in parallel propagation and that the tangent to a geodesic is always parallel (see Ex. 9.24.1 and 2). These are all properties which obviously obtain in Euclidean space and are as much as we can hope to retain in the curved surface.

We should notice that the vector obtained at a second point, by parallel propagation from the first, generally depends on the path. For example, the only nonvanishing Christoffel symbols for the sphere are

$$\begin{Bmatrix} 1 \\ 2 \quad 2 \end{Bmatrix} = -\sin u^1 \cos u^1$$

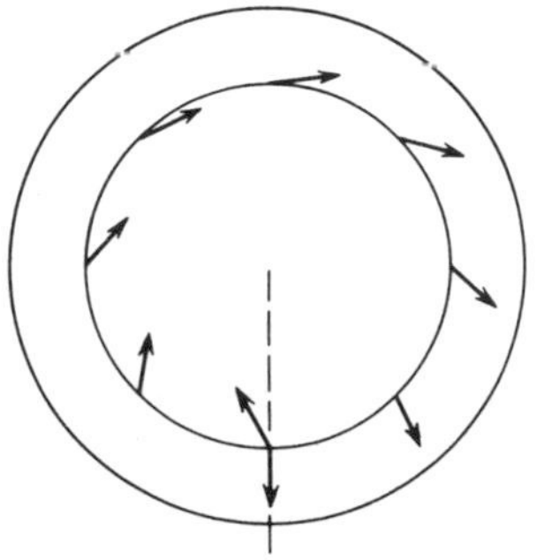

Fig. 9.4

and $$\begin{Bmatrix} 2 \\ 1 \quad 2 \end{Bmatrix} = \begin{Bmatrix} 2 \\ 2 \quad 1 \end{Bmatrix} = \cot u^1.$$

Thus for parallel propagation

$$\frac{dA^1}{dt} - \frac{1}{2}\sin 2u^1 A^2 \frac{du^2}{dt} = 0$$

$$\frac{dA^2}{dt} + \cot u^1 \left\{A^2 \frac{du^1}{dt} + A^1 \frac{du^2}{dt}\right\} = 0$$

showing the dependence on the path. If A^α is propagated around a great circle of longitude $u^2 =$ constant, $u^1 = t$, we have

$$\frac{dA^1}{dt} = 0 \qquad \frac{dA^2}{dt} + \cot u^1 A^2 = 0.$$

If $A^1 = A_0^1$, $A^2 = A_0^2$ at the equator $u^1 = \pi/2$, then

$$A^1 = A_0^1, \qquad A^2 = A_0^2 \csc u^1.$$

In contrast, if A^α is propagated around a circle of constant latitude, $u^1 = \theta =$ constant, $u^2 = t$, then

$$\frac{dA^1}{dt} - \frac{1}{2}\sin 2\theta\, A^2 = 0, \qquad \frac{dA^2}{dt} + \cot\theta\, A^1 = 0.$$

From this we have

$$\frac{d^2A^\alpha}{dt^2} + \cos^2\theta A^\alpha = 0,$$

so if $A^\alpha = A_0^\alpha$ when $t = 0$,

$$A^1 = A_0^1 \cos(t\cos\theta) + A_0^2 \sin\theta \sin(t\cos\theta)$$

$$A^2 = A_0^2 \cos(t\cos\theta) - A_0^1 \csc\theta \sin(t\cos\theta).$$

When a great circle of latitude has been completed, $t = 2\pi$ and it is evident that, in general, $A^\alpha \neq A_0^\alpha$. Thus we cannot always expect a vector propagated parallel to itself round a closed curve to be the same when it returns to the starting point. Figure 9.4 shows a view from the north pole of a vector with $A_0^2 = 0$ being propagated around the circle $\theta = \pi/4$.

If $A_\beta = a_{\alpha\beta}A^\alpha$ is the associated covariant vector of a parallel vector, we have

$$0 = \frac{\delta A^\alpha}{\delta t} = \frac{d}{dt}(a^{\alpha\beta}A_\beta) + \begin{Bmatrix} \alpha \\ \beta \ \ \gamma \end{Bmatrix} a^{\beta\epsilon}A_\epsilon \frac{du^\gamma}{dt}$$

$$= a^{\alpha\beta}\frac{dA_\beta}{dt} + \left[\frac{\partial a^{\alpha\beta}}{\partial u^\gamma}A_\beta + \begin{Bmatrix} \alpha \\ \beta \ \ \gamma \end{Bmatrix} a^{\beta\epsilon}A_\epsilon\right]\frac{du^\gamma}{dt}$$

$$= a^{\alpha\beta}\left[\frac{dA_\beta}{dt} - \begin{Bmatrix} \epsilon \\ \beta \ \ \gamma \end{Bmatrix} A_\epsilon \frac{du^\gamma}{dt}\right]$$

by substitution from Ex. 9.21.4. Hence a parallel covariant vector satisfies

$$\frac{\delta A_\beta}{\delta t} = \frac{dA_\beta}{dt} - \begin{Bmatrix} \epsilon \\ \beta \ \ \gamma \end{Bmatrix} A_\epsilon \frac{du^\gamma}{dt} = 0. \tag{9.24.4}$$

Exercise 9.24.1. Show that the angle between two vectors A^α and B^α remains constant in parallel propagation.

Exercise 9.24.2. Show that the tangent to a geodesic is parallel to itself and deduce that a vector propagated along a geodesic makes a constant angle with it.

Exercise 9.24.3. Why is the parallel propagation of a vector between two points independent of the path in Euclidean [3] space?

Exercise 9.24.4. If s is the arc length along a curve in the surface, the unit tangent is $\tau^\alpha = du^\alpha/ds$. Show that the intrinsic derivative $\delta\tau^\alpha/\delta s$ is perpendicular to the curve. If ν^α is the unit vector perpendicular to τ^α in the positive direction (that is, $\varepsilon_{\alpha\beta}\tau^\alpha\nu^\beta = +1$) and $\delta\tau^\alpha/\delta s = \sigma\nu^\alpha$, σ is called the geodesic curvature of the curve. Show that $\delta\nu^\alpha/\delta s = -\sigma\tau^\alpha$.

Exercise 9.24.5. Show that a geodesic has zero geodesic curvature.

Exercise 9.24.6. Find the geodesic curvature of the coordinate lines of an orthogonal coordinate system.

9.25. Covariant surface differentiation

The method of obtaining a derivative with the proper transformation properties of a tensor is exactly similar to the case of three dimensions treated in Section 7.55. Given a tensor field (say A_β^α) defined on a given region of a surface we select an arbitrary curve $[u^\gamma(t)]$ on the surface and two arbitrary parallel vector fields (B_α and C^β) on the curve. Then $E = A_\beta^\alpha B_\alpha C^\beta$ is a scalar and its derivative with respect to the parameter t along the curve is a scalar However, using Eqs. (9.24.3 and 4),

$$\frac{dE}{dt} = \left[\frac{\partial A^\alpha_\beta}{\partial u^\gamma} + \left\{\begin{matrix} & \alpha & \\ \delta & & \gamma\end{matrix}\right\} A^\delta_\beta - \left\{\begin{matrix} & \delta & \\ \beta & & \gamma\end{matrix}\right\} A^\alpha_\delta\right] B_\alpha C^\beta \frac{du^\gamma}{dt},$$

we see that the expression in the brackets must be a third order surface tensor which we write

$$A^\alpha_{\beta,\gamma} = \frac{\partial A^\alpha_\beta}{\partial u^\gamma} + \left\{\begin{matrix} & \alpha & \\ \delta & & \gamma\end{matrix}\right\} A^\delta_\beta - \left\{\begin{matrix} & \delta & \\ \beta & & \gamma\end{matrix}\right\} A^\alpha_\delta. \tag{9.25.1}$$

At the pole of a geodesic coordinate system this covariant derivative reduces to the partial derivative, just as it does everywhere in a Cartesian system. The generalization to a tensor of any order is immediate.

The metric tensor acts as a constant with respect to covariant differentiation. This may be proved by direct substitution or by the observation that $a_{\alpha\beta,\gamma} = 0$ at the pole of a geodesic system and so must vanish for all coordinate systems.

Exercise 9.25.1. Show that $a^{\alpha\beta}$, a, $\varepsilon^{\alpha\beta}$, and $\varepsilon_{\alpha\beta}$ act as constants with respect to covariant differentiation.

Exercise 9.25.2. Show that

$$\frac{\partial a}{\partial u^\alpha} = 2a\left\{\begin{matrix} & \beta & \\ \alpha & & \beta\end{matrix}\right\}.$$

Exercise 9.25.3. Show that

$$A^\alpha_{,\alpha} = \frac{1}{a^{1/2}} \frac{\partial}{\partial u^\alpha}(a^{1/2}A^\alpha).$$

9.26. The Gaussian or total curvature of a surface

We now turn to consider what can be said about the curvature of the surface without viewing it from outside of itself. We want an intrinsic measure of the departure of the surface from being a Euclidean two-dimensional space. Here again we can build on experience for we have encountered (Section 7.6) a tensor, the Riemann-Christoffel tensor, that vanishes identically in Euclidean space. Let us recall the analog of Eq. 7.6.2,

$$A_{\alpha,\beta\gamma} - A_{\alpha,\gamma\beta} = R^\delta_{\alpha\beta\gamma} A_\delta, \tag{9.26.1}$$

where

$$R^\delta_{\alpha\beta\gamma} = \frac{\partial}{\partial u^\beta}\left\{\begin{matrix} & \delta & \\ \gamma & & \alpha\end{matrix}\right\} - \frac{\partial}{\partial u^\gamma}\left\{\begin{matrix} & \delta & \\ \alpha & & \beta\end{matrix}\right\} + \left\{\begin{matrix} & \epsilon & \\ \alpha & & \gamma\end{matrix}\right\}\left\{\begin{matrix} & \delta & \\ \epsilon & & \beta\end{matrix}\right\} - \left\{\begin{matrix} & \epsilon & \\ \alpha & & \beta\end{matrix}\right\}\left\{\begin{matrix} & \delta & \\ \epsilon & & \gamma\end{matrix}\right\}. \tag{9.26.2}$$

This certainly vanishes for a plane or for a surface in which the metric tensor is a constant, but we cannot expect it to vanish in general. Even when geodesic coordinates are introduced at a point and the Christoffel symbols vanish their partial derivatives will not necessarily vanish.

The associated tensor

$$R_{\lambda\alpha\beta\gamma} = a_{\lambda\delta} R^{\delta}_{\alpha\beta\gamma} \tag{9.26.3}$$

is also expressible in terms of the Christoffel symbols and their derivatives (cf. Ex. 7.6.2) and is antisymmetric with respect to the pairs of indices λ, α and β, γ (cf. Ex. 9.26.1). Thus, such components as $R_{11\beta\gamma}$ and $R_{\lambda\alpha 22}$ are zero and of the 16 components only four survive, namely, $R_{1212} = -R_{2112} = -R_{1221} = R_{2121}$. The constant

$$K = \tfrac{1}{4}\varepsilon^{\lambda\alpha}\varepsilon^{\beta\gamma} R_{\lambda\alpha\beta\gamma} \tag{9.26.4}$$

is a scalar, namely,

$$K = \frac{R_{1212}}{a}. \tag{9.26.5}$$

This quantity is called the Gaussian or total curvature of the surface. It is an intrinsic measure calculable from the metric alone. We shall come across an illuminating interpretation of K when we take a look at the surface from outside, which we will now begin to do.

Exercise 9.26.1. The antisymmetry with respect to λ and α of $R_{\lambda\alpha\beta\gamma}$ is not quite so obvious as its antisymmetry with respect to β and γ. Prove the identity

$$R_{\lambda\alpha\beta\gamma} = \frac{1}{2}\left\{\frac{\partial^2 a_{\lambda\gamma}}{\partial u^\alpha\,\partial u^\beta} + \frac{\partial^2 a_{\alpha\beta}}{\partial u^\lambda\,\partial u^\gamma} - \frac{\partial^2 a_{\lambda\beta}}{\partial u^\alpha\,\partial u^\gamma} - \frac{\partial^2 a_{\alpha\gamma}}{\partial u^\lambda\,\partial u^\beta}\right\}$$
$$+\, a^{\mu\nu}\{[\alpha\beta,\mu][\lambda\gamma,\nu] - [\lambda\beta,\mu][\alpha\gamma,\nu]\},$$

from which the first antisymmetry is obvious.

Exercise 9.26.2. Show that the Gaussian curvature of a sphere of radius r is $K = r^{-2}$.

Exercise 9.26.3. Show that $K = 0$ for a cylinder.

Exercise 9.26.4. If the coordinate system is orthogonal, $a_{11} = b_1^2$, $a_{12} = 0$, $a_{22} = b_2^2$, show that

$$K = -\frac{1}{b_1 b_2}\left[\frac{\partial}{\partial u^1}\left(\frac{1}{b_1}\frac{\partial b_2}{\partial u^1}\right) + \frac{\partial}{\partial u^2}\left(\frac{1}{b_2}\frac{\partial b_1}{\partial u^2}\right)\right].$$

Exercise 9.26.5. In the notation of Section 9.21, show that for a surface $y^1 = u^1$, $y^2 = u^2$, $y^3 = f(u^1, u^2)$

$$K = \frac{f_{11}f_{22} - f_{12}^2}{(1 + f_1^2 + f_2^2)^2}.$$

Exercise 9.26.6. Show that $R_{\lambda\alpha\beta\gamma} = K\varepsilon_{\lambda\alpha}\varepsilon_{\beta\gamma}$.

9.31. The surface in space

Up to this point we have been viewing the surface intrinsically, working with coordinates, vectors and tensors in the surface itself. Now we wish to link together quantities which belong to the surface and the space and indeed form some hybrid quantities. We started from the assumption of a surface in Euclidean space and so could set up Cartesian coordinates. The equation of the surface is then

$$y^i = y^i(u^1, u^2)$$

where u^1 and u^2 are the surface coordinates. Actually there is no need to assume a Cartesian coordinate system and from now on we will take a general coordinate system x^i in space and write the equations of the surface

$$x^i = x^i(u^1, u^2). \tag{9.31.1}$$

Notice that we cannot invert these equations and write $u^\alpha = u^\alpha(x^1, x^2, x^3)$ for such an equation is only meaningful if the point $\mathbf{x}$ is on the surface.

The partial derivatives $\partial x^i/\partial u^\alpha$ are assumed to exist and will be so frequently used as to justify a special symbol; we write

$$t^i_\alpha = \frac{\partial x^i}{\partial u^\alpha}. \tag{9.31.2}$$

If we were to transform the space coordinates to some other system $\bar{x}^i$, the quantity t^i_α would become

$$\bar{t}^i_\alpha = \frac{\partial \bar{x}^i}{\partial u^\alpha} = \frac{\partial \bar{x}^i}{\partial x^j}\frac{\partial x^j}{\partial u^\alpha} = \frac{\partial \bar{x}^i}{\partial x^j} t^j_\alpha. \tag{9.31.3}$$

Thus t^i_α is a contravariant space vector for fixed α. Similarly, if the surface coordinates were transformed to $\tilde{u}^\alpha$, t^i_α would become

$$\tilde{t}^i_\alpha = \frac{\partial x^i}{\partial \tilde{u}^\alpha} = \frac{\partial x^i}{\partial u^\beta}\frac{\partial u^\beta}{\partial \tilde{u}^\alpha} = \frac{\partial u^\beta}{\partial \tilde{u}^\alpha} t^i_\beta, \tag{9.31.4}$$

showing that t^i_α is a covariant surface vector for any i. Thus the hybrid tensor t^i_α forms a link between the two coordinate systems.

t^i_α can be used to associate spatial and surface tensors. If A^α is a contravariant surface vector, then

$$A^i = t^i_\alpha A^\alpha \tag{9.31.5}$$

is an associated space vector. Since A^α is in the surface, A^i is necessarily tangent to the surface. Similarly, if A_i is a covariant space vector,

$$A_\alpha = t^i_\alpha A_i \tag{9.31.6}$$

is a covariant surface vector.

A tensor $T^{i\ldots j,\ \gamma\ldots\delta}_{k\ldots m,\ \alpha\ldots\beta}$ which transforms as a space tensor with respect to the italic indices and as a surface tensor with respect to the Greek indices, will be called a hybrid tensor.

Exercise 9.31.1. Prove the following example of the quotient rule. The 18 quantities $A^i_{j\alpha}$ are related to a covariant space vector D_j by the relation $A^i_{j\alpha}B^\alpha C_i = D_j$. B^α is a contravariant surface vector and C_i a covariant space vector, both being independent of $A^i_{j\alpha}$. Then $A^i_{j\alpha}$ is a hybrid tensor, second order and mixed in space and first order covariant in the surface.

9.32. The first fundamental form of the surface

The line element in the three-dimensional space is given by the metric

$$ds^2 = g_{ij}\,dx^i\,dx^j. \tag{9.32.1}$$

If this element lies in the surface, then dx^i corresponds to differentials in the surface coordinates such that

$$dx^i = t^i_\alpha\,du^\alpha. \tag{9.32.2}$$

Hence, the line element in the surface is

$$\begin{aligned} ds^2 &= g_{ij}t^i_\alpha t^j_\beta\,du^\alpha\,du^\beta \\ &= a_{\alpha\beta}\,du^\alpha\,du^\beta. \end{aligned} \tag{9.32.3}$$

Thus the metric tensor of the surface is linked to the metric tensor of the space by

$$a_{\alpha\beta} = g_{ij}t^i_\alpha t^j_\beta. \tag{9.32.4}$$

The quadratic form (9.32.3) is often called the first fundamental form of the surface.

The relation between Eqs. (9.31.5 and 6) is now apparent for

$$A_\alpha = t^i_\alpha A_i = t^i_\alpha g_{ij}A^j = t^i_\alpha g_{ij}t^j_\beta A^\beta = a_{\alpha\beta}A^\beta.$$

Thus the covariant surface vector associated with a covariant space vector is the associated surface vector of the contravariant surface vector associated with the associated contravariant space vector of the original covariant space vector—no less!

Exercise 9.32.1. Show that the length of the vector $A^i = t^i_\alpha A^\alpha$ in space is the same as the length of A^α in the surface.

Exercise 9.32.2. If the surface is taken to be the coordinate surface $x^3 = 0$ and the x^3 coordinate lines are a family of curves normal to the surface, show that $g_{13} = g_{23} = 0$. If, in addition, the lines u^1 and u^2 constant are the intersection of the surface with the surfaces x^1 and x^2 constant, show that $a_{\alpha\beta} = g_{\alpha\beta}$, $g = g_{33}a$, $a^{\alpha\beta} = g^{\alpha\beta}$, $g^{33} = 1/g_{33}$.

Exercise 9.32.3. Show that the element of area in the surface can be written $dS = a^{1/2}\,du^1\,du^2$.

9.33. The normal to the surface

If du^α is an infinitesimal surface vector, the infinitesimal space vector $dx^i = t^i_\alpha \, du^\alpha$ must be tangent to the surface. Since any vector A^α can be regarded as a very large multiple of an infinitesimal vector, it follows that $A^i = t^i_\alpha A^\alpha$ must be tangent to the surface, and may be called a tangent vector of the surface.

The cross product of two vectors A^j and B^k which are tangent to the surface is

$$\varepsilon_{ijk}A^jB^k = \varepsilon_{ijk}t^j_\alpha t^k_\beta A^\alpha B^\beta. \tag{9.33.1}$$

This is a vector of magnitude $|A|\,|B| \sin\theta$ and perpendicular to the surface. Let n_i be the covariant vector normal to the surface. By Exs. 9.14.2 and 9.32.1 we see that

$$|A|\,|B| \sin\theta = \varepsilon_{\alpha\beta}A^\alpha B^\beta \tag{9.33.2}$$

so that

$$n_i\varepsilon_{\alpha\beta}A^\alpha B^\beta = \varepsilon_{ijk}t^j_\alpha t^k_\beta A^\alpha B^\beta.$$

This must be true for all vectors A^α and B^β so that

$$n_i\varepsilon_{\alpha\beta} = \varepsilon_{ijk}t^j_\alpha t^k_\beta$$

or

$$n_i = \tfrac{1}{2}\varepsilon^{\alpha\beta}\varepsilon_{ijk}t^j_\alpha t^k_\beta. \tag{9.33.3}$$

We notice that $n_i t^i_\gamma = 0$ for γ must equal either α or β and $\varepsilon_{ijk}t^i_\gamma t^j_\alpha t^k_\beta$ would be a determinant with two rows the same. Thus the covariant surface vector $A_\alpha = t^i_\alpha A_i$ associated with the covariant space vector A_i is really a kind of projection of this vector on the surface. For, if A_i is written $\bar{A}_i + \alpha n_i$, where $\bar{A}_i$ is tangent to the surface,

$$A_\alpha = t^i_\alpha A_i = t^i_\alpha \bar{A}_i$$

and the component normal to the surface has been destroyed. Since $\alpha = n^jA_j$, we can write

$$\bar{A}_i = A_i - n_i(n^jA_j)$$

for the projection of A_i on the tangent plane. The length of this projection is

$$g^{ij}\bar{A}_i\bar{A}_j = g^{ij}A_iA_j - g^{ij}n_in_j(n^kA_k)^2 = (g^{ij} - n^in^j)A_iA_j$$
$$= a^{\alpha\beta}A_\alpha A_\beta$$

by Ex. 9.33.2. Thus the length in the surface of the covariant surface vector associated with a covariant space vector is the length of its projection on the tangent plane.

Exercise 9.33.1. In the special coordinate system of Ex. 9.32.2, show that $t^i_\alpha = 1$, $i = \alpha$, $t^i_\alpha = 0$, $i \neq \alpha$, and hence

$$n_1 = n_2 = 0, \qquad n_3 = g_{33}^{1/2}.$$

Exercise 9.33.2. Prove that

$$a^{\alpha\beta}t^i_\alpha t^j_\beta = g^{ij} - n^i n^j$$

by considering this relation first in the special coordinate system used above.

Exercise 9.33.3. Show in two distinct ways that

$$g_{ij}a^{\alpha\beta}t^i_\alpha t^j_\beta = 2.$$

Exercise 9.33.4. Show that the element of area defined by the increments du^1 and du^2 in the two coordinates is $dS = a^{1/2}\, du^1\, du^2$.

9.34. Covariant differentiation of hybrid tensors

We must now ask how the covariant surface derivative of a hybrid tensor should be taken. It will be sufficient to consider the tensor A^i_α and our method of treating it will follow the same course as before. In the surface $x^i = x^i(u^1, u^2)$ we take a curve $u^\alpha = u^\alpha(t)$. Its equation in space is $x^i = x^i(u^1(t), u^2(t))$ and

$$\frac{dx^i}{dt} = \frac{\partial x^i}{\partial u^\alpha}\frac{du^\alpha}{dt} = t^i_\alpha \frac{du^\alpha}{dt}. \tag{9.34.1}$$

On this we construct two vector fields by parallel propagation. The first is a contravariant surface vector field B^α, which accordingly satisfies the equation

$$\frac{dB^\alpha}{dt} + \begin{Bmatrix} \alpha \\ \beta \quad \gamma \end{Bmatrix} B^\beta \frac{du^\gamma}{dt} = 0, \tag{9.34.2}$$

and the second a covariant space vector C_i satisfying

$$\frac{dC_i}{dt} - \begin{Bmatrix} k \\ i \quad j \end{Bmatrix} C_k \frac{dx^j}{dt} = 0. \tag{9.34.3}$$

Now $A^i_\alpha B^\alpha C_i$ is a scalar and so its derivative with respect to t is a scalar. However, this is

$$\frac{d}{dt}(A^i_\alpha B^\alpha C_i) = \frac{dA^i_\alpha}{dt} B^\alpha C_i - A^i_\alpha \begin{Bmatrix} \alpha \\ \beta \quad \gamma \end{Bmatrix} B^\beta \frac{du^\gamma}{dt} C_i + A^i_\alpha B^\alpha \begin{Bmatrix} k \\ i \quad j \end{Bmatrix} C_k \frac{dx^j}{dt}.$$

By changing the dummy suffixes this can be written $(\delta A^i_\alpha/\delta t)B^\alpha C_i$, where

$$\frac{\delta A^i_\alpha}{\delta t} = \frac{dA^i_\alpha}{dt} - \begin{Bmatrix} \beta \\ \alpha \quad \gamma \end{Bmatrix} A^i_\beta \frac{du^\gamma}{dt} + \begin{Bmatrix} i \\ j \quad k \end{Bmatrix} A^j_\alpha \frac{dx^k}{dt}. \tag{9.34.4}$$

The quotient rule shows this to be a hybrid tensor of the same kind and we call it the intrinsic derivative of A^i_α along the curve. Using Eq. (9.34.1) we may write it as

$$\frac{\delta A^i_\alpha}{\delta t} = A^i_{\alpha,\beta}\frac{du^\beta}{dt} = \left[\frac{\partial A^i_\alpha}{\partial u^\beta} - \begin{Bmatrix} \gamma \\ \alpha \quad \beta \end{Bmatrix} A^i_\gamma + \begin{Bmatrix} i \\ j \quad k \end{Bmatrix} A^j_\alpha t^k_\beta\right]\frac{du^\beta}{dt}. \tag{9.34.5}$$

Thus $A^i_{\alpha,\beta}$ is a hybrid tensor of one higher order in surface covariance. The generalization of this formula is immediate. No confusion should be caused by the two kinds of Christoffel symbols since the Greek one is constructed from the $a_{\alpha\beta}$ exactly as the Italic one from the g_{ij} and both are evaluated on the surface. If the hybrid tensor is really invariant in one or other of the systems, then this derivative reduces to the covariant derivative defined before. It follows that $a_{\alpha\beta}$, g_{ij}, and all the tensors associated with them act as constants under this differentiation.

9.35. The second fundamental form of the surface

Consider the covariant surface derivative of t^i_α

$$t^i_{\alpha,\beta} = \frac{\partial t^i_\alpha}{\partial u^\beta} + \begin{Bmatrix} & i & \\ j & & k \end{Bmatrix} t^j_\alpha t^k_\beta - \begin{Bmatrix} & \gamma & \\ \alpha & & \beta \end{Bmatrix} t^i_\gamma \tag{9.35.1}$$

$$= \frac{\partial^2 x^i}{\partial u^\alpha \, \partial u^\beta} + \begin{Bmatrix} & i & \\ j & & k \end{Bmatrix} \frac{\partial x^j}{\partial u^\alpha} \frac{\partial x^k}{\partial u^\beta} - \begin{Bmatrix} & \gamma & \\ \alpha & & \beta \end{Bmatrix} \frac{\partial x^i}{\partial u^\gamma}.$$

This is symmetric in α and β. However, we also have $a_{\alpha\beta} = g_{ij} t^i_\alpha t^j_\beta$, which on differentiation gives

$$a_{\alpha\beta,\gamma} = 0 = g_{ij}\{t^i_{\alpha,\gamma} t^j_\beta + t^i_\alpha t^j_{\beta,\gamma}\}.$$

Now by the symmetry of the $t^i_{\alpha,\gamma}$, etc.,

$$\tfrac{1}{2}(a_{\gamma\alpha,\beta} + a_{\beta\gamma,\alpha} - a_{\alpha\beta,\gamma}) = g_{ij} t^i_{\alpha,\beta} t^j_\gamma = 0. \tag{9.35.2}$$

Since t^j_γ is tangent to the surface this equation implies that $t^i_{\alpha,\beta}$ is normal both to t^j_1 and t^j_2 and so normal to the surface. It is therefore proportional to n^i and so we may write

$$t^i_{\alpha,\beta} = b_{\alpha\beta} n^i \tag{9.35.3}$$

or

$$b_{\alpha\beta} = t^i_{\alpha,\beta} n_i = \left[\frac{\partial^2 x^i}{\partial u^\alpha \, \partial u^\beta} + \begin{Bmatrix} & i & \\ j & & k \end{Bmatrix} \frac{\partial x^j}{\partial u^\alpha} \frac{\partial x^k}{\partial u^\beta}\right] n_i \tag{9.35.4}$$

since the product of n_i and the last term of Eq. (9.35.1) vanishes identically. If the coordinates are Cartesian coordinates y^i, then

$$b_{\alpha\beta} = \frac{\partial^2 y^i}{\partial u^\alpha \, \partial u^\beta} n_i = \frac{1}{2} \varepsilon^{\rho\sigma} \varepsilon_{ijk} \frac{\partial^2 y^i}{\partial u^\alpha \, \partial u^\beta} \frac{\partial y^j}{\partial u^\rho} \frac{\partial y^k}{\partial u^\sigma}. \tag{9.35.5}$$

The quadratic form

$$b_{\alpha\beta} \, du^\alpha \, du^\beta \tag{9.35.6}$$

is called the second fundamental form of the surface. It can be shown that if two surfaces have the same first two fundamental forms, then they are intrinsically the same; that is, they can be brought into coincidence by a rigid motion. It is clear that the identity of the first fundamental form is not sufficient, for, if we write the equation of a cylinder as

$$y^1 = a\cos\frac{u^1}{a}, \qquad y^2 = a\sin\frac{u^1}{a}, \qquad y^3 = u^2,$$

we have $a_{11} = a_{22} = 1$, $a_{12} = 0$, which is the same as the metric of the Euclidean plane. However, the coefficients of the second fundamental form is $b_{11} = -1/a$, $b_{12} = b_{22} = 0$, whereas that of the plane is identically zero.

Exercise 9.35.1. Show that if the surface is given by $y^3 = f(y^1, y^2)$ in Cartesian coordinates, then

$$b_{\alpha\beta} = \frac{\dfrac{\partial^2 f}{\partial y^\alpha\,\partial y^\beta}}{\left\{1 + \left(\dfrac{\partial f}{\partial y^1}\right)^2 + \left(\dfrac{\partial f}{\partial y^2}\right)^2\right\}^{1/2}}.$$

Exercise 9.35.2. Show that for this surface

$$a^{\alpha\beta}b_{\alpha\beta} = \frac{[f_{11}(1 + f_2^2) - 2f_1f_2f_{12} + f_{22}(1 + f_1^2)]}{[1 + f_1^2 + f_2^2]^{3/2}},$$

where $f_{\alpha\beta} = \partial^2 f/\partial y^\alpha\,\partial y^\beta$, etc.

Exercise 9.35.3. Show that in any coordinate system

$$b_{\alpha\beta} = \tfrac{1}{2}\varepsilon^{\rho\sigma}\varepsilon_{ijk}t^i_{\alpha,\beta}t^j_\rho t^k_\sigma$$

Exercise 9.35.4. Calculate the second fundamental form for the sphere.

Exercise 9.35.5. Show that

$$\varepsilon^{\lambda\alpha}b_{\alpha\beta}\varepsilon^{\beta\gamma}b_{\gamma\mu} = -K\delta^\lambda_\mu.$$

9.36. The third fundamental form

The first fundamental form is connected with distance in the surface, the second with the rate of change of the tangent. The third fundamental form is connected with the rate of change of the normal. Since n^i is a unit vector $n^i_{,\alpha}$ is perpendicular to it and so must lie in the tangent plane. However, if it is tangential to the surface, then it must be expressible as a combination of the t^i_β

$$n^i_{,\alpha} = \nu^\beta_\alpha t^i_\beta. \tag{9.36.1}$$

However, $g_{ij}t^i_\beta n^j = 0$ so that differentiating convariantly with respect to u^α gives

$$\begin{aligned} g_{ij}t^i_{\beta,\alpha}n^j + g_{ij}t^i_\beta n^j_{,\alpha} &= g_{ij}b_{\beta\alpha}n^i n^j + g_{ij}t^i_\beta \nu^\gamma_\alpha t^j_\gamma \\ &= b_{\alpha\beta} + \nu^\gamma_\alpha a_{\gamma\beta} \;= 0. \end{aligned}$$

Thus multiplying through $a^{\beta\epsilon}$ we have

$$\nu_{\alpha}^{\epsilon} = -b_{\alpha\beta}a^{\beta\epsilon}, \tag{9.36.2}$$

which gives the surface tensor relating the rate of change of normal to the tangent vectors. Substituting back in Eq. 9.36.1 we have

$$n^{i}_{,\alpha} = -b_{\alpha\beta}a^{\beta\gamma}t^{i}_{\gamma}, \tag{9.36.3}$$

a set of equations known as Weingarten's formulae.

The symmetric tensor

$$c_{\alpha\beta} = g_{ij}n^{i}_{,\alpha}n^{j}_{,\beta} \tag{9.36.4}$$

is used to form the third fundamental form of the surface

$$c_{\alpha\beta}\, du^{\alpha}\, du^{\beta}. \tag{9.36.5}$$

This form is not independent of the other two, for substituting for $n^{i}_{,\alpha}$ and $n^{i}_{,\beta}$ in Eq. (9.36.4), we have

$$\begin{aligned} c_{\alpha\beta} &= g_{ij}b_{\alpha\delta}a^{\delta\gamma}t^{i}_{\gamma}b_{\beta\rho}a^{\rho\sigma}t^{j}_{\sigma} = a_{\gamma\sigma}b_{\alpha\delta}a^{\delta\gamma}b_{\beta\rho}a^{\rho\sigma} \\ &= b_{\alpha\sigma}a^{\sigma\rho}b_{\rho\beta}. \end{aligned} \tag{9.36.6}$$

9.37. The relation between the three fundamental forms—Gauss-Codazzi equations

We have seen that the third fundamental form is related to the first two. There is a further relation between the three forms and the two measures of curvature. The total or Gaussian curvature K has been defined intrinsically and we shall find it is related to the second fundamental form. The mean curvature H is defined by

$$2H = a^{\alpha\beta}b_{\alpha\beta}. \tag{9.37.1}$$

It is clearly zero for the plane but its interpretation as a curvature must wait until we consider curves in the surface.

Let us use Cartesian coordinates and geodesic surface coordinates so that the Christoffel symbols can all be made to vanish at a particular point. The derivatives of the Christoffel symbols in space will vanish but not those in the surface. In this system

$$t^{i}_{\alpha,\beta\gamma} = \frac{\partial^3 x^i}{\partial u^\alpha\, \partial u^\beta\, \partial u^\gamma} - \left[\frac{\partial}{\partial u^\gamma}\begin{Bmatrix} \epsilon \\ \alpha \quad \beta \end{Bmatrix}\right] t^{i}_{\epsilon}.$$

Thus

$$\begin{aligned} t^{i}_{\alpha,\beta\gamma} - t^{i}_{\alpha,\gamma\beta} &= \left[\frac{\partial}{\partial u^\beta}\begin{Bmatrix} \epsilon \\ \gamma \quad \alpha \end{Bmatrix} - \frac{\partial}{\partial u^\gamma}\begin{Bmatrix} \epsilon \\ \alpha \quad \beta \end{Bmatrix}\right] t^{i}_{\epsilon} \\ &= R^{\epsilon}{}_{.\alpha\beta\gamma}t^{i}_{\epsilon}, \end{aligned} \tag{9.37.2}$$

by definition of the Riemann-Christoffel tensor.

Also, since $t^i_{\alpha,\beta} = b_{\alpha\beta}n^i$, we have

$$t^i_{\alpha,\beta\gamma} = b_{\alpha\beta,\gamma}n^i + b_{\alpha\beta}n^i_{,\gamma}$$
$$= b_{\alpha\beta,\gamma}n^i - b_{\alpha\beta}b_{\gamma\delta}a^{\delta\epsilon}t^i_\epsilon. \qquad (9.37.3)$$

If we form the expression on the left-hand side of Eq. 9.37.2 from this last formula, we have

$$R^\epsilon_{\alpha\beta\gamma}t^i_\epsilon = (b_{\alpha\beta,\gamma} - b_{\alpha\gamma,\beta})n^i + (b_{\gamma\alpha}b_{\beta\delta} - b_{\alpha\beta}b_{\gamma\delta})a^{\delta\epsilon}t^i_\epsilon. \qquad (9.37.4)$$

Multiplying this by n_i we see that the first and last terms vanish since $n_i t^i_\epsilon = 0$ and

$$b_{\alpha\beta,\gamma} = b_{\alpha\gamma,\beta}. \qquad (9.37.5)$$

This being so, and Eq. 9.37.4 true for any t^i_ϵ ,it can be written by lowering the suffix

$$R_{\lambda\alpha\beta\gamma} = b_{\lambda\beta}b_{\alpha\gamma} - b_{\alpha\beta}b_{\lambda\gamma}. \qquad (9.37.6)$$

The first of these equations is associated with the name of Codazzi, the second with Gauss. In particular, the only independent term of $R_{\lambda\alpha\beta\gamma}$ is R_{1212} and

$$R_{1212} = b_{11}b_{22} - b_{12}b_{21} = b. \qquad (9.37.7)$$

Hence,

$$K = \frac{R_{1212}}{a} = \frac{b}{a}. \qquad (9.37.8)$$

Now

$$c_{\alpha\beta} - 2Hb_{\alpha\beta} = b_{\alpha\gamma}a^{\gamma\delta}b_{\delta\beta} - (a^{\gamma\delta}b_{\gamma\delta})b_{\alpha\beta}$$
$$= a^{\gamma\delta}[b_{\alpha\gamma}b_{\beta\delta} - b_{\alpha\beta}b_{\gamma\delta}]$$
$$= -a_{\alpha\beta}\frac{b}{a},$$

so that

$$c_{\alpha\beta} - 2Hb_{\alpha\beta} + Ka_{\alpha\beta} = 0. \qquad (9.37.9)$$

Exercise 9.37.1. If τ^α is a unit vector in the surface, show that

$$K = 2H\tau^\alpha b_{\alpha\beta}\tau^\beta - \tau^\alpha b_{\alpha\beta}a^{\beta\gamma}b_{\gamma\delta}\tau^\delta.$$

Exercise 9.37.2. Show that

$$b_{\alpha\beta}a^{\beta\gamma}b_{\gamma\delta}a^{\delta\alpha} = 4H^2 - 2K.$$

9.38. Curves in the surface

Another way of looking at the curvature of the surface is to regard a curve in the surface as a space curve and look at its properties. If $u^\alpha = u^\alpha(s)$ is a curve in the surface and s the arc length along it, then

$$a_{\alpha\beta}\dot{u}^\alpha\dot{u}^\beta = 1 \qquad (9.38.1)$$

and

$$\tau^\alpha = \dot{u}^\alpha \qquad (9.38.2)$$

is a unit tangent vector to the curve. (A dot denotes intrinsic differentiation with respect to s.) The equation of the curve in space is

$$x^i = x^i(u^1(s), u^2(s)) \qquad \text{and} \qquad \tau^i = \dot{x}^i = t^i_\alpha \dot{u}^\alpha = t^i_\alpha \tau^\alpha. \tag{9.38.3}$$

We may translate the Serret-Frenet formulae, which were derived in Section 3.12 in Cartesian coordinates, into general coordinates. Then

$$\begin{aligned} \dot{\tau}^i &= \kappa\nu^i \\ \dot{\nu}^i &= -\kappa\tau^i + \mu\beta^i \\ \dot{\beta}^i &= -\mu\nu^i \end{aligned} \tag{9.38.4}$$

Here ν^i is the unit principal normal defined by the first of these equations and β^i is the binormal orthogonal to both the tangent and principal normal. κ is the curvature and μ* the torsion. The derivation of these equations follows precisely along the lines of Section 3.12.

Differentiating Eq. 9.38.3 and comparing with the first of the Serret-Frenet formulae we have

$$\dot{\tau}^i = t^i_{\alpha,\beta}\dot{u}^\alpha\dot{u}^\beta + t^i_\alpha\ddot{u}^\alpha = b_{\alpha\beta}\dot{u}^\alpha\dot{u}^\beta n^i + t^i_\alpha\ddot{u}^\alpha = \kappa\nu^i.$$

Now $\ddot{u}^\alpha$ is a surface vector and $t^i_\alpha\ddot{u}^\alpha$ is therefore tangential to the surface. It follows that

$$\kappa_{(n)} = \kappa n_i \nu^i = b_{\alpha\beta}\dot{u}^\alpha\dot{u}^\beta = \frac{b_{\alpha\beta}\, du^\alpha\, du^\beta}{a_{\alpha\beta}\, du^\alpha\, du^\beta}. \tag{9.38.5}$$

However, $\kappa n_i \nu^i$ is the projection on the normal to the surface of a vector of length κ along the principal normal to the curve. We see that this projected curvature is independent of the precise curve chosen in the surface provided it is tangent to the direction τ^α. It is called the normal curvature of the surface in the direction τ^α, and Eq. 9.38.5 shows that it is the ratio of the first two fundamental forms. Figure 9.5 shows the curve C in the surface S whose principal normal does not coincide with the normal to the surface. Only if C is a geodesic will the two coincide.

We may now ask in what direction the normal curvature has its maximum and minimum values. That is, we require the stationary values of $b_{\alpha\beta}\tau^\alpha\tau^\beta$ subject to $a_{\alpha\beta}\tau^\alpha\tau^\beta = 1$. This is the familiar problem of finding the characteristic roots and vectors of the tensor $b_{\alpha\beta}$. The two values of κ will be given by the quadratic

$$\begin{vmatrix} b_{11} - \kappa a_{11} & b_{12} - \kappa a_{12} \\ b_{21} - \kappa a_{21} & b_{22} - \kappa a_{22} \end{vmatrix} = 0.$$

* μ was previously written $1/\sigma$, but here σ will be reserved for the geodesic curvature.

However, this gives
$$b - 2aH\kappa + a\kappa^2 = 0$$
since
$$a_{11}b_{22} - a_{12}b_{21} - a_{21}b_{12} + a_{22}b_{11} \\ = a(a^{22}b_{22} + a^{21}b_{21} + a^{12}b_{12} + a^{11}b_{11}) = 2aH.$$
Hence, dividing through by a we have
$$\kappa^2 - 2H\kappa + K = 0. \tag{9.38.6}$$
It follows that if κ_1 and κ_2 are the greatest and least curvatures
$$\tfrac{1}{2}(\kappa_1 + \kappa_2) = H, \qquad \kappa_1\kappa_2 = K. \tag{9.38.7}$$
These are called the principal curvatures at the point and the corresponding

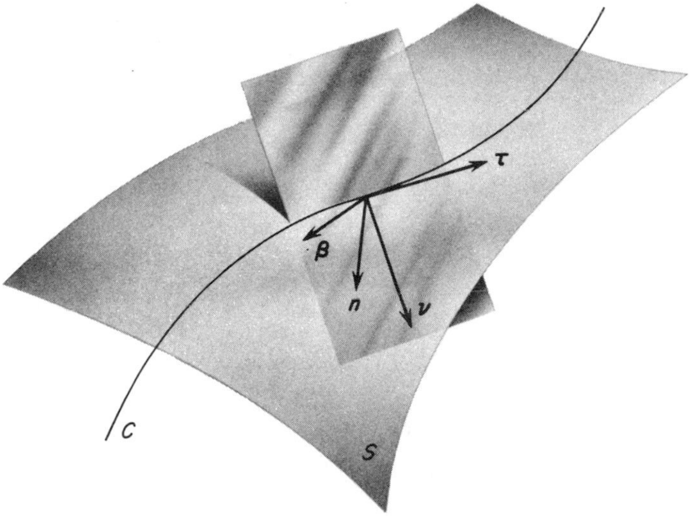

direction the directions of principal curvature. This affords a very pleasing interpretation of the mean and total curvature.

A surface for which $K = 0$ is called *developable* since it may be rolled out on a plane without stretching. The cylinder and cone are developable and we can obviously introduce a Cartesian metric by taking u^1 to be the distance along a generator and u^2 the distance along the orthogonal trajectory. All other developable surfaces can be obtained from a space curve whose tangents will be the generators of a surface (cf. Ex. 9.38.3).

Exercise 9.38.1. Derive Eq. 9.38.4 in the manner of Section 3.12.

Exercise 9.38.2. Show that the normal curvature of a surface in a given direction is the curvature of the geodesic in that direction.

Exercise 9.38.3. A surface is generated from the tangents of a curve $\xi^i(s)$. A point on the surface is thus

$$x^i = \xi^i(s) + (t - s)\frac{d\xi^i}{ds}.$$

If s is the arc length along the curve and τ^i, ν^i, β^i are the unit tangent normal and binormal to the curve, show that with $u^1 - s$, $u^2 = t$ we have

$$a_{11} = \kappa^2(u^2 - u^1)^2, \qquad a_{12} = 0, \qquad a_{22} = 1, \qquad b_{11} = -\kappa\mu(u^2 - u^1),$$

$$b_{12} = b_{22} = 0, \qquad H = \frac{-\mu}{2\kappa(u^2 - u^1)}, \qquad K = 0, \qquad n^i = \beta^i.$$

(κ and μ are defined by Eq. 9.38.4.)

9.41. Differential operators in a surface

If $\varphi(u^1, u^2)$ is a scalar function defined in the surface $\varphi_{,\alpha} = \partial\varphi/\partial u^\alpha$ is a covariant vector or surface gradient of φ. If A^α is a contravariant surface vector $A^\alpha_{,\alpha}$ is a scalar which we may call the surface divergence of A^α. The form of divergence most useful for calculation has been mentioned above (Ex. 9.25.2 and 3), namely,

$$A^\alpha_{,\alpha} = \frac{1}{a^{1/2}}\frac{\partial}{\partial u^\alpha}(a^{1/2}A^\alpha). \tag{9.41.1}$$

The surface divergence of a covariant vector can be obtained by first raising the index,

$$a^{\alpha\beta}A_{\alpha,\beta} = \frac{1}{a^{1/2}}\frac{\partial}{\partial u^\beta}(a^{1/2}a^{\alpha\beta}A_\alpha). \tag{9.41.2}$$

If the covariant vector is the gradient of a scalar we have the surface Laplacian

$$a^{\alpha\beta}\varphi_{,\alpha\beta} = \frac{1}{a^{1/2}}\frac{\partial}{\partial u^\beta}\left(a^{1/2}a^{\alpha\beta}\frac{\partial\varphi}{\partial u^\alpha}\right). \tag{9.41.3}$$

The length of the surface gradient

$$a^{\alpha\beta}\varphi_{,\alpha}\varphi_{,\beta} \tag{9.41.3}$$

is also known as Beltrami's first differential parameter of and the Laplacian as his second. There is also Beltrami's mixed differential parameter of φ and ψ, namely,

$$a^{\alpha\beta}\varphi_{,\alpha}\psi_{,\beta}. \tag{9.41.4}$$

By analogy with three dimensions we might define the surface curl as

$$\varepsilon^{\alpha\beta}A_{\beta,\alpha} = a^{-1/2}\left(\frac{\partial A_2}{\partial u^1} - \frac{\partial A_1}{\partial u^2}\right). \tag{9.41.5}$$

The surface divergence and curl of a space vector can also be defined. The surface divergence of a covariant space vector is

$$a^{\alpha\beta}t^i_\alpha A_{i,\beta} \tag{9.41.6}$$

and that of a contravariant space vector

$$g_{ij}a^{\alpha\beta}t^i_\alpha A^j_{,\beta}. \tag{9.41.7}$$

The surface divergence of the normal vector is

$$\begin{aligned} g_{ij}a^{\alpha\beta}t^i_\alpha n^j_{,\beta} &= -g_{ij}a^{\alpha\beta}b_{\beta\gamma}a^{\gamma\delta}t^j_\delta t^i_\alpha \\ &= -a^{\alpha\beta}b_{\beta\gamma}a^{\gamma\delta}a_{\delta\alpha} \\ &= -2H. \end{aligned} \tag{9.41.8}$$

A space vector may be represented as the sum of a normal and a tangential part

$$A^i = A_{(n)}n^i + A^\alpha_{(t)}t^i_\alpha.$$

For since $n_i t^i_\alpha = 0$ and $g_{ij}n^i t^j_\beta = 0$,

$$\begin{aligned} A_{(n)} &= A^i n_i, \\ A^\alpha_{(t)} &= a^{\alpha\beta}g_{ij}t^j_\beta A^i. \end{aligned} \tag{9.41.9}$$

Then, using Eq. (9.41.8) and Weingarten's formula, we have, after some manipulation, that the surface divergence of A^i is

$$A^\alpha_{(t),\alpha} - 2HA_{(n)}. \tag{9.41.10}$$

The surface curl of a space vector is similarly defined as

$$\varepsilon_{ijk}a^{\alpha\beta}A^j_{,\alpha}t^k_\beta. \tag{9.41.11}$$

Exercise 9.41.1. Show that $a^{\alpha\beta}\varphi_{,\alpha}\varphi_{,\beta} = g_{ij}\varphi^i\varphi^j$, where $\varphi^i = a^{\alpha\beta}t^i_\alpha\varphi_{,\beta}$ is the space vector associated with the surface gradient.

Exercise 9.41.2. If $\bar{A}_i = A_i - (A_j n^j)n_i$ is the "projection" of A_i on the tangent plane to the surface, show that

$$a^{\alpha\beta}t^i_\alpha\bar{A}_{i,\beta} = a^{\alpha\beta}(t^i_\alpha A_i)_{,\beta}.$$

Exercise 9.41.3. Show that the surface curl of n^i is zero.

9.42. Green's and Stokes' theorems in a surface

For any suitable surface S and curve C on it, we have Stokes' theorem in space (cf. Eq. 7.57.4),

$$\iint_S \varepsilon^{ijk}A_{j,i}n_k\,dS = \oint_C A_j t^j\,ds, \tag{9.42.1}$$

where n_k is the unit normal to the surface and t^j the unit tangent to the curve. Let us now take the special coordinate system in which the surface is $x^3 = 0$ and the x^3 coordinate line are normal to the surface. In the surface we take its intersections with the surfaces x^1 and x^2 constant to be the u^1 and u^2 coordinate lines. Then

$$\begin{aligned} t^i_\alpha &= \delta^i_\alpha, \qquad i, \alpha = 1, 2, \qquad t^3_\alpha = 0, \\ a_{\alpha\beta} &= g_{\alpha\beta}, \qquad g_{3\alpha} = g_{\alpha 3} = 0, \qquad g = g_{33}a. \end{aligned} \tag{9.42.2}$$

Moreover, $n_1 = n_2 = 0$, $n_3 = g_{33}^{1/2}$, and so

$$\varepsilon^{ijk} n_k = \epsilon^{ij3}\left(\frac{g_{33}}{g}\right)^{1/2} = \frac{\epsilon^{ij3}}{a^{1/2}} = \varepsilon^{\alpha\beta}$$

if $i = \alpha$, $j = \beta$. If we write $A_\alpha = t^i_\alpha A_i$, then,by Eq. (9.42.2), $A_\alpha = A_i$, $i = \alpha = 1, 2$ and $A_{j,i} = A_{\beta,\alpha}$ for $i = \alpha$, $j = \beta$. Also, if $\tau_\beta = t^j_\beta t_j$ is the unit surface vector tangent to the curve, $\tau_\beta = t_j$, $j = \beta = 1, 2$, $t_3 = 0$ and $A_{,j}t^j = A^j t_j = A^\beta \tau_\beta = A_\beta \tau^\beta$. Thus Eq. 9.42.1 becomes

$$\iint_S \varepsilon^{\alpha\beta} A_{\beta,\alpha}\, dS = \oint_C A_\beta \tau^\beta\, ds, \tag{9.42.3}$$

which is an equation entirely in surface vectors. However, it is a valid tensor equation and so holds not only in the special coordinate system described but also in any system of surface coordinates. It evidently relates the surface integral of the curl of the surface vector to the circulation around the bounding curve, and may be called Stokes' theorem for the surface.

Now let

$$A_\beta = \varepsilon_{\gamma\beta} B^\gamma \tag{9.42.4}$$

so that

$$\varepsilon^{\alpha\beta} A_{\beta,\alpha} = \varepsilon^{\alpha\beta}\varepsilon_{\gamma\beta} B^\gamma_{,\alpha} = \delta^\alpha_\gamma B^\gamma_{,\alpha} = B^\alpha_{,\alpha}.$$

Also $\varepsilon_{\gamma\beta}\tau^\beta = \nu_\gamma$ the outward unit normal to the curve in the surface (cf. Ex. 9.14.1). Hence, by substitution in Eq. (9.42.3),

$$\iint_S B^\alpha_{,\alpha}\, dS = \oint_C B^\alpha \nu_\alpha\, ds. \tag{9.42.5}$$

This is clearly the surface form of Green's theorem. If $B^\alpha = a^{\alpha\beta}\varphi,_\beta\psi$, then

$$\iint_S \psi a^{\alpha\beta}\varphi_{,\alpha\beta}\, dS + \iint_S a^{\alpha\beta}\psi_{,\alpha}\varphi_{,\beta}\, dS = \oint_C \psi a^{\alpha\beta}\varphi_{,\beta}\nu_\alpha\, ds \tag{9.42.6}$$

and, in particular, if $\psi = 1$,

$$\iint_S a^{\alpha\beta}\varphi_{,\alpha\beta}\, dS = \oint_C \varphi_{,\alpha}\nu^\alpha\, ds \tag{9.42.7}$$

relating the surface integral of the Laplacian to the line integral of the normal derivative.

Exercise 9.42.1. Show that if $a^{\alpha\beta}\varphi_{,\alpha\beta} = -1$ and on S and $\varphi_{,\alpha}\nu^{\alpha} = 0$ on C

$$\iint\limits_S \psi \, dS = \iint\limits_S a^{\alpha\beta}\psi_{,\alpha}\varphi_{,\beta} \, dS.$$

Exercise 9.42.2. Show that

$$\iint\limits_S a^{\alpha\beta}[\psi\varphi_{,\alpha\beta} - \varphi\psi_{,\alpha\beta}] \, dS = \oint\limits_C [\psi\varphi_{,\alpha} - \varphi\psi_{,\alpha}]\nu^{\alpha} \, ds.$$

Exercise 9.42.3. Using the decomposition $A^i = A_{(n)}n^i + A^{\alpha}_{(t)}t^i_{\alpha}$, show that

$$\iint\limits_S a^{\alpha\beta}t^i_{\alpha}A_{i,\beta} = \oint\limits_C A^{\alpha}_{(t)}\tau_{\alpha} \, ds - 2\iint\limits_S A_{(n)}H \, dS.$$

Exercise 9.42.4. Relate Eq. (9.42.5) to Green's theorem in space by considering a family of closed surfaces that shrink down onto the two sides of S within the curve C.

The Equations of Surface Flow

10

We propose here to give some account of the equations governing fluid motion in a surface or interface. The importance of this subject is shown by the growing interest in the behavior of surfaces and Scriven (in the paper listed at the end of this chapter and referred to constantly in it) gives many references to diverse applications. Its interest here lies not only in these applications but also in the fact that it is itself an application of the methods and ideas we have developed for non-Euclidean space. As has been remarked before, Cartesian tensors really suffice for three-dimensional flows; for the space of everyday life, being Euclidean, always admits of a Cartesian frame of reference. However, the surface is a two-dimensional non-Euclidean space and from the outset demands a full tensorial treatment. Again, the flow in a surface is not just flow in a two-dimensional space whose governing equations will be immediate analogs of the three-dimensional ones. For, in contrast with the three-dimensional space, this surface is a two-dimensional space that can move within a space of higher dimensions, namely, the three-dimensional space surrounding it. Moreover, it may be the region of contact of two bulk fluids. This is again a new feature, for a bulk fluid can never be the interface of two four-dimensional fluids. We shall sometimes be thinking intrinsically and at others be seeing the connection of the surface with the space around it.

10.11. Velocity in a surface

Consider the surface at some arbitrary time which, for convenience, we take to be $t = 0$. A system of surface coordinates u^1, u^2 can be introduced and particles can be labelled by these coordinates. A system of convected coordinates can thus be set up in which a particle retains its coordinate label. We will distinguish convected coordinates by capital Greek indices; thus u^Γ = constant is a material line in the surface and moving with its flow. The element of length in the surface will be given by a metric tensor

$$ds^2 = a_{\Gamma\Delta}\, du^\Gamma\, du^\Delta, \tag{10.11.1}$$

where $a_{\Gamma\Delta}$ is a function of the surface coordinates and the time.

It is not possible to set up a fixed coordinate system in a moving surface by intrinsic considerations alone. Indeed, any other system of surface coordinates would provide a reference frame and the velocity of a particle would be velocity relative to this frame. If we look outside the surface, however, we can say that a point is fixed in the surface if its velocity in space is wholly normal to the surface. Let lower case Greek indices be used to denote the coordinates thus "fixed" in the surface. Then the motion in the surface is a continuous transformation in two-dimensional space

$$u^\alpha = u^\alpha(u^\Gamma, t), \qquad \alpha, \Gamma = 1, 2. \tag{10.11.2}$$

This relation has an inverse

$$u^\Gamma = u^\Gamma(u^\alpha, t), \qquad \alpha, \Gamma = 1, 2, \qquad (10.11.2)$$

which gives the convected coordinates of the particle at u^α. We have a metric tensor in the fixed coordinate system

$$ds^2 = a_{\alpha\beta}\, du^\alpha\, du^\beta, \tag{10.11.3}$$

where

$$a_{\alpha\beta} = \frac{\partial u^\Gamma}{\partial u^\alpha}\frac{\partial u^\Delta}{\partial u^\beta}\, a_{\Gamma\Delta}, \tag{10.11.4}$$

and unless the surface is stationary this also will be a function of time.

The surface velocity at a point is defined by

$$V^\alpha = \frac{du^\alpha}{dt}, \tag{10.11.5}$$

where the differentiation is with u^Γ held constant. The acceleration at a point is similarly defined as the material derivative of V^α. Thus,

$$A^\alpha = \frac{dV^\alpha}{dt} = \frac{\partial V^\alpha}{\partial t} + V^\beta V^\alpha_{,\beta}. \tag{10.11.6}$$

We shall need certain identities later, of which one will be proved here and others given as exercises. Since the space is curved, second covariant derivatives do not commute, but

$$V^{\mu}_{,\beta\gamma} - V^{\mu}_{,\gamma\beta} = a^{\alpha\mu}R_{\lambda\alpha\beta\gamma}V^{\lambda} = Ka^{\alpha\mu}\varepsilon_{\lambda\alpha}\varepsilon_{\beta\gamma}V^{\lambda}.$$

If we contract on μ and γ and multiply by $a^{\beta\nu}$,

$$a^{\beta\nu}(V^{\mu}_{,\beta\mu} - V^{\mu}_{,\mu\beta}) = KV^{\nu} \tag{10.11.7}$$

since

$$a^{\beta\nu}a^{\alpha\mu}\varepsilon_{\beta\mu}\varepsilon_{\lambda\alpha} = \delta^{\nu}_{\lambda} \tag{10.11.8}$$

(cf. Ex. 9.13.8).

Thus, as soon as second derivatives appear, we are liable also to have the Gaussian or total curvature.

Exercise 10.11.1. Recalling that $\varepsilon^{\alpha\beta}a_{\beta\gamma} = a^{\alpha\beta}\varepsilon_{\beta\gamma}$, show that

$$\varepsilon^{\lambda\mu}\varepsilon^{\alpha\beta}a_{\beta\gamma}V^{\gamma}_{,\alpha\lambda} = a^{\alpha\lambda}V^{\mu}_{,\alpha\lambda} - a^{\alpha\mu}V^{\lambda}_{,\alpha\lambda}.$$

Hence, show that

$$a^{\lambda\beta}V^{\alpha}_{,\lambda\beta} - a^{\alpha\beta}V^{\lambda}_{,\lambda\beta} = KV^{\alpha} - \varepsilon^{\alpha\beta}\varepsilon^{\lambda\mu}a_{\mu\gamma}V^{\gamma}_{,\lambda\beta}\,.$$

Compare this with the three-dimensional formula

$$\nabla^2\mathbf{V} - \nabla(\nabla\cdot\mathbf{V}) = -\nabla\wedge(\nabla\wedge\mathbf{V}).$$

10.12. Operations with a time dependent metric

It will be of great convenience later if we notice first some formulae connected with a time dependent metric. A dot will be used to denote the derivative with respect to time, that is,

$$\dot{a}_{\alpha\beta} = \frac{\partial}{\partial t}\,a_{\alpha\beta}(u^1, u^2, t). \tag{10.12.1}$$

Since

$$a^{\alpha\beta}a_{\beta\gamma} = \delta^{\alpha}_{\gamma},$$

we have

$$\dot{a}^{\alpha\beta}a_{\beta\gamma} = -a^{\alpha\beta}\dot{a}_{\beta\gamma}\,.$$

Multiplying both sides by $a^{\gamma\delta}$ gives

$$\dot{a}^{\alpha\delta} = -a^{\alpha\beta}a^{\delta\gamma}\dot{a}_{\beta\gamma}. \tag{10.12.2}$$

This shows that the raising and lowering of indices does not commute with the differentiation.

Again, the determinant

$$a = a_{11}a_{22} - a_{12}a_{21},$$

so

$$\begin{aligned}\dot{a} &= \dot{a}_{11}a_{22} + a_{11}\dot{a}_{22} - \dot{a}_{12}a_{21} - a_{12}\dot{a}_{21}\\ &= a(a^{11}\dot{a}_{11} + a^{22}\dot{a}_{22} + a^{12}\dot{a}_{12} + a^{21}\dot{a}_{21}).\end{aligned}$$

Thus

$$\frac{\dot{a}}{a} = a^{\alpha\beta}\dot{a}_{\alpha\beta}. \tag{10.12.3}$$

If we take the identity (cf. Ex. 9.13.7)

$$\varepsilon^{\alpha\lambda}\varepsilon^{\beta\mu} + \varepsilon^{\alpha\mu}\varepsilon^{\beta\lambda} = 2a^{\alpha\beta}a^{\lambda\mu} - a^{\alpha\mu}a^{\beta\lambda} - a^{\alpha\lambda}a^{\beta\mu}, \tag{10.12.4}$$

and multiply it by $\dot{a}_{\lambda\mu}$, we have

$$2\varepsilon^{\alpha\lambda}\varepsilon^{\beta\mu}\dot{a}_{\lambda\mu} = 2a^{\alpha\beta}a^{\lambda\mu}\dot{a}_{\lambda\mu} - 2a^{\alpha\lambda}a^{\beta\mu}\dot{a}_{\lambda\mu}$$

or

$$\varepsilon^{\alpha\lambda}\varepsilon^{\beta\mu}\dot{a}_{\lambda\mu} = a^{\alpha\beta}\frac{\dot{a}}{a} + \dot{a}^{\alpha\beta}. \tag{10.12.5}$$

Another form in which we shall need this identity is

$$\varepsilon^{\alpha\lambda}\varepsilon^{\beta\mu}(\dot{a}_{\lambda\mu})_{,\beta} = (\dot{a}^{\alpha\beta})_{,\beta} + 2a^{\alpha\beta}\left(\frac{\dot{a}}{2a}\right)_{,\beta}. \tag{10.12.6}$$

The element of area defined by an increment du^1 in one coordinate and du^2 in another is

$$dS = a^{1/2}\,du^1\,du^2. \tag{10.12.7}$$

If Γ is any function of position on the surface and time, and S is a material part of the surface,

$$\iint_S \Gamma\,dS = \iint_S \Gamma a^{1/2}\,du^1\,du^2 = \iint_\Sigma \Gamma A^{1/2}\,du^{\mathrm{I}}\,du^{\mathrm{II}},$$

where Σ is the part of surface from which S has been convected, A is the determinant of the metric tensor, and du^{I} and du^{II} are increments of the convected coordinates. Now $A = J^2 a$ where $J = |\partial u^\alpha/\partial u^\Gamma|$ is the Jacobian of the transformation from fixed to convected coordinates, hence, we can write the integral as

$$\iint_\Sigma \Gamma J a^{1/2}\,du^{\mathrm{I}}\,du^{\mathrm{II}}.$$

If now we take the material or convected derivative of this, we have

$$\begin{aligned}
\frac{d}{dt}\iint_S \Gamma\,dS &= \iint_\Sigma \frac{d}{dt}(\Gamma J a^{1/2})\,du^{\mathrm{I}}\,du^{\mathrm{II}} \\
&= \iint_\Sigma \left[\frac{d\Gamma}{dt}Ja^{1/2} + \Gamma\frac{dJ}{dt}a^{1/2} + \Gamma J\frac{da^{1/2}}{dt}\right]du^{\mathrm{I}}\,du^{\mathrm{II}} \\
&= \iint_\Sigma \left[\frac{d\Gamma}{dt} + \Gamma V^\alpha_{,\alpha} + \Gamma\frac{\dot{a}}{2a}\right]Ja^{1/2}\,du^{\mathrm{I}}\,du^{\mathrm{II}} \\
&= \iint_S \left[\dot{\Gamma} + \Gamma V^\alpha_{,\alpha} + \Gamma\frac{\dot{a}}{2a}\right]dS.
\end{aligned} \tag{10.12.9}$$

Here we have used the fact that $\dot{J} = V^\alpha_{,\alpha} J$. This is the analog of Reynolds' transport theorem in the surface and we see that there is a new term arising from the time dependence of the metric. In a fixed surface the last term will vanish.

The equation may be manipulated in the usual way by the use of Green's theorem and gives

$$\frac{d}{dt}\iint_S \Gamma\, dS = \iint_S \left[\frac{\partial \Gamma}{\partial t} + \Gamma \frac{\dot{a}}{2a} + (\Gamma V^\alpha)_{,\alpha}\right] dS$$

$$= \iint_S \left(\frac{\partial \Gamma}{\partial t} + \Gamma \frac{\dot{a}}{2a}\right) dS + \oint_C \Gamma V^\alpha m_\alpha\, dS \qquad (10.12.10)$$

where m_α is the unit surface vector normal to the bounding curve C. The corollary to this theorem again holds. If γ is any conserved scalar quantity, that is,

$$\frac{d}{dt}\iint_S \gamma\, dS = 0,$$

then

$$\frac{d}{dt}\iint_S \gamma\Gamma\, dS = \iint_S \gamma \frac{d\Gamma}{dt}\, dS. \qquad (10.12.11)$$

Exercise 10.12.1. Show that

$$\dot{a}_{\alpha\beta} = \frac{da_{\alpha\beta}}{dt} = \frac{\partial a_{\alpha\beta}}{\partial t}.$$

10.21. Strain in the surface

The element of length in convected coordinates is

$$ds^2 = a_{\Gamma\Delta}\, du^\Gamma\, du^\Delta. \qquad (10.21.1)$$

If we take the material or convected derivative of this, we have

$$2ds \frac{d(ds)}{dt} = \dot{a}_{\Gamma\Delta}\, du^\Gamma\, du^\Delta$$

or

$$\frac{1}{ds}\frac{d}{dt}(ds) = \frac{S_{\Gamma\Delta}\, du^\Gamma\, du^\Delta}{a_{\Gamma\Delta}\, du^\Gamma\, du^\Delta}, \qquad (10.21.2)$$

where

$$S_{\Gamma\Delta} = \tfrac{1}{2}\dot{a}_{\Gamma\Delta}. \qquad (10.21.3)$$

In the fixed coordinate system the rate of strain or deformation tensor $S_{\Gamma\Delta}$ has components $S_{\alpha\beta}$. In the nomenclature of Section 8.32 $S_{\alpha\beta}$ is the convected derivative $(d_c/dt)(\tfrac{1}{2}a_{\alpha\beta})$ and, using the equations given there, becomes

$$S_{\alpha\beta} = \tfrac{1}{2}\dot{a}_{\alpha\beta} + \tfrac{1}{2}(V_{\alpha,\beta} + V_{\beta,\alpha}). \qquad (10.21.4)$$

The deformation tensor $S_{\alpha\beta}$ is susceptible of the usual interpretations in terms of the principal axes of stretching.

The dilation of area is

$$\frac{1}{dS}\frac{d_c}{dt}(dS) = \frac{\dot{a}}{2a} + V^{\alpha}_{,\alpha} = a^{\alpha\beta}S_{\alpha\beta}. \tag{10.21.5}$$

Exercise 10.21.1. Show that the rate of change of the angle between elements du^{α}, dv^{β}, originally orthogonal to one another, is

$$\frac{-2S_{\alpha\beta}\,du^{\alpha}\,dv^{\beta}}{\{a_{\alpha\beta}\,du^{\alpha}\,du^{\beta}\,a_{\gamma\delta}\,dv^{\gamma}\,dv^{\delta}\}^{1/2}}.$$

Exercise 10.21.2. Show that the rate of strain given by Ex. (10.21.1) has two stationary values which are roots of the quadratic

$$\lambda^2 - (a^{\alpha\beta}S_{\alpha\beta})\lambda + (\varepsilon^{\alpha\lambda}\varepsilon^{\beta\mu}S_{\alpha\beta}S_{\lambda\mu}) = 0.$$

Exercise 10.21.3. Establish Eq. 10.21.5 (cf. Section 10.12).

10.22. Stress in the surface

The surface stress is a force per unit length of line element. If m_{α} is the unit normal to a line element we can show by an argument on an elementary triangle, which is entirely analogous to the spatial argument on the tetrahedron, that the stress on the element is given by a tensor

$$T^{\alpha\beta}m_{\alpha} = t^{\beta}_{(m)}. \tag{10.22.1}$$

By parallel propagation along the coordinate lines we can construct an arbitrary parallel field l_{β} over any finite part of the surface. If C is the bounding curve of this part of the surface S, then the total component of stress in the direction of l_{β} is

$$\oint_C t^{\beta}_{(m)}l_{\beta}\,ds = \oint_C T^{\alpha\beta}m_{\alpha}l_{\beta}\,ds = \iint_S T^{\alpha\beta}_{,\alpha}l_{\beta}\,dS. \tag{10.22.2}$$

In a state of rest we suppose that the stress is normal to the line element and independent of its orientation. Then, if it is true for any m_{α} that

$$T^{\alpha\beta}m_{\alpha} = \sigma m^{\beta} = \sigma a^{\alpha\beta}m_{\alpha},$$

it follows that

$$T^{\alpha\beta} = \sigma a^{\alpha\beta} \tag{10.22.3}$$

in a state of rest. σ is called the interfacial or surface tension and corresponds to the pressure in bulk fluids. If there are no forces acting within the surface and it is in equilibrium, then the net stress on any finite part of the surface must be zero. Thus, by Eq. (10.22.2) $T^{\alpha\beta}_{,\alpha} = 0$ and since $a^{\alpha\beta}_{,\alpha} = 0$ this implies that $\sigma_{,\alpha} = 0$ or σ is constant over the surface.

10.23. Constitutive relations for the surface

The arguments which have previously been adduced in constructing the constitutive relation for the Stokesian fluid can be readily modified to the present case. If the surface is isotropic, the difference $(T^{\alpha\beta} - \sigma a^{\alpha\beta})$ of the stress from its equilibrium value must be an isotropic function of $S_{\alpha\beta}$ which vanishes with $S_{\alpha\beta}$. It must therefore have the form

$$T^{\alpha\beta} - \sigma a^{\alpha\beta} = A a^{\alpha\beta} + B^{\alpha\beta\lambda\mu} S_{\lambda\mu} \tag{10.23.1}$$

where A is a scalar and $B^{\alpha\beta\lambda\mu}$ an isotropic tensor function of the invariants of S. There is no product term in the tensor $S_{\alpha\beta}$ as in the three-dimensional case, for since a tensor satisfies its own characteristic equation, a square or higher power could always be expressed in terms of the first power and the invariants. There are two invariants, the coefficients in the quadratic characteristic equation, namely,

$$a^{\alpha\beta} S_{\alpha\beta} \quad \text{and} \quad \varepsilon^{\alpha\lambda}\varepsilon^{\beta\mu} S_{\alpha\beta} S_{\lambda\mu}. \tag{10.23.2}$$

A Newtonian surface fluid is a Stokesian surface fluid in which the stress depends only linearly on the rate of strain. Thus A must be proportional to the first invariant and $B^{\alpha\beta\lambda\mu}$ a constant isotropic tensor. By precisely the same arguments as before, the symmetric isotropic fourth order tensor is a linear combination of $a^{\alpha\beta}a^{\lambda\mu}$ and $(a^{\alpha\lambda}a^{\beta\mu} + a^{\alpha\mu}a^{\beta\lambda})$. For the Newtonian surface fluid we can therefore write

$$T^{\alpha\beta} = (\sigma + \kappa a^{\lambda\mu} S_{\lambda\mu}) a^{\alpha\beta} + \epsilon E^{\alpha\beta\lambda\mu} S_{\lambda\mu}, \tag{10.23.3}$$

where

$$E^{\alpha\beta\lambda\mu} = a^{\alpha\lambda}a^{\beta\mu} + a^{\alpha\mu}a^{\beta\lambda} - a^{\alpha\beta}a^{\lambda\mu}. \tag{10.23.4}$$

There is some convenience in this form since the trace $a_{\alpha\beta}E^{\alpha\beta\lambda\mu} = 0$. κ and ϵ are called the coefficients of dilatational and shear surface viscosity respectively. In what follows they will be assumed to be constant.

Exercise 10.23.1. Show that

$$a^{\lambda\mu} S_{\lambda\mu} = \frac{\dot{a}}{2a} + V^{\lambda}_{,\lambda}$$

and

$$E^{\alpha\beta\lambda\mu} S_{\lambda\mu} = -\left\{\dot{a}^{\alpha\beta} + a^{\alpha\beta}\frac{\dot{a}}{2a}\right\} + \{a^{\alpha\lambda}V^{\beta}_{,\lambda} + a^{\beta\lambda}V^{\alpha}_{,\lambda} - a^{\alpha\beta}V^{\lambda}_{,\lambda}\}.$$

10.31. Intrinsic equations of surface motion

If γ is the surface density (mass per unit area) and the surface has no interchange with its surroundings, we can obtain a continuity equation from the principle of the conservation of mass. Thus, if the mass of any portion S of the surface is constant,

$$\frac{d}{dt}\iint_S \gamma \, dS = 0.$$

Applying the transport theorem we have

$$\iint_S \left[\frac{d\gamma}{dt} + \gamma V^{\alpha}_{,\alpha} + \gamma \frac{\dot{a}}{2a}\right] dS = 0$$

and since this is true for an arbitrary part of the surface,

$$\frac{d\gamma}{dt} + \gamma V^{\alpha}_{,\alpha} + \gamma \frac{\dot{a}}{2a} = 0. \tag{10.31.1}$$

Now let an arbitrary parallel field of covariant surface vectors l_α be defined on the surface and let F^α be the external force per unit area. If we make a balance of linear momentum in the direction of the parallel field,

$$\frac{d}{dt}\iint_S \gamma V^\alpha l_\alpha \, dS = \iint_S \gamma A^\alpha l_\alpha \, dS = \iint_S F^\alpha l_\alpha \, dS + \oint_C t^{\alpha}_{(m)} l_\alpha \, ds.$$

Using Eq. (10.22.2) to reduce the last integral we have

$$\iint_S [\gamma A^\alpha - F^\alpha - T^{\beta\alpha}_{,\beta}] l_\alpha \, dS$$

and, since both S and l_α are arbitrary,

$$\gamma A^\alpha = F^\alpha + T^{\beta\alpha}_{,\beta}. \tag{10.31.2}$$

If angular momentum is conserved

$$\iint_S \gamma \varepsilon_{\alpha\beta} A^\alpha u^\beta \, dS = \iint_S \varepsilon_{\alpha\beta} F^\alpha u^\beta \, dS + \oint_C \varepsilon_{\beta\gamma} t^{\beta}{}_{(m)} u^\gamma \, dS$$

Substituting on the left-hand side from Eq. (10.31.2) and using Eq. (10.22.1)

$$\iint_S \varepsilon_{\alpha\beta} T^{\gamma\alpha}_{,\gamma} u^\beta \, dS = \oint_C \varepsilon_{\beta\gamma} T^{\alpha\beta} u^\gamma m_\alpha \, dS$$

$$= \iint_S (\varepsilon_{\beta\gamma} T^{\alpha\beta} u^\gamma)_{,\alpha} \, dS$$

by Green's theorem. However, since $u^{\gamma}_{,\alpha} = \delta^{\gamma}_{\alpha}$, this gives

$$\varepsilon_{\beta\gamma} T^{\gamma\beta} = 0, \tag{10.31.3}$$

and hence the stress tensor is symmetric. The last term in Eq. (10.31.2) could thus be written $T^{\alpha\beta}_{,\beta}$.

10.32. Intrinsic equations for a Newtonian surface fluid

For the Newtonian surface fluid we have the constitutive equation (10.23.3)

$$T^{\alpha\beta} = (\sigma + \kappa a^{\lambda\mu} S_{\lambda\mu}) a^{\alpha\beta} + \epsilon E^{\alpha\beta\lambda\mu} S_{\lambda\mu}. \tag{10.32.1}$$

To calculate $T^{\gamma\beta}_{,\beta}$ for the equation of motion we recall the assumption that κ and ϵ are constant. Then

$$T^{\alpha\beta}_{,\beta} = \sigma_{,\beta} a^{\alpha\beta} + \kappa a^{\alpha\beta}(a^{\lambda\mu} S_{\lambda\mu})_{,\beta} + \epsilon(E^{\alpha\beta\lambda\mu} S_{\lambda\mu})_{,\beta}. \tag{10.32.2}$$

The first term is just the gradient of the surface tension and using Ex. 10.23.1 the second term is

$$\kappa a^{\alpha\beta} a^{\lambda\mu} S_{\lambda\mu,\beta} = \kappa a^{\alpha\beta} \left(\frac{\dot{a}}{2a}\right)_{,\beta} + \kappa a^{\alpha\beta} V^{\lambda}_{,\lambda\beta}. \tag{10.32.3}$$

The third term requires a little more manipulation. From Ex. 10.23.1 we have

$$E^{\alpha\beta\lambda\mu} S_{\lambda\mu,\beta} = -\left\{\dot{a}^{\alpha\beta} + a^{\alpha\beta} \frac{\dot{a}}{2a}\right\}_{,\beta} + \{a^{\alpha\lambda} V^{\beta}_{,\lambda\beta} + a^{\beta\lambda} V^{\alpha}_{,\lambda\beta} - a^{\alpha\beta} V^{\lambda}_{,\lambda\beta}\}.$$

The first and third terms in this last bracket can be reduced by Eq. (10.11.7) to give

$$E^{\alpha\beta\lambda\mu} S_{\lambda\mu,\beta} = -\left\{\dot{a}^{\alpha\beta} + a^{\alpha\beta} \frac{\dot{a}}{2a}\right\}_{,\beta} + \{a^{\beta\lambda} V^{\alpha}_{,\lambda\beta} + K V^{\alpha}\}. \tag{10.32.4}$$

We may now substitute from Eqs. (10.32.3 and 4) into Eq. (10.32.2) and thence into the equation of motion (10.31.2). This gives

$$\gamma A^{\alpha} - F^{\alpha} = a^{\alpha\beta}\sigma_{,\beta} + \kappa a^{\alpha\beta}\left[V^{\lambda}_{,\lambda\beta} + \left(\frac{\dot{a}}{2a}\right)_{,\beta}\right]$$
$$+ \epsilon\left[a^{\lambda\beta} V^{\alpha}_{,\lambda\beta} + K V^{\alpha} - (\dot{a}^{\alpha\beta})_{,\beta} - a^{\alpha\beta}\left(\frac{\dot{a}}{2a}\right)_{,\beta}\right].$$

Now write $\kappa = (\kappa + \epsilon) - \epsilon$ and absorb the term

$$-\epsilon a^{\alpha\beta}\left[V^{\lambda}_{,\lambda\beta} + \left(\frac{\dot{a}}{2a}\right)_{,\beta}\right]$$

in the last bracket. This becomes

$$a^{\lambda\beta} V^{\alpha}_{,\lambda\beta} - a^{\alpha\beta} V^{\lambda}_{,\lambda\beta} + K V^{\alpha} - (\dot{a}^{\alpha\beta})_{,\beta} - a^{\alpha\beta}\left(\frac{\dot{a}}{a}\right)_{,\beta}.$$

Using Ex. 10.11.1 on the first two terms and Eq. (10.12.6) on the last two, this is

$$2KV^\alpha - \varepsilon^{\alpha\beta}\varepsilon^{\lambda\mu}a_{\mu\gamma}V^\gamma_{,\lambda\beta} - \varepsilon^{\alpha\lambda}\varepsilon^{\beta\mu}(\dot{a}_{\lambda\mu})_{,\beta} = 2KV^\alpha - \varepsilon^{\alpha\beta}[\varepsilon^{\lambda\mu}V_{\mu,\lambda}]_{,\beta} - \varepsilon^{\alpha\lambda}\varepsilon^{\beta\mu}(\dot{a}_{\lambda\mu})_{,\beta}.$$

The second term here is a curl of the surface curl of the velocity. It corresponds to the term $\nu\nabla \wedge (\nabla \wedge \mathbf{v}) = \nu\nabla \wedge \mathbf{w}$ in the Navier-Stokes equations (Eq. 6.11.8).

The intrinsic equations of motion of a Newtonian surface fluid may be written in unconventional but explanatory form as follows:

γA^α	acceleration or inertial force	
$= F^\alpha$	body force	
$+ a^{\alpha\beta}\sigma_{,\beta}$	surface tension gradient	
$+ (\kappa + \epsilon)a^{\alpha\beta}\left[V^\lambda_{,\lambda} + \dfrac{\dot{a}}{2a}\right]_{,\beta}$	viscous resistance to dilatation	(10.32.4)
$- \epsilon\varepsilon^{\alpha\beta}[\varepsilon^{\lambda\mu}V_{\mu,\lambda}]_{,\beta}$	viscous resistance to shear	
$+ 2\varepsilon KV^\alpha$	force due to intrinsic curvature	
$- \epsilon\varepsilon^{\alpha\lambda}\varepsilon^{\beta\mu}(\dot{a}_{\lambda\mu})_{,\beta}$	force due to movement of the surface.	

If the surface is stationary, the terms in $\dot{a}$ and $\dot{a}_{\lambda\mu}$ vanish. If the surface is plane or developable ($K = 0$) and

$$\gamma A^\alpha = F^\alpha + a^{\alpha\beta}\sigma_{,\beta} + (\kappa + \epsilon)a^{\alpha\beta}V^\lambda_{,\lambda\beta} - \epsilon\varepsilon^{\alpha\beta}[\varepsilon^{\lambda\mu}V_{\mu,\lambda}]_{,\beta}$$

or

$$\gamma A^\alpha = F^\alpha + a^{\alpha\beta}\sigma_{,\beta} + \kappa a^{\alpha\beta}V^\lambda_{,\lambda\beta} + \epsilon a^{\lambda\beta}V^\alpha_{,\lambda\beta}. \tag{10.32.5}$$

These are clearly the Navier-Stokes equations and it is interesting to note that no essentially new terms are involved so long as the Gaussian curvature is zero.

10.41. The continuity of the surface and its surroundings

Up to this point we have confined attention to the surface itself. An important application, however, is to a surface which is the interface of two bulk fluids, for such interfaces exhibit the behavior of a two-dimensional fluid.

If x^i, $i = 1, 2, 3$, are coordinates in space, the equation of the surface at time t has the form

$$x^i = x^i(u^1, u^2, t). \tag{10.41.1}$$

The hybrid tensors

$$t^i_\alpha = \frac{\partial x^i}{\partial u^\alpha} \tag{10.41.2}$$

will again come in to link surface and space vectors but now it must be remembered that they are also functions of time. The contravariant space velocity of a fluid particle in the surface is

$$U^i = \frac{dx^i}{dt} = \frac{\partial x^i}{\partial u^\alpha}\frac{du^\alpha}{dt} + \frac{\partial x^i}{\partial t} = t^i_\alpha V^\alpha + \dot{x}^i, \tag{10.41.3}$$

where

$$\dot{x}^i = \frac{\partial x^i}{\partial t}.$$

If W^i and $\hat{W}^i$ are the velocities of the bulk fluids on either side of the surface, their components tangential to the surface in the direction of u^α are $g_{ij}t^i_\alpha W^j$ and $g_{ij}t^i_\alpha \hat{W}^j$, respectively. If there is no slip at the surface, these velocities must be equal and equal to the tangential component of U^i, namely, $g_{ij}t^i_\alpha U^j$. Thus, the condition for no slip is

$$V_\alpha = g_{ij}t^i_\alpha(W^j - \dot{x}^j) = g_{ij}t^i_\alpha(\hat{W}^j - \dot{x}^j). \tag{10.41.4}$$

If n_i is the normal to the surface pointing from the bulk fluid $\hat{B}$ to the bulk fluid B on the other side, the relative velocities of the two fluids toward the surface are $(\hat{W}^i - U^i)n_i$ and $(U^i - W^i)n_i$, respectively. If $\hat{\rho}$ and ρ are their densities the net rate of transport of mass to the surface is

$$[\hat{\rho}(\hat{W}^i - U^i) + \rho(U^i - W^i)]n_i$$

per unit area. Thus

$$\frac{d}{dt}\iint_S \gamma\, dS = \iint_S [\hat{\rho}(\hat{W}^i - U^i) + \rho(U^i - W^i)]n_i\, dS,$$

and applying the transport theorem and the usual arguments

$$\frac{d\gamma}{dt} + \gamma V^\alpha_{,\alpha} + \gamma\frac{\dot{a}}{2a} = [\hat{\rho}\hat{W}^i + (\rho - \hat{\rho})U^i - \rho W^i]n_i. \tag{10.41.5}$$

This is the continuity equation.

If there is no interchange, we recover the continuity equation (10.31.1) by equating the normal components of the velocities

$$W^i n_i = \hat{W}^i n_i = U^i n_i = \dot{x}^i n_i. \tag{10.41.6}$$

The last equality follows from the fact that $n_i t^i_\alpha = 0$. Since there is also no slip

$$U^i = W^i = \hat{W}^i. \tag{10.41.7}$$

Exercise 10.41.1. Show that

$$V^\alpha_{,\alpha} + \frac{\dot{a}}{2a} = a^{\alpha\beta}(U_j t^j_\alpha)_{,\beta} - 2Hn^j U_j.$$

10.42. Connection between surface strain and the surroundings

The intrinsic rate of strain in the surface is given by the tensor

$$S_{\alpha\beta} = \tfrac{1}{2}\dot{a}_{\alpha\beta} + \tfrac{1}{2}(V_{\alpha,\beta} + V_{\beta,\alpha}). \tag{10.42.1}$$

If, for simplicity, we continue with the case where there is no interchange of material between the surface and bulk phases, then

$$U^i = W^i = \hat{W}^i \quad \text{and} \quad n_i W^i = n_i \dot{x}^i. \tag{10.42.2}$$

Thus

$$\begin{aligned} V_{\alpha,\beta} &= g_{ij}[t^i_\alpha(W^j - \dot{x}^j)]_{,\beta} \\ &= g_{ij}n^i b_{\alpha\beta}(W^j - \dot{x}^j) + g_{ij}t^i_\alpha(W^j_{,\beta} - \dot{x}^j_{,\beta}) \\ &= t^i_\alpha(W_{i,\beta} - \dot{x}_{i,\beta}) \end{aligned} \tag{10.42.3}$$

since $g_{ij}n^i(W^j - \dot{x}^j) = 0$. We also have

$$\dot{a}_{\alpha\beta} = \frac{\partial}{\partial t}(g_{ij}t^i_\alpha t^j_\beta) = \frac{\partial g_{ij}}{\partial t} t^i_\alpha t^j_\beta + g_{ij}\left[\frac{\partial t^i_\alpha}{\partial t} t^j_\beta + t^i_\alpha \frac{\partial t^j_\beta}{\partial t}\right],$$

where this differentiation with respect to time is with u^α held constant. Although g_{ij} is independent of time when regarded as a function of space variables x^i, yet because the surface is moving through space it becomes a function of time for fixed u^α by Eq. (10.41.1). Thus,

$$\frac{\partial g_{ij}}{\partial t} = \frac{\partial g_{ij}}{\partial x^k}\dot{x}^k = \left[\begin{Bmatrix} m \\ k \quad i \end{Bmatrix} g_{mj} + \begin{Bmatrix} m \\ j \quad k \end{Bmatrix} g_{im}\right]\dot{x}^k,$$

the last term following from the fact that $g_{ij,k} = 0$. Also,

$$\frac{\partial t^i_\alpha}{\partial t} = \frac{\partial}{\partial t}\left(\frac{\partial x^i}{\partial u^\alpha}\right) = \frac{\partial}{\partial u^\alpha}\left(\frac{\partial x^i}{\partial t}\right) = \frac{\partial \dot{x}^i}{\partial u^\alpha},$$

so that on substitution and with change of dummy affixes

$$\dot{a}_{\alpha\beta} = g_{ij}[\dot{x}^i_{,\alpha}t^j_\beta + t^i_\alpha \dot{x}^j_{,\beta}] \tag{10.42.4}$$

and hence

$$S_{\alpha\beta} = \tfrac{1}{2}[t^i_\alpha W_{i,\beta} + W_{i,\alpha}t^i_\beta]. \tag{10.42.5}$$

By symmetry

$$\begin{aligned} a^{\lambda\mu}S_{\lambda\mu,\beta} &= a^{\lambda\mu}(t^i_\lambda W_{i,\mu})_{,\beta} \\ &= a^{\lambda\mu}b_{\lambda\beta}n^i W_{i,\mu} + a^{\lambda\mu}t^i_\lambda W_{i,\mu\beta} \\ &= (a^{\lambda\mu}t^i_\lambda W_{i,\mu})_{,\beta}. \end{aligned} \tag{10.42.6}$$

This we recognize as the gradient of the surface divergence (cf. Eq. 9.41.6).*

* I am indebted to Mr J. D. Eliassen for the elucidation of this section.

10.43. Dynamical connection between the surface and its surroundings

If m_β is the surface vector normal to the boundary C of a part of the surface S, the surface stress acting on an element of C is $T^{\alpha\beta}m_\beta\,ds$. This has components in space

$$T^i\,ds = t^i_\alpha T^{\alpha\beta} m_\beta\,ds.$$

The surface acceleration A^α is a tangential component of the surface's acceleration A^i. There is also a normal component of acceleration n_jA^j so that we may write

$$A^i = t^i_\alpha A^\alpha + n^i n_j A^j. \tag{10.43.1}$$

Similarly, for the body force F^i, which may have a normal component though *in* the surface it is manifest only in its tangential components F^α,

$$F^i = t^i_\alpha F^\alpha + n^i n_j F^j. \tag{10.43.2}$$

If l_i is a parallel covariant vector field, a balance for the linear momentum in this direction gives

$$\frac{d}{dt}\iint_S \gamma U^i l_i\,dS = \iint_S \gamma A^i l_i\,dS = \iint_S F^i l_i\,dS + \oint_C T^i l_i\,ds.$$

With the usual arguments about the arbitrariness of S and l_i and an application of Green's theorem to the last integral we have

$$\gamma A^i = F^i + (t^i_\alpha T^{\alpha\beta})_{,\beta}. \tag{10.43.3}$$

However,

$$(t^i_\alpha T^{\alpha\beta})_{,\beta} = t^i_{\alpha,\beta}T^{\alpha\beta} + t^i_\alpha T^{\alpha\beta}_{,\beta} = n^i b_{\alpha\beta}T^{\alpha\beta} + t^i_\alpha T^{\alpha\beta}_{,\beta},$$

and substituting both this and the resolutions (10.43.1 and 2) in Eq. (10.43.3) gives

$$t^i_\alpha\{\gamma A^\alpha - F^\alpha - T^{\alpha\beta}_{,\beta}\} + n^i\{\gamma n_j A^j - n_j F^j - b_{\alpha\beta}T^{\alpha\beta}\} = 0. \tag{10.43.4}$$

This is a resolution of the equations into tangential and normal parts. If we multiply through by $g_{ij}t^j_\gamma$, the second term disappears since $g_{ij}n^i t^j_\gamma = 0$ and we recover the intrinsic equations of motion in the surface

$$\gamma A^\alpha = F^\alpha + T^{\alpha\beta}_{,\beta}. \tag{10.43.5}$$

If we multiply through by n_i, the first terms disappear and we have

$$\gamma n_j A^j = n_j F^j + b_{\alpha\beta}T^{\alpha\beta}. \tag{10.43.6}$$

This shows very prettily how the second fundamental tensor $b_{\alpha\beta}$ interacts with the stress tensor to give a normal component. For a static film $A^j = 0$ and $T^{\alpha\beta} = \sigma a^{\alpha\beta}$; thus,

$$-n_j F^j = \sigma a^{\alpha\beta} b_{\alpha\beta} = 2H\sigma. \tag{10.43.7}$$

There must be a difference in normal force (that is, pressure) on either side of a film equal to 2σ times the mean curvature if the film is to be kept stationary.

To reduce these equations further we must plunge into some rather heavy algebra and for clarity we will give the final result immediately and then go back to sketch the intermediate steps. The final equation may be written

$$\gamma \frac{dW^i}{dt} = F^i + t^i_\alpha a^{\alpha\beta}\sigma_{,\beta} + (\kappa + \epsilon)t^i_\alpha a^{\alpha\beta}(a^{\lambda\mu}t^j_\lambda W_{j,\mu})_{,\beta}$$
$$+ \epsilon t^i_\alpha\{2Ka^{\alpha\beta}(t^j_\beta W_j) - \varepsilon^{\alpha\beta}[\varepsilon^{\lambda\mu}(t^j_\mu W_j)_{,\lambda}]_{,\beta} - 2\varepsilon^{\alpha\lambda}b_{\lambda\beta}\varepsilon^{\beta\mu}(n^j W_j)_{,\mu}\}$$
$$+ n^i\{2H\sigma + 2H(\kappa + \epsilon)(t^j_\lambda a^{\lambda\mu}W_{j,\mu}) + 2\epsilon t^j_\lambda \varepsilon^{\lambda\alpha}b_{\alpha\beta}\varepsilon^{\beta\mu}W_{j,\mu}\}. \qquad (10.43.7)$$

To see the meaning of each of these terms we notice that certain combinations occur frequently. Thus, if L_β is a surface vector, $t^i_\alpha a^{\alpha\beta}L_\beta$ is simply its ith space component. If we denote the surface gradient $M_{,\beta}$ by grad M, it has space components $t^i_\alpha a^{\alpha\beta}M_{,\beta}$. Similarly,

$$a^{\lambda\mu}t^j_\lambda W_{j,\mu} = \text{div } W_s \quad \text{and} \quad \varepsilon^{\lambda\mu}(t^j_\mu W_j)_{,\lambda} = \text{curl } W_s,$$

where W_s is velocity in the surface. Using the vertical form we have, for Eq. (10.43.7),

(i) $\gamma \dfrac{dW^i}{dt} =$	inertial force
(ii) F^i	external force
(iii) $+t^i_\alpha a^{\alpha\beta}\sigma_{,\beta}$	gradient of surface tension
(iv) $+(\kappa + \epsilon)t^i_\alpha a^{\alpha\beta}(a^{\lambda\mu}t^j_\lambda W_{j,\mu})_{,\beta}$	grad (div W_s), dilatational force
(v) $+2\epsilon K t^i_\alpha a^{\alpha\beta}t^j_\beta W_j$	force due to W_s and total curvature
(vi) $-\epsilon t^i_\alpha \varepsilon^{\alpha\beta}[\varepsilon^{\lambda\mu}(t^j_\mu W_j)_{,\lambda}]_{,\beta}$	curl (curl W_s)
(vii) $-2\epsilon t^i_\alpha \varepsilon^{\alpha\lambda}b_{\lambda\beta}\varepsilon^{\beta\mu}(n^j W_j)_{,\mu}$	effect of varying normal velocity
(viii) $+n^i 2H\sigma$	normal force due to surface tension
(ix) $+n^i 2H(\kappa + \epsilon)(t^j_\lambda a^{\lambda\mu}W_{j,\mu})$	normal force due to dilatation
(x) $+n^i 2\epsilon t^j_\lambda \varepsilon^{\lambda\alpha}b_{\alpha\beta}\varepsilon^{\beta\mu}W_{j,\mu}$	normal force due to shear

We notice that $t^j_\beta W_j$ is the tangential component of W_j, so that $t^i_\alpha a^{\alpha\beta}t^j_\beta W_j$ is merely this surface velocity built back up again into a space vector.

To fill in the algebraic gap that we have left we will outline the necessary manipulations in such a way that the reader should be able to follow the steps and find that the details left for him to supply are quite trivial ones such as the rearrangement of dummy indices, raising or lowering of indices, or taking notice of a symmetry. We list first some identities which have appeared and which will be needed denoting them by capital letters (A), . . . , (K). A capital letter enclosed in parentheses following any term of an equation introduces this letter as an abbreviation for that term. If the term is itself a parenthesis with other terms within it these will be referred by a suffix. For example, $(t^i_\alpha V^\alpha + \dot{x}^i)$(Y) means the whole bracket would be referred to as Y and if needed $t^i_\alpha V^\alpha$ would be referred to as Y_1 and $\dot{x}^i$ as Y_2; the example is merely illustrative, the notation being useful only for more complicated expressions.

The identities we have are:

A. $2a^{\alpha\beta}a^{\lambda\mu} - a^{\alpha\lambda}a^{\beta\mu} - a^{\alpha\mu}a^{\beta\lambda} = \varepsilon^{\alpha\lambda}\varepsilon^{\beta\mu} + \varepsilon^{\alpha\mu}\varepsilon^{\beta\lambda}$,

B. $\dfrac{\dot{a}}{a} = a^{\alpha\beta}\dot{a}_{\alpha\beta}$,

C. $\dot{a}^{\alpha\beta} = -a^{\alpha\lambda}a^{\beta\mu}\dot{a}_{\lambda\mu}$,

D. $2a^{\alpha\beta}\left(\dfrac{\dot{a}}{2a}\right)_{,\beta} + (\dot{a}^{\alpha\beta})_{,\beta} = \varepsilon^{\alpha\lambda}\varepsilon^{\beta\mu}(\dot{a}_{\lambda\mu})_{,\beta}$,

E. $\varepsilon^{\lambda\alpha}b_{\alpha\beta}\varepsilon^{\beta\gamma}b_{\gamma\mu} = -K\delta^\lambda_\mu$,

F. $a^{\alpha\beta}(V^\mu_{,\beta\mu} - V^\mu_{,\mu\beta}) = KV^\alpha$,

G. $a^{\beta\mu}V^\alpha_{,\beta\mu} - a^{\alpha\beta}V^\lambda_{,\lambda\beta} = KV^\alpha - \varepsilon^{\alpha\beta}\varepsilon^{\lambda\mu}a_{\gamma\mu}V^\gamma_{,\lambda\beta}$,

H. $\dot{a}_{\alpha\beta} = g_{ij}\{t^i_\alpha\dot{x}^j_{,\beta} + \dot{x}^i_{,\alpha}t^j_\beta\} = \{t^i_\alpha\dot{x}_{i,\beta} + \dot{x}_{i,\alpha}t^i_\beta\}$

I. $V_\alpha = t^j_\alpha(W_j - \dot{x}_j)$,

J. $\dot{x}_{j,\alpha\beta} = \dot{x}_{j,\beta\alpha}$,

K. $n^j_{,\mu} = -t^j_\nu a^{\nu\lambda}b_{\lambda\mu}$.

To get from Eq. (10.43.4) to (10.43.7) we write the stress tensor in the form

$$T^{\alpha\beta} = \sigma a^{\alpha\beta} + (\kappa + \epsilon)a^{\lambda\mu}S_{\lambda\mu}a^{\alpha\beta} + \epsilon(E^{\alpha\beta\lambda\mu} - a^{\alpha\beta}a^{\lambda\mu})S_{\lambda\mu}(\text{P}).$$

Using (A) and Eq. (10.42.5), this last term (P) will contribute to $T^{\alpha\beta}b_{\alpha\beta}$ a term $-2\epsilon\varepsilon^{\alpha\lambda}\varepsilon^{\beta\mu}b_{\alpha\beta}t^i_\lambda W_{i,\mu}$. Hence, again using Eq. (10.42.5) for $S_{\alpha\beta}$, we have

$$b_{\alpha\beta}T^{\alpha\beta} = 2H\sigma + (\kappa + \epsilon)2Ha^{\lambda\mu}t^i_\lambda W_{i,\mu} - 2\epsilon\varepsilon^{\alpha\lambda}\varepsilon^{\beta\mu}t^i_\lambda W_{i,\mu}.$$

We notice that this accounts for the terms (viii), (ix), and (x) of Eq. (10.43.7). For the term $T^{\alpha\beta}_{,\beta}$ we already have, from Eq. (10.32.4),

$$T^{\alpha\beta}_{,\beta} = a^{\alpha\beta}\sigma_{,\beta} + (\kappa + \epsilon)a^{\lambda\mu}S_{\lambda\mu,\beta}a^{\alpha\beta}(\text{Q}) + \epsilon[2KV^\alpha - \varepsilon^{\alpha\beta}\varepsilon^{\lambda\mu}V_{\mu,\lambda\beta} - \varepsilon^{\alpha\lambda}\varepsilon^{\beta\mu}(\dot{a}_{\lambda\mu})_{,\beta}](\text{R}).$$

The first of these terms will clearly become (iii) of Eq. (10.43.7) when multiplied by t^i_α. Substituting for $a^{\lambda\mu}S_{\lambda\mu,\beta}$ in (Q) from Eq. (10.42.6) it will give (iv) of Eq. (10.43.7) when multiplied by t^i_α. The terms (i) and (ii) carry straight over from the original equation and so present no difficulty. The pièce de résistance is (R) which must evidently account for the terms (v), (vi), and (vii).

In demonstrating this we may drop the factor of ϵ which is common to all terms.

Now

$$t^i_\alpha(\mathrm{R}) = 2Kt^i_\alpha V^\alpha(\mathrm{L}) - t^i_\alpha \varepsilon^{\alpha\beta}\varepsilon^{\lambda\mu}V^\mu_{,\lambda\beta}(\mathrm{M}) - \varepsilon^{\alpha\lambda}\varepsilon^{\beta\mu}t^i_\alpha(\dot a_{\lambda\mu})_{,\beta}(\mathrm{N}).$$

Also by (I)

$$(\mathrm{L}) = 2Ka^{\alpha\beta}t^i_\alpha t^j_\beta W_j(\mathrm{L}_1) - 2Ka^{\alpha\beta}t^i_\alpha t^j_\beta \dot x_j(\mathrm{L}_2),$$

and

$$\begin{aligned}(\mathrm{M}) &= -t^i_\alpha \varepsilon^{\alpha\beta}\varepsilon^{\lambda\mu}[t^i_\mu(W_j - \dot x_j)]_{,\lambda\beta} \\ &= -t^i_\alpha \varepsilon^{\alpha\beta}[\varepsilon^{\lambda\mu}(t^j_\mu W_j)_{,\lambda}]_{,\beta}(\mathrm{M}_1) + t^i_\alpha \varepsilon^{\alpha\beta}[\varepsilon^{\lambda\mu}(t^j_\mu \dot x_j)_{,\lambda}]_{,\beta}(\mathrm{M}_2).\end{aligned}$$

Using (II) and $t^i_{\alpha,\beta} = b_{\alpha\beta}n^i$ we have

$$\begin{aligned}(\mathrm{N}) &= -\varepsilon^{\alpha\lambda}\varepsilon^{\beta\mu}t^i_\alpha[t^j_\lambda \dot x_{j,\mu} + t^j_\mu \dot x_{j,\lambda}]_{,\beta} \\ &= -\varepsilon^{\alpha\lambda}\varepsilon^{\beta\mu}t^i_\alpha[b_{\lambda\beta}n^j\dot x_{j,\mu} + t^j_\lambda \dot x_{j,\mu\beta} + b_{\mu\beta}n^j\dot x_{j,\lambda} + t^j_\mu \dot x_{j,\lambda\beta}](\mathrm{N}).\end{aligned}$$

However, (N_2) and (N_3) are both zero under the influence of $\varepsilon^{\beta\mu}$ because of (J) and the symmetry of $b_{\mu\beta}$. Leaving (N_1) for the moment,

$$\begin{aligned}(\mathrm{N}_4) &= -\varepsilon^{\alpha\lambda}\varepsilon^{\beta\mu}t^i_\alpha t^j_\mu \dot x_{j,\beta\lambda}, \text{ by (J)}, \\ &= -\varepsilon^{\alpha\lambda}\varepsilon^{\beta\mu}t^i_\alpha[(t^j_\mu \dot x_{j,\beta})_{,\lambda} - b_{\mu\lambda}n^j\dot x_{j,\beta}](\mathrm{S}).\end{aligned}$$

Further

$$(\mathrm{S}_1) = -t^i_\alpha \varepsilon^{\alpha\lambda}\varepsilon^{\beta\mu}(t^j \dot x_j)_{,\beta\lambda} + t^i_\alpha \varepsilon^{\alpha\lambda}(\varepsilon^{\beta\mu}t^j_{\mu,\beta}\dot x_j)_{,\lambda}$$

and the last term of this vanishes by the symmetry of $t^j_{\mu,\beta} = b_{\mu\beta}n^j$ while the first term is (M_2) with a negative sign and the indices β and λ interchanged. Thus (S_1) vanishes and takes (M_2) with it. By interchanging the suffixes β and μ we see that (N_1) and (S_2) are the same and their sum is

$$2(\mathrm{N}_1) = -2t^i_\alpha \varepsilon^{\alpha\lambda}b_{\lambda\beta}\varepsilon^{\beta\mu}[(n^j\dot x_j)_{,\mu} - n^j_{,\mu}\dot x_j](\mathrm{T}).$$

However, using (K) we have

$$(\mathrm{T}_2) = -2t^i_\alpha \varepsilon^{\alpha\lambda}b_{\lambda\beta}\varepsilon^{\beta\mu}b_{\mu\sigma}a^{\sigma\nu}t^j_\nu \dot x_j = 2Kt^i_\alpha a^{\alpha\nu}t^j_\nu \dot x_j$$

by (E), and this is equal but opposite in sign to (L_2). Hence, the only terms that survive (mirabile dictu) are L_1, M_1, and T_1, and since $n^j W_j = n^j \dot x_j$ these are the required terms (v), (vi), and (vii) of Eq. (10.43.7).

10.51. Surface equations as boundary conditions at an interface

Surface body forces may arise from some intrinsic force or due to the drag of the adjacent bulk fluid. An intrinsic force, such as that due to the weight of the surface, will be denoted by $\gamma\mathscr{F}^i$, while the traction of the bulk phases are $\mathscr{T}^i$ and $\hat{\mathscr{T}}^i$. Thus

$$F^i = \gamma\mathscr{F}^i + \mathscr{T}^i - \hat{\mathscr{T}}^i. \tag{10.51.1}$$

If T^{ij} is the stress tensor in the bulk fluid

$$\mathscr{T}^i = T^{ij}n_j \tag{10.51.2}$$

and for a Newtonian fluid this is

$$\begin{aligned}\mathscr{T}^i &= (-p + \lambda W^k_{,k})n^i + \mu(g^{jq}W^i_{,q} + g^{ip}W^j_{,p})n_j \\ &= -pn^i + (\lambda + 2\mu)W^k_{,k}n^i - \mu g_{mn}n_j\varepsilon^{mjk}\varepsilon^{nil}(W_{k,l} + W_{l,k}).\end{aligned} \tag{10.51.3}$$

When these equations are substituted back into Eq. (10.43.7) they provide the full boundary conditions at the interface of two fluids when there is no exchange of material.

If we choose the space coordinates in such a way that the interface is $x^3 =$ constant and take the surface coordinates u^1 and u^2 the relations between the metrics are simple. In particular, if all coordinates are orthogonal

$$\begin{aligned}&t^i_\alpha = \delta^i_\alpha, \qquad a_{\alpha\beta} = g_{\alpha\beta}, \\ &g_{ij} = 0, \qquad i \neq j, \qquad g_{33} = 1, \qquad n^i = (0, 0, 1).\end{aligned} \tag{10.51.4}$$

If the surface is stationary in space

$$W_3 = n^iW_i = 0. \tag{10.51.5}$$

10.52. The plane interface

Let the plane be $x^3 = 0$ in Cartesian coordinates. We will use x, y, z in place of x^1, x^2, x^3 and A_x will denote the physical component $A(1)$. We assume that the density of the interface is so small that the inertial and intrinsic body forces can be neglected. Then, since

$$g_{11} = g_{22} = g_{33} = 1, \qquad b_{\alpha\beta} = H = K = 0, \tag{10.52.1}$$

the equations simplify enormously and give

$$\hat{\mathscr{T}}_x - \mathscr{T}_x = \frac{\partial\sigma}{\partial x} + (\kappa + \epsilon)\frac{\partial}{\partial x}\left(\frac{\partial W_x}{\partial x} + \frac{\partial W_y}{\partial y}\right) + \epsilon\frac{\partial}{\partial y}\left(\frac{\partial W_x}{\partial y} - \frac{\partial W_y}{\partial x}\right) \tag{10.52.2}$$

$$\hat{\mathscr{T}}_y - \mathscr{T}_y = \frac{\partial\sigma}{\partial y} + (\kappa + \epsilon)\frac{\partial}{\partial y}\left(\frac{\partial W_x}{\partial x} + \frac{\partial W_y}{\partial y}\right) - \epsilon\frac{\partial}{\partial x}\left(\frac{\partial W_x}{\partial y} - \frac{\partial W_y}{\partial x}\right) \tag{10.52.3}$$

$$\hat{\mathscr{T}}_z - \mathscr{T}_z = 0 \tag{10.25.4}$$

If there is no velocity or variation in the y-direction we have

$$\hat{\mathscr{T}}_x - \mathscr{T}_x = \frac{\partial\sigma}{\partial x} + (\kappa + \epsilon)\frac{\partial^2 W_x}{\partial x^2}. \tag{10.52.5}$$

10.53. The cylindrical interface

If the interface is the cylinder $\rho = R$ and ϕ, z denote the other cylindrical polar coordinates, we may take $x^1 = \phi$, $x^2 = z$, $x^3 = \rho$ and have

$$\begin{aligned} g_{11} &= R^2, \qquad g_{22} = g_{33} = 1, \\ b_{11} &= -R, \qquad b_{12} = b_{22} = 0; \\ K &= 0, \qquad H = -1/2R. \end{aligned} \tag{10.53.1}$$

Then, with the same assumptions,

$$\hat{\mathscr{T}}_\rho - \mathscr{T}_\rho = -\frac{\sigma}{R} - \frac{\kappa + \epsilon}{R}\left(\frac{1}{R}\frac{\partial W_\phi}{\partial \phi} + \frac{\partial W_z}{\partial z}\right) + \frac{2\epsilon}{R}\frac{\partial W_z}{\partial z}, \tag{10.53.2}$$

$$\hat{\mathscr{T}}_\phi - \mathscr{T}_\phi = \frac{1}{R}\frac{\partial \sigma}{\partial \phi} + \frac{\kappa + \epsilon}{R}\frac{\partial}{\partial \phi}\left(\frac{1}{R}\frac{\partial W_\phi}{\partial \phi} + \frac{\partial W_z}{\partial z}\right) + \epsilon\frac{\partial}{\partial z}\left(\frac{\partial W_\phi}{\partial z} - \frac{1}{R}\frac{\partial W_z}{\partial \phi}\right), \tag{10.53.3}$$

$$\hat{\mathscr{T}}_z - \mathscr{T}_z = \frac{\partial \sigma}{\partial z} + (\kappa + \epsilon)\frac{\partial}{\partial z}\left(\frac{1}{R}\frac{\partial W_\phi}{\partial \phi} + \frac{\partial W_z}{\partial z}\right) - \frac{\epsilon}{R}\frac{\partial}{\partial \phi}\left(\frac{\partial W_\phi}{\partial z} - \frac{1}{R}\frac{\partial W_z}{\partial \phi}\right). \tag{10.53.4}$$

A generalization of the cylindrical interface is the developable surface. If we introduce the metric in the form of Ex. 9.38.3, the normal component of the equations gives

$$n_3(\hat{\mathscr{T}}^3 - \mathscr{T}^3) = b_{11}T^{11} = 2H\{\sigma + \kappa(W^1_{,1} + W^2_{,2}) + \epsilon(W^1_{,1} - W^2_{,2})\}. \tag{10.53.5}$$

There are also certain simplifications in the other two equations but they are still too complicated to be worth giving in detail.

10.54. The spherical interface

If the interface is the sphere $r = R$, we may take spherical polar coordinates

$$x^1 = \theta, \qquad x^2 = \phi, \qquad x^3 = r = R.$$

Then

$$\begin{aligned} &g_{11} = R^2, \qquad g_{22} = R^2\sin^2\theta, \qquad g_{33} = 1, \qquad b_{11} = -R,\ b_{12} = 0, \\ &b_{22} = -R\sin^2\theta, \qquad H = -\frac{1}{R}, \qquad K = \frac{1}{R^2}. \end{aligned} \tag{10.54.1}$$

The equations for a surface of constant radius R become

$$\hat{\mathscr{T}}_r - \mathscr{T}_r = -\frac{2\sigma}{R} - \frac{2\kappa}{R}\left\{\frac{1}{R\sin\theta}\left[\frac{\partial(W_\theta\sin\theta)}{\partial\theta} + \frac{\partial W_\phi}{\partial\phi}\right]\right\}, \tag{10.54.2}$$

$$\begin{aligned}\hat{\mathscr{T}}_\theta - \mathscr{T}_\theta &= \frac{1}{R}\frac{\partial\sigma}{\partial\theta} + (\kappa+\epsilon)\frac{1}{R}\frac{\partial}{\partial\theta}\left\{\frac{1}{R\sin\theta}\left[\frac{\partial(W_\theta\sin\theta)}{\partial\theta} + \frac{\partial W_\phi}{\partial\phi}\right]\right\} \\ &+ \epsilon\left[\frac{2W_\theta}{R^2} + \frac{1}{R\sin\theta}\frac{\partial}{\partial\phi}\left\{\frac{1}{R\sin\theta}\left[\frac{\partial W_\theta}{\partial\phi} - \frac{\partial(W_\phi\sin\theta)}{\partial\theta}\right]\right\}\right],\end{aligned} \tag{10.54.3}$$

$$\begin{aligned}\hat{\mathscr{T}}_\phi - \mathscr{T}_\phi &= \frac{1}{R\sin\theta}\frac{\partial\sigma}{\partial\phi} + \frac{\kappa+\epsilon}{R\sin\theta}\frac{\partial}{\partial\phi}\left\{\frac{1}{R\sin\theta}\left[\frac{\partial(W_\theta\sin\theta)}{\partial\theta} + \frac{\partial W_\phi}{\partial\phi}\right]\right\} \\ &+ \epsilon\left[\frac{2W_\phi}{R^2} - \frac{1}{R}\frac{\partial}{\partial\theta}\left\{\frac{1}{R\sin\theta}\left[\frac{\partial W_\theta}{\partial\phi} - \frac{\partial(W_\phi\sin\theta)}{\partial\theta}\right]\right\}\right].\end{aligned} \tag{10.54.4}$$

If the sphere is expanding or contracting, its radius being a function of time $R(t)$, then the metric is as given in Eq. (10.54.1) but $W_3 = W_r = \dot{R}(t)$. Assuming the velocity to be purely radial we have

$$\hat{\mathscr{T}}_r - \mathscr{T}_r = -\frac{2\sigma}{R} - \frac{4\kappa\dot{R}}{R^2} \tag{10.54.5}$$

and $\hat{\mathscr{T}}_\theta = \mathscr{T}_\theta$, $\hat{\mathscr{T}}_\phi = \mathscr{T}_\phi$. The second term is the extra pressure needed to overcome the resistance to dilatation.

BIBLIOGRAPHY

This chapter is in the nature of a somewhat extended gloss on L. E. Scriven's paper

Dynamics of a fluid interface, Chem. Engng. Sci. *12* (1960), pp. 98–108.

Further references will be found in this paper.

Equations for Reacting Fluids 11

In this final chapter we shall consider the equations of motion and energy in a system of reacting chemical species. Our treatment is somewhat brief for the foundations of this topic in continuum mechanics are not yet thoroughly established and the discussion adds little as an example of the application of tensor methods. It is, however, of such importance that its entire omission would be unpardonable. Our treatment follows the paper by H. J. Merk referred to in the bibliography and references for further reading will also be found there.

The basic idea is that the various chemical species composing the system can be regarded as interpenetrating continua. We thus imply that these component species are entirely miscible and that there is no change of phase during reaction. Except where explicitly required, we shall avoid the use of tensor components and use bold face letters for vectors and tensors. Ordinary Cartesian coordinates are to be understood throughout and the suffixes will refer to the components of the system (that is, chemical species) and not to tensor components.

11.11. The conservation of matter

In a mixture of N chemical species A_i we let the suffix i denote properties of the ith species, $i = 1, \ldots, N$. If m_i is the mass of one mole of A_i and n_i its molar concentration (moles per unit volume) the mass concentration of A_i is

$$\rho_i = m_i n_i. \tag{11.11.1}$$

The velocity $\mathbf{v}_i$ of the species A_i may be defined as if this were the only continuum present and $\rho_i \mathbf{v}_i$ is the mass flux in the fixed Cartesian coordinate system.

In constructing a mass balance we have to consider that the species is being created or destroyed at every point. Let r_i be the rate at which A_i is being created in units of mass per unit volume per unit time. It is a little difficult to think of a material volume of one species since this is being created and destroyed, so that in setting up a mass balance we use a fixed control volume. Then the rate of change of mass of A_i is due to its net flux in through the surface and its change by reaction. Thus, if V is this fixed volume with surface S,

$$\frac{\partial}{\partial t} \iiint_V \rho_i \, dV = \iiint_V \frac{\partial \rho_i}{\partial t} \, dV = -\iint_S \rho_i \mathbf{v}_i \cdot \mathbf{n} \, dS + \iiint_V r_i \, dV,$$

where $\mathbf{n}$ is the outward normal. Transforming the surface integral by Green's theorem and making the usual deduction from the arbitrariness of V, we have

$$\frac{\partial \rho_i}{\partial t} + \nabla \cdot (\rho_i \mathbf{v}_i) = r_i. \tag{11.11.2}$$

The total mass per unit volume is the density ρ of the mixture,

$$\rho = \Sigma \, \rho_i. \tag{11.11.3}$$

(All summations will be understood to run from 1 to N.) We can define an average velocity, called the mean mass velocity $\mathbf{v}$, by

$$\rho \mathbf{v} = \Sigma \, \rho_i \mathbf{v}_i. \tag{11.11.4}$$

Moreover, in a reaction, although the mass of each species changes, the total mass remains constant,

$$\Sigma \, r_i = 0. \tag{11.11.5}$$

If we add Eqs. (11.11.2) we obtain an over-all continuity equation

$$\frac{\partial \rho}{dt} + \nabla \cdot (\rho \mathbf{v}) = \frac{d\rho}{dt} + \rho \nabla \cdot \mathbf{v} = 0 \tag{11.11.6}$$

Since there is no net creation of mass and mean velocity has been calculated as a mass average, this is the familiar continuity equation.

The mass fraction, mass of A_i per unit mass of mixture, is given by

$$c_i = \frac{\rho_i}{\rho}. \tag{11.11.7}$$

The mean velocity is thus

$$\mathbf{v} = \Sigma \, c_i \mathbf{v}_i. \tag{11.11.8}$$

11.12. Mass fluxes

The mass flux $\rho_i \mathbf{v}_i$ is the mass flux of A_i with respect to the fixed coordinate frame. Similarly, the mean mass flux $\rho \mathbf{v}$ is the total mass flux with respect to fixed coordinates. The velocity of the species A_i relative to the mean is $\mathbf{v}_i - \mathbf{v}$ and so the flux of A_i relative to coordinates moving with the mean velocity is

$$\mathbf{j}_i = \rho_i(\mathbf{v}_i - \mathbf{v}). \tag{11.12.1}$$

This is called the barycentric mass flux of A_i and will be the principal one that we shall use. It is not the only mass flux that can be defined and it is important in any discussion of multicomponent systems to specify carefully the nature of the flux used.

We might use the molar flux $n_i \mathbf{v}_i$ with respect to fixed axes to define a mean molar velocity

$$\mathbf{v}^* = \frac{\Sigma\, n_i \mathbf{v}_i}{\Sigma\, n_i} = \Sigma\, c_i^* \mathbf{v}_i, \tag{11.12.2}$$

where

$$n = \Sigma\, n_i \tag{11.12.3}$$

is the total number of moles per unit volume and

$$c_i^* = \frac{n_i}{n} \tag{11.12.4}$$

is the mole fraction of A_i. The molar flux of A_i with respect to coordinates moving with mean molar velocity is then

$$\mathbf{j}_i^* = n_i(\mathbf{v}_i - \mathbf{v}^*). \tag{11.12.5}$$

Since each of this fluxes is defined relative to its own mean

$$\Sigma\, \mathbf{j}_i = \Sigma\, \mathbf{j}_i^* = 0. \tag{11.12.6}$$

Other fluxes such as $\rho_i(\mathbf{v}_i - \mathbf{v}^*)$ are defined relative to a different mean; in this case we have a mass flux relative to the molar average velocity. Here $\Sigma\, \rho_i(\mathbf{v}_i - \mathbf{v}^*) = \rho(\mathbf{v} - \mathbf{v}^*)$.

The continuity equation may be expressed in terms of the mass concentration and barycentric flux. For

$$\rho \frac{dc_i}{dt} = \rho \frac{d}{dt}\left(\frac{\rho_i}{\rho}\right) = \frac{d\rho_i}{dt} - \frac{\rho_i}{\rho}\frac{d\rho}{dt}$$

$$= r_i - \rho_i \nabla \cdot \mathbf{v}_i + \frac{\rho_i}{\rho}\rho \nabla \cdot \mathbf{v} + (\mathbf{v} - \mathbf{v}_i) \cdot \nabla \rho_i,$$

so that

$$\rho \frac{dc_i}{dt} + \nabla \cdot \mathbf{j}_i = r_i. \tag{11.12.7}$$

Exercise 11.12.1. If m is the mean molar weight, ρ/n, show that

$$c_i = \frac{m_i}{m} c_i^*.$$

Exercise 11.12.2. Show that for a binary mixture

$$\rho_1(\mathbf{v}_1 - \mathbf{v}^*) = \frac{m}{m_2} \mathbf{j}_1.$$

11.13. Stoichiometric and kinetic relations

Since the chemical species are being created and destroyed, there is a definite relation between the rates r_i and the reactions taking place. Suppose that there is only one reaction and that this can be written

$$\Sigma \alpha_i A_i = 0. \tag{11.13.1}$$

The numbers α_i are the stoichiometric coefficients of the reaction. If the reaction is to be regarded as going in a particular direction it is convenient to give positive stoichiometric coefficients to the products which are formed. The rate of formation of A_i in moles per unit volume per unit time is r_i/m_i, and by the stoichiometry of the reaction this must be proportional to α_i. Hence, we can write

$$r_i = \alpha_i m_i r, \tag{11.13.2}$$

where r is the rate of the reaction. With the convention used, r is positive for the forward reaction.

When there are M simultaneous reactions they may be written

$$\sum_{i=1}^{N} \alpha_i^j A_i = 0, \qquad j = 1, \ldots, M. \tag{11.13.3}$$

Then a rate of reaction r^j may be defined for each reaction and

$$r_i = m_i \sum_{j=1}^{M} \alpha_i^j r^j. \tag{11.13.4}$$

Note that the r_i are mass rates per unit volume whereas the reaction rates r, r^j are mole rates per unit volume.

The net change in mass is zero so that

$$\sum_{i=1}^{N} \alpha_i m_i = \sum_{i=1}^{N} \alpha_i^j m_i = 0. \tag{11.13.5}$$

Substituting Eq. (11.13.4) in Eq. (11.11.2) and dividing through by m_i, we have

$$\frac{\partial n_i}{\partial t} + \nabla \cdot (n_i \mathbf{v}_i) = \Sigma\, \alpha_i^j r^j. \tag{11.13.6}$$

Hence
$$\frac{\partial n}{\partial t} + \nabla \cdot (n\mathbf{v}^*) = \sum_{i=1}^{N} \sum_{j=1}^{M} \alpha_i^j r^j. \tag{11.13.7}$$

The rates of reaction r or r^j are supposed to be obtainable from considerations of chemical kinetics as functions of the composition temperature and pressure. Thus,
$$r^j = r^j(c_1, \ldots, c_N, T, p). \tag{11.13.8}$$
In many cases these expressions are not fully established. Sometimes the reaction is regarded as being extremely rapid so that the concentrations obey an equilibrium condition
$$r^j(c_1, \ldots, c_N, T, p) = 0. \tag{11.13.9}$$
Such equilibrium relations are often known when the reaction rate itself is quite unknown.

Exercise 11.13.1. Show that
$$n\frac{\partial c_i^*}{\partial t} + n(\mathbf{v}^* \cdot \nabla)c_i^* + \nabla \cdot \mathbf{j}_i^* = r_i^*,$$
where, for a single reaction,
$$r_i^* = \left(\alpha_i - c_i^* \sum_{k=1}^{N} \alpha_k\right) r.$$

11.2. The conservation of momentum

It is assumed that a viscous stress tensor can be defined for the mixture so that if p is the pressure the stress tensor is
$$\mathbf{T} = -p\mathbf{I} + \mathbf{P}, \tag{11.2.1}$$
where $\mathbf{I}$ is the unit tensor and $\mathbf{P}$ the viscous stress tensor. If the mixture is Newtonian, we would write
$$\mathbf{P} = \lambda(\nabla \cdot \mathbf{v})\mathbf{I} + \mu\{(\nabla\mathbf{v}) + (\nabla\mathbf{v})'\}. \tag{11.2.2}$$
The coefficients λ and μ will depend on composition, temperature, and pressure. More general expressions could be written down for non-Newtonian mixtures the coefficients being functions of these variables. Theoretically the rates of reaction should affect the stress tensor, but this effect is usually neglected.

The balance of linear momentum and usual reductions will give
$$\rho\frac{d\mathbf{v}}{dt} = -\nabla p + \Sigma\, \rho_i \mathbf{F}_i + \nabla \cdot \mathbf{P}, \tag{11.2.3}$$

the familiar equations of motion. Here it must be supposed that each species may be acted on by different external forces, and that $\mathbf{F}_i$ is the external force per unit mass acting on A_i. Substitution of Eq. (11.2.2) into this last equation gives a form of the Navier-Stokes equation.

11.31. The conservation of energy

We consider the total energy of the fluid to be the sum of its kinetic and internal energies and denote the specific internal energy by e. If $\mathbf{j}_q$ is the total heat flux vector, the usual balance by the first law of thermodynamics gives

$$\rho \frac{d}{dt}(e + \tfrac{1}{2}v^2) = -\nabla \cdot \mathbf{j}_q - \nabla \cdot (p\mathbf{v}) + \Sigma\, \rho_i \mathbf{v}_i \cdot \mathbf{F}_i + \nabla \cdot (\mathbf{P} \cdot \mathbf{v}). \quad (11.31.1)$$

This can be manipulated in the same way as was done in Section 6.3 to give

$$\rho \frac{de}{dt} = -\nabla \cdot \mathbf{j}_q - p\nabla \cdot \mathbf{v} + \Sigma\, \mathbf{j}_i \cdot \mathbf{F}_i + \Upsilon, \quad (11.31.2)$$

where $\Upsilon = \mathbf{P} : (\nabla \mathbf{v})$ is the irreversible dissipation of energy by viscous forces. $p\nabla \cdot \mathbf{v}$ is a reversible interchange with the strain energy and $\Sigma\, \mathbf{j}_i \cdot \mathbf{F}_i$ the net work by the external forces. The heat flux vector is the sum of the heat flux by the diffusion of heat and the heat transported with the mass fluxes. Let h_i be the specific enthalpy of A_i; then we write

$$\mathbf{j}_q = \mathbf{q} + \Sigma\, h_i \mathbf{j}_i. \quad (11.31.3)$$

The specific enthalpy of the mixture is

$$h = \Sigma\, c_i h_i \quad (11.31.4)$$

and is related to the specific internal energy by

$$e = h - \frac{p}{\rho}. \quad (11.31.5)$$

Substituting these into Eq. (11.31.2) gives

$$\rho \frac{dh}{dt} + \nabla \cdot (\Sigma\, h_i \mathbf{j}_i) = -\nabla \cdot \mathbf{q} + \frac{dp}{dt} + \Sigma\, \mathbf{j}_i \mathbf{F}_i + \Upsilon. \quad (11.31.6)$$

The enthalpy of the mixture changes with temperature, pressure, and composition. Thus,

$$\frac{dh}{dt} = \frac{d}{dt}(\Sigma\, c_i h_i) = \Sigma \frac{dc_i}{dt} h_i + \Sigma\, c_i \frac{dh_i}{dt}.$$

By Eq. (11.12.7) the first of these terms is

$$\Sigma\, h_i \frac{dc_i}{dt} = \Sigma \frac{h_i}{\rho}(r_i - \nabla \cdot \mathbf{j}_i)$$
$$= \frac{1}{\rho}\{\Sigma\, h_i r_i - \Sigma\, h_i (\nabla \cdot \mathbf{j}_i)\}. \quad (11.31.7)$$

Here the first term is $-Q/\rho$, where Q is the rate of generation of heat per unit volume by reaction. It is

$$Q = -\Sigma\, h_i r_i = -r\, \Sigma\, \alpha_i m_i h_i = (-\Delta H) r$$

and ΔH is the heat of reaction (negative for an exothermic one). For simultaneous reactions

$$Q = -\Sigma\, h_i r_i = \Sigma\, r^j(-\Delta H)^j, \tag{11.31.8}$$

where $(\Delta H)^j$ is the heat of the jth reaction. The second term in dh/dt is

$$\left[\Sigma\, c_i\left(\frac{\partial h_i}{\partial T}\right)\right]\frac{dT}{dt} + \left[\Sigma\, c_i\left(\frac{\partial h_i}{\partial p}\right)\right]\frac{dp}{dt} + \Sigma\left[\Sigma\, c_i\,\frac{\partial h_i}{\partial c_j}\right]\frac{dc_j}{dt}.$$

However, enthalpy is an extensive property and

$$\rho h = \Sigma\, \rho_j h_j = \Sigma\, \rho_j \frac{\partial(\rho h)}{\partial \rho_j} = \Sigma\, \rho_j\left\{h_j + \Sigma\, \rho_i \frac{\partial h_i}{\partial \rho_j}\right\}$$

showing that $\Sigma\, c_i(\partial h_i/\partial \rho_j) = 0$. For ideal gases $\partial h_i/\partial p = 0$ but more generally we may write

$$\rho\, \Sigma\, c_i\,\frac{\partial h_i}{\partial p} = \rho l + 1, \tag{11.31.9}$$

where l is of the nature of a latent heat. Combining all these equations with Eq. (11.31.6) and writing $c_p = \Sigma\, c_i(\partial h_i/\partial T)$ for the specific heat at constant pressure, we have

$$\rho c_p \frac{dT}{dt} + \Sigma\, \mathbf{j}_i \cdot \nabla h_i = -\nabla\cdot\mathbf{q} - \rho l\,\frac{dp}{dt} + Q + \Sigma\, \mathbf{j}_i \cdot \mathbf{F}_i + \Upsilon. \tag{11.31.10}$$

11.32. The diffusion of heat and matter

The heat flux $\mathbf{q}$ and the mass fluxes $\mathbf{j}_i$ have yet to be related to driving forces. These driving forces are the gradients of temperature and chemical potential. From the principles of irreversible thermodynamics, these fluxes should be linear combinations of all the driving forces. The driving forces may be written

$$\mathbf{x}_0 = -\frac{\nabla T}{T} \tag{11.32.1}$$

$$\mathbf{x}_i = -T\nabla\frac{\mu_i}{T} - h_i\,\frac{\nabla T}{T} + \mathbf{F}_i, \tag{11.32.2}$$

where μ_i is the chemical potential. In terms of these, the fluxes are

$$\mathbf{q} = \beta_{00}\mathbf{x}_0 + \Sigma\, \beta_{0j}\mathbf{x}_j \tag{11.32.3}$$

$$\mathbf{j}_i = \beta_{i0}\mathbf{x}_0 + \Sigma\, \beta_{ij}\mathbf{x}_j. \tag{11.32.4}$$

The β_{ij} are coefficients which will depend on the thermodynamic state and composition. They are not independent for $\beta_{ij} = \beta_{ji}$ and $\sum_{j=1}^{N} \beta_{ij} = 0$. β_{00}/T is the thermal conductivity. The $\beta_{i0} = D_i^T$ are thermal diffusion coefficients. In the remaining terms several effects may be distinguished. We will omit the detailed manipulations, which may be found in Merk's paper, but the final result can be expressed as

$$\mathbf{j}_i = \frac{n^2}{\rho} \sum_{j=1}^{N} m_i m_j D_{ij} \mathbf{d}_j - D_i^T \frac{\nabla T}{T}, \tag{11.32.5}$$

where, for an ideal gas,

$$\mathbf{d}_j = (\nabla c_j^*) + (1 - c_j)\nabla \ln p - \frac{c_j}{p} (\rho \mathbf{F}_j - \Sigma\, \rho_i \mathbf{F}_i). \tag{11.32.6}$$

The first term in d_j is the concentration gradient which, with the multicomponent diffusion coefficients D_{ij}, gives the flux due to the material diffusion. The second term is a pressure diffusion and the third a diffusion brought on by the unequal external forces. The second term in Eq. (11.33.5) is the thermal diffusion effect.

11.33. Transport in binary mixtures

By ordinary diffusion we mean the diffusion that takes place in the absence of thermal, pressure, or forced diffusion. In binary mixtures the expression for the flux by ordinary diffusion is relatively simple. If D_{12} is the binary diffusion coefficient of species 1 in species 2,

$$\mathbf{j}_1 = -\frac{n^2}{\rho} m_1 m_2 D_{12} \nabla c_1^*. \tag{11.33.1}$$

If the subscripts are interchanged,

$$\mathbf{j}_2 = -\frac{n^2}{\rho} m_1 m_2 D_{21} \nabla c_2^*. \tag{11.33.2}$$

However, $\mathbf{j}_1 + \mathbf{j}_2 = 0$ and $c_1^* + c_2^* = 1$, hence

$$D_{12} = D_{21} = D \tag{11.33.3}$$

is the common diffusion coefficient. For nonideal systems the activity a should be used rather than the mole fraction c^* and the diffusivity

$$D^* = \frac{D}{(\partial \ln a/\partial \ln c^*)_{p,T}} \tag{11.33.4}$$

is held to be much less dependent on concentration.

If all effects are taken into account the flux $\mathbf{j}_1 = -\mathbf{j}_2$ for the binary case becomes

$$\mathbf{j}_1 = -\frac{n^2}{\rho} m_1 m_2 D^* \left[\left(\frac{\partial \ln a_1}{\partial \ln c_1^*} \right) \nabla c_1^* - \frac{\rho_1 \rho_2}{n\rho RT} (\mathbf{F}_1 - \mathbf{F}_2) + \frac{\rho_2 - \rho_1}{n\rho RT} \nabla p + k_T \nabla \ln T \right]. \quad (11.33.5)$$

The various elements of ordinary, forced, pressure, and thermal diffusion can clearly be distinguished here.

$$k_T = \frac{\rho D_1^T}{n^2 m_1 m_2 D^*}$$

is the so-called thermal diffusion ratio. If it is positive, component 1 moves toward the cold region.

BIBLIOGRAPHY

The presentation of this chapter has been largely influenced by a paper of H. J. Merk

The macroscopic equations for simultaneous heat and mass transfer in isotropic, continuous and closed systems, App. Sci. Res. **A8** (1959) pp. 73-99.
The original derivation of the equations for reacting fluids is in the work of Hirschfelder and Curtiss. The exposition and references given in Hirschfelder, J. O., C. F. Curtiss and R. B. Bird, Molecular theory of gases and liquids (New York: John Wiley, 1954) will lead the reader most readily to the original sources.

A very good review with numerous further references is

Fredrickson, A.G., and R. B. Bird, Transport phenomena in multicomponent systems, in Handbook of Fluid Dynamics. New York: McGraw-Hill, 1961.

See also

Bird, R. B., W. E. Stewart, and E. N. Lightfoot, Transport phenomena. New York: Wiley, 1960.

The subject of thermal diffusion is discussed fully in the monograph

Grew, K. E., and T. L. Ibbs, Thermal diffusion in gases. Cambridge: Cambridge University Press, 1952.

Appendix A

Résumé of Three-dimensional Coordinate Geometry and Matrix Theory

This appendix is designed to link on with the undergraduate preparation of the engineer and to provide a convenient summary of the material used in the text. It contains much that the reader may already know in another context or notation and may serve as a refresher or an enlightener as needed. It is convenient to intertwine the threads of geometry and algebra though each has a continuity of its own.

A.1. Cartesian coordinate systems

The ordinary space of everyday life is a three-dimensional Euclidean space and any point of it may be labeled by the device of introducing a Cartesian frame of reference. Any point O and any line $O1$ through it is taken and two other lines $O2$ and $O3$ at right angles to $O1$ and to each other are constructed. The point O is called the origin of coordinates and the lines $O1$, $O2$, and $O3$ the coordinate axes. If they have the mutual orientation shown in Fig. A.1, the frame of reference is said to be right-handed since the order $O1$, $O2$, $O3$ corresponds to the orientation of the thumb, first, and second fingers of the right hand. If the $O1$ and $O2$ axes were the same but the $O3$ axis pointed downwards, this would give a left-handed frame. Any two right-handed frames can be brought into coincidence by bringing the origins together and then rotating one of them, but a right-handed frame can only be turned into a left-handed frame by reflection of one of the coordinate axes in the plane of the other two.

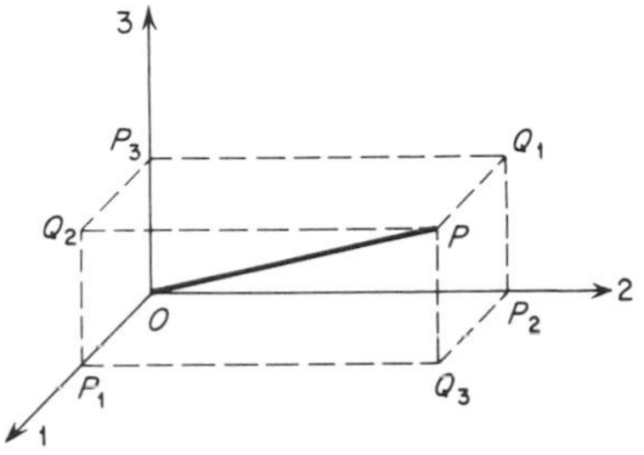

Fig. A.1

From any point P perpendiculars PQ_1, PQ_2, and PQ_3 can be dropped to the three coordinate planes $O23$, $O31$, $O12$ so that OP is the diagonal of a parallelepiped with sides parallel to the coordinate axes. The lengths of the sides of this parallelepiped are called the coordinates of the point P relative to the given frame of reference. Thus, we write x_1, x_2, x_3 for the lengths of PQ_1, PQ_2, and PQ_3, respectively. The lengths of OP_1, OP_2, and OP_3 are also x_1, x_2, x_3 and these are the projections of OP on to the three coordinate axes. (To see this observe that the line PP_3 lies in the plane $P_3Q_1PQ_2$, which is parallel to the O_{12} plane, and so is perpendicular to O_3.) The coordinates of a point P may thus be regarded as the lengths of the perpendiculars dropped from P on to the coordinate planes or as the lengths of the projections of OP on the coordinate axes.

If the length of OP is r, then by the theorem of Pythagoras,

$$r^2 = x_1^2 + x_2^2 + x_3^2.$$

Let angles POP_1, POP_2, and POP_3 be respectively α_1, α_2, and α_3, then

$$x_i = r \cos \alpha_i = r\lambda_i, \qquad i = 1, 2, 3, \tag{A.1.1}$$

and the three numbers λ_1, λ_2, λ_3 are called the direction cosines of the line OP. Substituting Eq. (A.1.1) in the expression for r we see that

$$\lambda_1^2 + \lambda_2^2 + \lambda_3^2 = 1 \tag{A.1.2}$$

so that only two direction cosines are arbitrary.

Exercise A.1.1. Interpret geometrically the surface for whose points $\lambda_1 =$ constant. Show that the two surfaces $\lambda_1 =$ constant, $\lambda_2 =$ constant will meet in two straight lines provided $\lambda_1^2 + \lambda_2^2 < 1$.

A.2. The projection of one line on another—Orthogonality

Consider two points P and Q, as shown in Fig. A.2, neither of which is at the origin, and let their coordinates be (x_1, x_2, x_3) and (y_1, y_2, y_3), respectively. If r and s are the lengths of OP and OQ and θ the angle between them, then, by the cosine rule for the plane triangle OPQ,

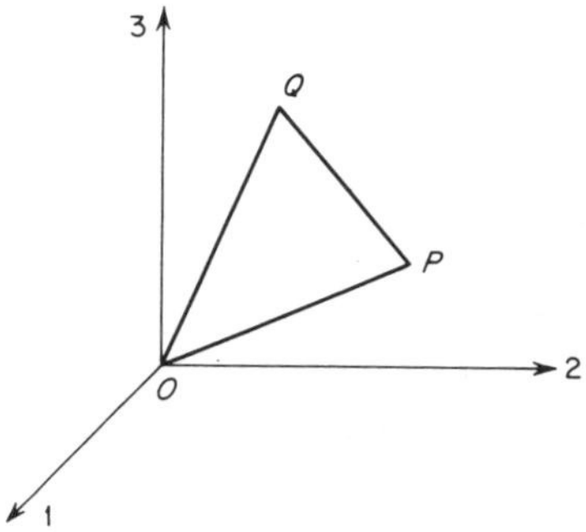

Fig. A.2

$$PQ^2 = r^2 + s^2 - 2rs\cos\theta. \quad \text{(A.2.1)}$$

However, the projections of PQ on to the three axes are $(y_1 - x_1)$, $(y_2 - x_2)$, and $(y_3 - x_3)$ respectively so that by Pythagoras' theorem

$$\begin{aligned} PQ^2 &= (y_1 - x_1)^2 \\ &\quad + (y_2 - x_2)^2 + (y_3 - x_2)^3 \\ &= (x_1^2 + x_2^2 + x_3^2) + (y_1^2 + y_2^2 + y_3^2) \\ &\quad - 2(x_1y_1 + x_2y_2 + x_3y_3). \quad \text{(A.2.2)} \end{aligned}$$

Comparing these two expressions and remembering that

$$r^2 = x_1^2 + x_2^2 + x_3^2$$

and

$$s^2 = y_1^2 + y_2^2 + y_3^2$$

we see that

$$rs\cos\theta = x_1y_1 + x_2y_2 + x_3y_3. \quad \text{(A.2.3)}$$

If λ_1, λ_2, λ_3 are the direction cosines of OP, then dividing by r we have

$$s\cos\theta = \lambda_1y_1 + \lambda_2y_2 + \lambda_3y_3 \quad \text{(A.2.4)}$$

which is the length of the projection of OQ on the line OP. If μ_1, μ_2, μ_3 are the direction cosines of OQ (that is, $y_i = s\mu_i$) then

$$\cos\theta = \lambda_1\mu_1 + \lambda_2\mu_2 + \lambda_3\mu_3 \quad \text{(A.2.5)}$$

which gives an expression for the angle between two directions in terms of their direction cosines.

If the angle between two directions is a right angle they are said to be orthogonal. In this case, $\cos\theta = 0$ and the condition for orthogonality is

$$\lambda_1\mu_1 + \lambda_2\mu_2 + \lambda_3\mu_3 = 0. \quad \text{(A.2.6)}$$

Exercise A.2.1. Show that in setting up a new coordinate system relative to the old, by first taking any line $O1'$ then any line $O2'$ at right angles and a third $O3'$ mutually orthogonal, only three numbers can be arbitrarily specified.

Exercise A.2.2. If 2θ is the angle between the two straight lines referred to in Ex. A.1.1, show that

$$\cos^2\theta = \lambda_1^2 + \lambda_2^2.$$

A.3. The line, plane, and surface

If $A(a_1, a_2, a_3)$ is a fixed point and $P(x_1, x_2, x_3)$ any point on a straight line through A, then

$$\frac{x_1 - a_1}{\lambda_1} = \frac{x_2 - a_2}{\lambda_2} = \frac{x_3 - a_3}{\lambda_3}, \tag{A.3.1}$$

for the differences in coordinates $(x_i - a_i)$ must always be proportional to the direction cosines. If the common ratio in (A.3.1) is s, then

$$x_i = a_i + \lambda_i s, \qquad i = 1, 2, 3 \tag{A.3.2}$$

and s is a parameter, namely, the distance from the fixed point A to the general point P.

In general, if x_i is specified as a function of a parameter s, the set of equations $x_i = f_i(s)$ defines a curve in space. Thus, for example,

$$x_1 = a \cos s, \quad x_2 = a \sin s, \quad x_3 = bs$$

shows that for all s the point P is at a constant distance $a = (x_1^2 + x_2^2)$ from the x_3 axis and that as s increases the point moves around this axis with constant angular velocity and also parallel to it with constant speed. The curve is thus a helix drawn on a cylinder of radius a whose axis is O_3.

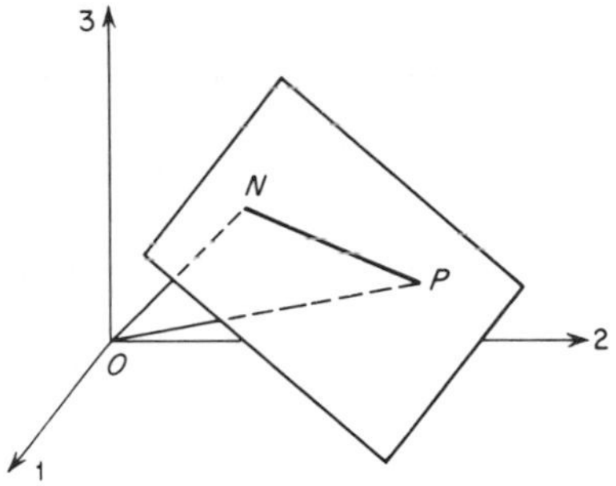

Fig. A.3

The plane has the property that any line lying in it (for example, PN in Fig. A.3) is perpendicular to the normal to the plane. If ON, the perpendicular from the origin to the plane, has direction cosines $(\lambda_1, \lambda_2, \lambda_3)$ and P is any point (x_1, x_2, x_3) in the plane, then, by (A.2.4), the length of ON is $\lambda_1 x_1 + \lambda_2 x_2 + \lambda_3 x_3$. However, this is a constant, say p, so that the equation of the plane is

$$\lambda_1 x_1 + \lambda_2 x_2 + \lambda_3 x_3 = p. \tag{A.3.3}$$

Multiplying this equation by an arbitrary constant we see that any linear equation

$$a_1 x_1 + a_2 x_2 + a_3 x_3 = b \tag{A.3.4}$$

is the equation of a plane; the direction cosines of the normal are

$$\lambda_i = \frac{a_i}{(a_1^2 + a_2^2 + a_3^2)^{1/2}} \quad \text{and} \quad p = \frac{b}{(a_1^2 + a_2^2 + a_3^2)^{1/2}}.$$

Any equation of the form

$$f(x_1, x_2, x_3) = 0 \tag{A.3.5}$$

represents a surface, for given x_1 and x_2, we can, in general, solve for the value or values of x_3. For example,

$$x_1^2 + x_2^2 + x_3^2 - r^2 = 0 \tag{A.3.6}$$

expresses the fact that the point $P(x_1, x_2, x_3)$ is at a distance r from the origin and so is the equation of the sphere. Given x_1, x_2 we find x_3 has two values $\pm(r^2 - x_1^2 - x_2^2)^{1/2}$ provided that $x_1^2 + x_2^2 \leq r^2$. It is sometimes convenient to represent a surface parametrically by writing

$$x_i = g_i(u_1, u_2). \tag{A.3.7}$$

Such a representation is not unique for the sphere might be given by

$$x_1 = u_1, \qquad x_2 = u_2, \qquad x_3 = \pm(r^2 - u_1^2 - u_2^2)^{1/2}$$

or by

$$x_1 = u_1 \cos u_2, \qquad x_2 = u_1 \sin u_2, \qquad x_3 = \pm(r^2 - u_1^2)^{1/2}$$

either of which will satisfy Eq. (A.3.6) identically. We can always return to the form (A.3.5) from the parametric form by eliminating u_1 and u_2 between the three equations (A.3.7).

If two surfaces intersect but do not coincide with one another, they intersect in a curve. Suppose that one surface is given parametrically by $x_i = g_i(u_1, u_2)$ and the other in the form $f(x_1, x_2, x_3) = 0$, then if we substitute for the x_i in f, we have an equation

$$f(g_1(u_1, u_2), g_2(u_1, u_2), g_3(u_1, u_2)) = 0$$

between u_1 and u_2. This defines u_2 as a function of u_1 so that on the intersection $x_i = g_i(u_1, u_2(u_1))$ is a function of only one parameter; this is a curve.

Exercise A.3.1. Show that the intersection of the plane (A.3.3) with the sphere (A.3.6) is a circle of radius $(r^2 - p^2)^{1/2}$ and that its projection on the 12 plane is an ellipse of area $\pi\lambda_3(r^2 - p^2)$.

A.4. Row and column vectors—change of origin and scale

The set of three coordinates (x_1, x_2, x_3) represents one entity, the point P. It will be convenient to have a notation and develop an algebra that reflects this unity. Let x with no suffix stand for the set of three numbers x_1, x_2, x_3 arranged as a column

$$x = \begin{bmatrix} x_1 \\ x_2 \\ x_3 \end{bmatrix}. \tag{A.4.1}$$

x is known as a *column vector* and, though it is unfortunate that the word vector should be used in a different sense here than in the title of the book, confusion may be avoided if we always add the word *column.* (The words *row* and *latent* will have the same purpose later.)

The three *elements*, as they are called, might also be arranged as a row and we write

$$x' = [x_1, x_2, x_3]. \tag{A.4.2}$$

x' is called a *row vector* and is the *transpose* of x, for the operation of transposition is to interchange rows for columns in any array.

Two row vectors or two column vectors may be added by adding the corresponding elements, but a row vector may not be added to a column vector for they are not arrayed in the same way. Geometrically the operation of addition is related to the translation of the origin of a frame of reference to a new position. If P and Q are the points (x_1, x_2, x_3) and (y_1, y_2, y_3) with respect to the frame of reference $O123$ and P has coordinates (z_1, z_2, z_3) with respect to a frame of reference $Q1'2'3'$, then, clearly, $x_i = y_i + z_i$. (Figure A.4 shows this for $i = 2$.) If the points are represented by column vectors, then

$$x = y + z \quad \text{or} \quad z = x - y,$$

or if by row vectors, then

$$x' = y' + z' \quad \text{or} \quad z' = x' - y'.$$

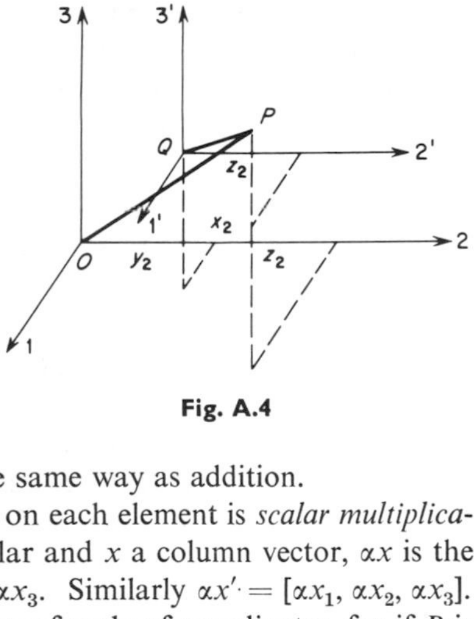

Fig. A.4

We see that subtraction is defined in the same way as addition.

Another operation which works alike on each element is *scalar multiplication.* If α is an ordinary number or scalar and x a column vector, αx is the column vector with elements αx_1, αx_2, αx_3. Similarly $\alpha x' = [\alpha x_1, \alpha x_2, \alpha x_3]$. Geometrically it corresponds to a change of scale of coordinates, for if P is (x_1, x_2, x_3) in the original system, and if the unit length in this system is made α units of length in a new coordinate system, then the new coordinates of P will be αx_1, αx_2, and αx_3.

A.5. Matrices and quadrics

The quadratic expression

$$ax_1^2 + bx_2^2 + cx_3^2 + 2fx_2x_3 + 2gx_3x_1 + 2hx_1x_2 = 1 \tag{A.5.1}$$

represents a surface since it is of the form (A.3.5). We shall see later that it is a quadric, that is an ellipsoid, hyperboloid, or some special form of these. We have met an example in the sphere where $a = b = c = r^{-2}$ and $f = g = h = 0$. The set of six numbers a, b, c, f, g, h represents a single entity, the quadric, just as the set of three coordinates represented a point. We write this set in a square array

$$A \equiv \begin{bmatrix} a & h & g \\ h & b & f \\ g & f & c \end{bmatrix} \tag{A.5.2}$$

and denote it by a capital letter. The position of the individual elements is determined by the suffixes of the quadratic term of which it is the coefficient. Thus a goes into the first row and first column because it is the coefficient of x_1^2 and the f's go into the second row and third column and the third row and second column respectively since $2f$ is the coefficient of x_2x_3.

We now define an operation of multiplication which can be performed on matrices and row or column vectors. It consists in taking elements from the rows of the first factor and multiplying by the corresponding elements of the columns in the second factor and adding them together. Thus

$$Ax \equiv \begin{bmatrix} ax_1 + hx_2 + gx_3 \\ hx_1 + bx_2 + fx_3 \\ gx_1 + fx_2 + cx_3 \end{bmatrix} \tag{A.5.3}$$

is a column vector. Naturally this is only possible if there are the same number of elements in the rows of the first factor as there are in the columns of the second, that is to say, the first factor has as many columns as the second has rows. The product Ax' is therefore meaningless but $x'A$ is the row vector which is the transpose of Eq. (A.5.3). Also we note that Eq. (A.5.2) can be written

$$x'Ax = 1 \tag{A.5.4}$$

so that the notation reflects the real unity of the quadric.

A more systematic notation for the elements of a matrix A is a_{ij}, $i, j = 1, 2, 3$, the first suffix referring to the row and the second to the column in which the element stands. Thus

$$A = \begin{bmatrix} a_{11} & a_{12} & a_{13} \\ a_{21} & a_{22} & a_{23} \\ a_{31} & a_{32} & a_{33} \end{bmatrix}. \tag{A.5.5}$$

Then the elements of the product Ax are $\sum_{j=1}^{3} a_{ij}x_j$, the summation being on adjacent indices and the remaining index showing that the product is a column vector. Similarly, the quadratic form $x'Ax$ is

$$\sum_{i=1}^{3}\sum_{j=1}^{3} x_i a_{ij} x_j.$$

The definition of matrix multiplication will be made more formal and general in Section A.7.

Two matrices A and B may be added if they have the same number of rows and columns. The elements of the sum are the sums of the corresponding elements. A matrix A may be multiplied by a scalar α to give a matrix whose elements are those of A, each multiplied by the number α.

If $a_{ij} = a_{ji}$ the matrix is said to be symmetric (the matrix A.5.2 is an example of this) and if $a_{ij} = -a_{ji}$ it is antisymmetric or skew-symmetric. Recalling that the operation of transposition is to interchange rows and columns we see that the i, j^{th} element of the transposed matrix A' is a_{ji}. Hence, a matrix is symmetric if $A' = A$ and antisymmetric if $A' = -A$. In general, a matrix is neither symmetric nor antisymmetric, but it is sometimes convenient to represent it as the sum of two matrices one of which is symmetric and the other antisymmetric. This may be done by writing

$$a_{ij} = \tfrac{1}{2}(a_{ij} + a_{ji}) + \tfrac{1}{2}(a_{ij} - a_{ji})$$

and so

$$A = \tfrac{1}{2}(A + A') + \tfrac{1}{2}(A - A'). \tag{A.5.6}$$

The matrix

$$I = \begin{bmatrix} 1 & \cdot & \cdot \\ \cdot & 1 & \cdot \\ \cdot & \cdot & 1 \end{bmatrix} \tag{A.5.7}$$

is called the unit or identity matrix, since it leaves unchanged any matrix it multiplies; thus $IA = A$, $Ix = x$, $x'I = x'$. It is the only matrix that does this, for if two matrices I and J give $Ix = x$ and $Jx = x$ for all x, then by subtraction $(I - J)x = 0$ a column vector all of whose elements are zero. However, if this is true for all x, then $I - J$ must be identically zero and $I = J$.

Exercise A.5.1. Show that in matrix notation the straight line through a with direction λ is given by $x = a + \lambda s$ and the plane with normal λ by $\lambda' x = p$.

Exercise A.5.2. If $x = a$ is any point on the quadric $x'Ax = 1$, show that the line $x = a + \lambda s$ is tangent to the quadric if $\lambda' Aa = 0$. Hence, show that the tangent plane at this point is $a'Ax = 1$.

Exercise A.5.3. Show that $x'A' = (Ax)'$.

A.6. Matrices and rotations of axes*

Let us consider the transformation of the coordinate frame by an arbitrary rigid rotation. Fig. A.5 shows the original frame $O123$ in which the coordinates of P are x_1, x_2, x_3 and the new frame $O\overline{123}$ in which they are $\bar{x}_1, \bar{x}_2, \bar{x}_3$. Let l_{ij} be the cosine of the angle between Oi and $O\bar{j}$, $i, j = 1, 2, 3$. Thus l_{1j}, l_{2j}, l_{3j} are the direction cosines of $O\bar{j}$ in the old system and l_{i1}, l_{i2}, l_{i3}, the direction cosines of Oi in the new. By definition the new coordinate $\bar{x}_j$ is the length of the projection of OP on the axis $O\bar{j}$, that is,

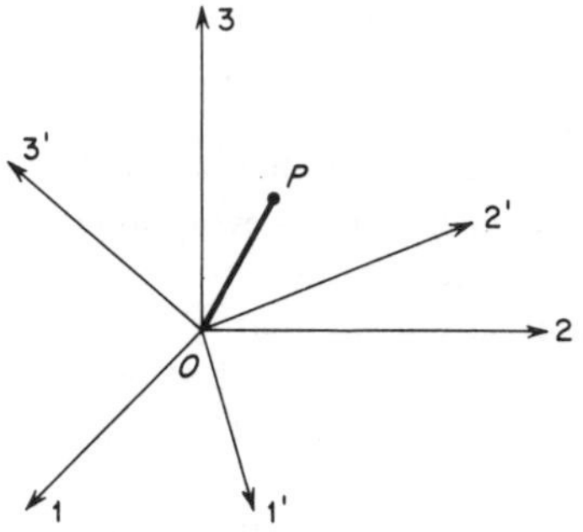

Fig. A.5

$$\bar{x}_j = l_{1j}x_1 + l_{2j}x_2 + l_{3j}x_3, \qquad j = 1, 2, 3. \tag{A.6.1}$$

Alternatively, working from the new system to the old, x_i is the projection of OP on to Oi and so

$$x_i = l_{i1}\bar{x}_1 + l_{i2}\bar{x}_2 + l_{i3}\bar{x}_3, \qquad i = 1, 2, 3. \tag{A.6.2}$$

If we form the numbers l_{ij} into a matrix

$$L \equiv \begin{bmatrix} l_{11} & l_{12} & l_{13} \\ l_{21} & l_{22} & l_{23} \\ l_{31} & l_{32} & l_{33} \end{bmatrix}$$

and the coordinates into column vectors x and $\bar{x}$ we see at once that Eq. (A.6.2) can be written

$$x = L\bar{x}. \tag{A.6.3}$$

Recalling that the transpose L' of L will have the element l_{ji} in the i, j^{th} position, Eq. (A.6.1) is evidently

$$\bar{x} = L'x \tag{A.6.4}$$

Substituting from Eq. (A.6.4) into Eq. (A.6.3) we have

$$x = LL'x = Ix,$$

and since I is the only matrix satisfying $x = Ix$, we must have $L'L = I$. By substituting from Eq. (A.6.3) into Eq. (A.6.4)

$$L'L = LL' = I. \tag{A.6.5}$$

A matrix L for which this relation holds is called orthogonal.

* An excellent discussion of the algebra of rotations is given by A. Mayer, Rotations and their algebra, SIAM Review, **2** (1960), pp. 77–122.

If the multiplication of the two matrices L and L' is performed in detail, we have various relations between the nine quantities l_{ij}. From the diagonal terms we have

$$l_{i1}^2 + l_{i2}^2 + l_{i3}^2 = 1$$

or (A.6.6)

$$l_{1j}^2 + l_{2j}^2 + l_{3j}^2 = 1$$

corresponding to the fact that these sets of numbers are direction cosines and so the sum of their squares is unity. From the off diagonal terms, we have

$$l_{i1}l_{p1} + l_{i2}l_{p2} + l_{i3}l_{p3} = 0, \qquad i \neq p,$$

or (A.6.7)

$$l_{1j}l_{1q} + l_{2j}l_{2q} + l_{3j}l_{3q} = 0, \qquad j \neq q,$$

expressing the fact that both old and new coordinates axes are mutually orthogonal.

Exercise A.6.1. Show that the matrix

$$Z(\theta) = \begin{bmatrix} \cos\theta & -\sin\theta & . \\ \sin\theta & \cos\theta & . \\ . & . & 1 \end{bmatrix}$$

represents a rotation through an angle θ about the z-axis.

Exercise A.6.2. The derivative of the matrix $Z(\theta)$ with respect to θ is obtained by differentiating each element separately. If this derivative is $\dot{Z}(\theta)$, show that $\dot{Z}(\theta) = Z(\theta)\dot{Z}(0)$.

Exercise A.6.3. Show that a rotation through an angle θ about a line with direction cosines $\lambda_1, \lambda_2, \lambda_3$ is given by an orthogonal matrix $L = I\cos\theta + (\lambda\lambda')(1 - \cos\theta) + \Lambda\sin\theta$ where

$$\Lambda = \begin{bmatrix} . & -\lambda_3 & \lambda_2 \\ \lambda_3 & . & -\lambda_1 \\ -\lambda_2 & \lambda_1 & . \end{bmatrix}.$$

A.7. The laws of matrix algebra

With the preceding discussion to give a feel for the manipulation of matrices with three rows and/or columns, we may now give a terse description of the algebra of matrices in a more general form.

Definition: An array of the mn numbers a_{ij}, $i = 1, \ldots, m$, $j = 1, \ldots, n$, in m rows and n columns is called a matrix A of order $m \times n$ with elements a_{ij},

$$A = \begin{bmatrix} a_{11} & a_{12} & \ldots & a_{1n} \\ a_{21} & a_{22} & \ldots & a_{2n} \\ & & \ldots & \\ a_{m1} & a_{m2} & \ldots & a_{mn} \end{bmatrix}.$$

The element in the ith row and j^{th} column may be called the ij^{th} element and denoted by $(A)_{ij} = a_{ij}$. An $m \times 1$ matrix is called a column vector and a $1 \times n$ matrix a row vector.

Scalar multiplication: If α is any number the scalar product, αA is the matrix obtained by multiplying each element of A by α, that is, $(\alpha A)_{ij} = \alpha a_{ij}$.

Addition: Two matrices can be added if they are of the same order and $(A + B)_{ij} = (A)_{ij} + (B)_{ij}$.

Subtraction: $A - B = A + (-1)B$.

Transposition: The matrix A' is obtained from A by interchanging rows and columns. It is therefore an $n \times m$ matrix with elements $(A')_{ij} = a_{ji}$.

Multiplication: Two matrices A, B are conformable for multiplication in the product AB if A has as many columns as B has rows. The ij^{th} element of the product is

$$(AB)_{ij} = \sum_k a_{ik} b_{kj}, \tag{A.7.1}$$

where the summation is over the common range of the suffix k. Note that $(AB)' = B'A'$.

For square matrices we have the following definitions. If

$$a_{ij} = 0, \qquad i \neq j,$$
$$a_{ij} = 1, \qquad i = j,$$

A is called the identity matrix I. For all conformable matrices $IB = BI = B$. If $A = A'$, it is symmetric; if $A = -A'$, it is antisymmetric. If $A'A = AA' = I$, it is said to be orthogonal. If a matrix B exists such that $AB = BA = I$, B is called the inverse of A and written A^{-1}. (The construction of the inverse will occupy us in the next section.)

Matrix algebra has the following laws:

$$A + B = B + A \tag{A.7.2}$$
$$(A + B) + C = A + (B + C) \tag{A.7.3}$$
$$(AB)C = A(BC) \tag{A.7.4}$$
$$A(B + C) = AB + AC \tag{A.7.5}$$

In general, $AB \neq BA$ for one of these need not be defined and even for square matrices they are not generally equal. If they are equal, the two matrices are said to commute.

Exercise A.7.1. Establish the laws (A.7.2-5) from the definitions and laws of the algebra of real numbers.

Exercise A.7.2. Show that $Z(\theta)$ and $Z(\phi)$ commute. See Ex. A.6.1 for definition.

A.8. Determinants—the inverse of a matrix

The determinant is a number calculated from all the elements of a square matrix as follows. If $ij \cdots p$ is a permutation of the first n integers $1, 2, \ldots, n$, it is called even or odd according as the natural order can be restored by an even or odd number of interchanges. Thus, 312 is an even permutation for it requires two interchanges (of 1 and 3 and then of 2 and 3 or of 1 and 2 and then of 1 and 3) to return it to the order 123, but 132 is odd since the single interchange of 2 and 3 suffices.

The determinant of an $n \times n$ matrix is

$$\det A = |A| = \Sigma \pm a_{1i}a_{2j} \cdots a_{np}, \tag{A.8.1}$$

where the summation is over all permutations $i, j, \ldots, p$ of $1, 2, \ldots, n$ and the sign accords with the parity of the permutation. Thus, for a 3×3 determinant, we have

$$\begin{aligned} A = {} & a_{11}a_{22}a_{33} + a_{12}a_{23}a_{31} + a_{13}a_{21}a_{32} \\ & -a_{12}a_{21}a_{33} - a_{11}a_{23}a_{32} - a_{13}a_{22}a_{31}. \end{aligned} \tag{A.8.2}$$

In each term of the sum there is one element from each row and one from each column but no two elements have their row or column in common. If $i, j, \ldots, p$ is rearranged into natural order $1, 2, \ldots, n$ and the same system of interchanges is performed on the set $1, 2, \ldots, n$ this produces a permutation that is said to be conjugate to the original permutation. Thus 312 requires the interchange of the first and second numbers followed by the interchange of the second and third, to restore the order 123. If we do this to 123, we obtain first 213 and then 231 which is therefore conjugate to 312. It is clear from the construction that conjugate permutations are of the same parity and it follows that the determinant can also be written

$$A = \Sigma \pm a_{i1}a_{j2} \ldots, a_{pn} \tag{A.8.3}$$

and hence that

$$|A'| = |A| . \tag{A.8.4}$$

We may also rearrange the sum of the $n!$ terms in Eq. (A.8.1) into n sums of $(n-1)!$ terms by taking out factors from the first row. Thus

$$\begin{aligned}\Sigma \pm a_{1i}a_{2j} \cdots a_{np} = a_{11} \Sigma \pm a_{2j} \ldots a_{np} \\ -a_{12} \Sigma \pm a_{2j} \ldots a_{np} \\ \ldots \\ +(-1)^{n+1}a_{1n} \Sigma \pm a_{2j} \ldots a_{np}.\end{aligned}$$

In the first term $j, \ldots, p$ is a permutation of $2, 3, \ldots, n$ and the parity of its permutation is the same as the parity of the permutation $1, j, \ldots, p$ of $1, 2, \ldots, n$. In the second term $j, \ldots, p$ is a permutation of $1, 3, \ldots, n$ and if it is an even permutation then $2, j, \ldots, p$ will be an even permutation of $2, 1, 3, \ldots, n$, that is, an odd permutation of $1, 2, 3, \ldots, n$. The negative sign is therefore associated with this term and the $\pm$ sign after the summation agrees with the parity of the permutation of the indices. In fact, the coefficient of $(-1)^{i+1}a_{1i}$ is $\Sigma \pm a_{2j} \ldots a_{np}$ where $j, \ldots, p$ is a permutation of $1, 2, \ldots, i-1, i+1, \ldots, n$ is evidently just the determinant of the $(n-1)x(n-1)$ matrix obtained by striking out the first row and i^{th} column of A. This is called the cofactor of a_{ij} in the determinant of A and may be denoted by A_{ij}. Thus

$$|A| = \sum_{i=1}^{n} a_{1i}A_{1i}, \tag{A.8.5}$$

and this is called the expansion of the determinant on the first row. A similar argument may be made about the expansion on any row or column and

$$|A| = \sum_{q=1}^{n} a_{iq}A_{iq} = \sum_{p=1}^{n} a_{pj}A_{pj}, \qquad i, j = 1, \ldots, n. \tag{A.8.6}$$

If we form a new matrix A^* by interchanging two rows or two columns we just switch the parity of the permutation in each term of Eq. (A.8.1) or Eq. (A.8.3) and so $|A^*| = -|A|$. It follows that if two rows or two columns of A are the same so that their interchange leaves A unchanged then $|A| = -|A| = 0$; the determinant vanishes and A is said to be singular. Also, we see from the definition that $|A|$ is linear in the elements of any one row or column so that if, for example, $a_{1i} = b_{1i} + c_{1i}$, then $|A| = |B| + |C|$ where B and C are matrices with b_{1i} and c_{1i} as their first row. It follows that we may replace any row or column by the sum of itself and some multiple of another row or column without changing the value of the determinant, for the determinant added is identically zero. Thus

$$\sum_{q=1}^{n} a_{iq}A_{jq} = \sum_{p=1}^{n} a_{pi}A_{pj} = 0, \qquad i \neq j, \tag{A.8.7}$$

a pair of formulas sometimes called expansion by *alien* cofactors.

The matrix constructed from A by replacing each element by its cofactor and transposing is called the adjugate of A; $(\text{adj } A)_{ij} = A_{ji}$. Consider now the product $A(\text{adj } A)$. Its ij^{th} element will be

$$\sum_{q=1}^{n} a_{iq}(\text{adj } A)_{qj} = \Sigma\, a_{iq} A_{jq},$$

which equals $|A|$ or zero according as $i = j$ or $i \neq j$. It follows that

$$A^{-1} = \frac{\text{adj } A}{|A|}, \qquad \text{(A.8.8)}$$

for then $AA^{-1} = I$. Similarly, $A^{-1}A = I$. The inverse will exist except when $|A| = 0$, that is, A is singular.

The linear equations

$$Ax = b, \qquad \text{(A.8.9)}$$

which written out in full would read

$$\begin{aligned} a_{11}x_1 + a_{12}x_2 + \dots + a_{1n}x_n &= b_1 \\ &\dots \\ a_{n1}x_1 + a_{n2}x_2 + \dots + a_{nn}x_n &= b_n, \end{aligned} \qquad \text{(A.8.10)}$$

can be solved if A is nonsingular, for multiplying through by A^{-1}

$$A^{-1}Ax = Ix = x = A^{-1}b. \qquad \text{(A.8.11)}$$

If A is nonsingular and b is identically zero, then the only solution is the trivial one $x = 0$, but if A is singular, there will be a nontrivial solution x. In the simplest case we may suppose that though $|A| = 0$, one of the cofactors is not zero and without loss of generality take this to be A_{nn}. Then the last equation is a linear combination of the first $n - 1$ and tells us nothing new. The first $(n - 1)$ can be rewritten

$$\begin{aligned} a_{11}\frac{x_1}{x_n} + a_{12}\frac{x_2}{x_n} + \dots + a_{1,n-1}\frac{x_{n-1}}{x_n} &= -a_{1n} \\ \dots \qquad\qquad & \dots \\ a_{n-1,1}\frac{x_1}{x_n} + a_{n-1,2}\frac{x_2}{x_n} + \dots + a_{n-1,n-1}\frac{x_{n-1}}{x_n} &= -a_{1,n-1} \end{aligned}$$

and these have a nontrivial solution for the ratios x_i/x_n, $i = 1, \dots, n - 1$. We shall not stop to go into the full details of the proof but record, for later use, the theorem. The homogeneous equations

$$Ax = 0 \qquad \text{(A.8.12)}$$

have a nontrivial solution provided that $|A| = 0$.

If the elements of the matrix are functions of some variable say $a_{ij} = a_{ij}(s)$, then the derivative of $|A|$ with respect to s is the sum of the n determinants obtained by replacing one row (or one column) by the derivatives of its elements. This is immediate from the definition of the determinant given in Eq. (A.8.1) for the derivative of each term is the sum of terms each with only one element differentiated.

Exercise A.8.1. Show that the ij^{th} component of A^{-1} is

$$\frac{\partial[\ln |A|]}{\partial a_{ji}}.$$

A.9. Partitioned matrices—Laplacian expansion—product of determinants

It is sometimes convenient to regard the matrix as composed not of scalar elements but of submatrices of smaller order. Such a matrix is called a partitioned matrix. For example, the matrix A may be regarded as composed of three column vectors

$$a^{(i)} = \begin{bmatrix} a_{1i} \\ a_{2i} \\ a_{3i} \end{bmatrix}, \qquad i = 1, 2, 3,$$

and we would write

$$A = [a^{(1)} \;\vdots\; a^{(2)} \;\vdots\; a^{(3)}]$$

or

$$A = \left[\begin{array}{cc:c} a_{11} & a_{12} & a_{13} \\ a_{21} & a_{22} & a_{23} \\ \hdashline a_{31} & a_{32} & a_{33} \end{array}\right] = \left[\begin{array}{c:c} A^{11} & A^{12} \\ \hdashline A^{21} & A^{22} \end{array}\right], \tag{A.9.1}$$

where the submatrices A^{ij} are of orders 2×2, 2×1, 1×2, and 1×1. If the partitioning of the columns of A is the same as that of the rows of B then the matrices are said to be conformally partitioned for the multiplication AB and the product can be written down by multiplying the submatrices as if they were elements. An example will suffice to show this. Suppose A is as given in Eq. (A.9.1) and

$$B = \left[\begin{array}{c:cc} b_{11} & b_{12} & b_{13} \\ b_{21} & b_{22} & b_{23} \\ \hdashline b_{31} & b_{32} & b_{33} \end{array}\right] = \left[\begin{array}{c:c} B^{11} & B^{12} \\ \hdashline B^{21} & B_{22} \end{array}\right],$$

Then the matrices are conformally partitioned and the product is the partitioned matrix

$$AB = \left[\begin{array}{c:c} A^{11}B^{11} + A^{12}B^{21} & A^{11}B^{12} + A^{12}B^{22} \\ \hdashline A^{21}B^{11} + A^{22}B^{21} & A^{21}B^{12} + A^{22}B^{22} \end{array}\right]$$

the submatrices being of order 2×1, 2×2, 1×1, 1×2.

It is clearly not possible to expand the determinant of a partitioned matrix by treating the submatrices as elements, but there are expansions other than on the elements of a single row or single column; such is the Laplacian expansion. Take any m rows $(m < n)$ of an $n \times n$ determinant which without loss of generality can be taken as the first m. From these $n!/m!(n - m)!$ determinants may be concocted by taking m of the n columns. Each of these has a cofactor, an $(n - m) \times (n - m)$ determinant composed of elements from the last rows and the remaining columns. An appropriate sign must be given to the product of each $m \times m$ determinants and its cofactor (see Ex. 9.1) and the sum of all these products is the determinant of the original matrix. In a special case the result is simple for suppose that A and B are square matrices which are submatrices in a partitioned matrix

$$D = \begin{bmatrix} A & \vdots & 0 \\ \cdots & \vdots & \cdots \\ C & \vdots & B \end{bmatrix}.$$

The upper right-hand submatrix is identically zero and C can be arbitrary. Then $|D| = |A|\,|B|$ for expanding on the first m rows ($m \times m$ being the order of A) all the determinants will vanish except that of the first m rows, that is, $|A|$. Similar relations hold when the identically zero submatrix is in other positions.

This type of expansion can be used to prove the rule for the product of matrices. If A and B are both $n \times n$ matrices $|AB| = |A|\,|B|$. Consider the following identity between three $2n \times 2n$ partitioned matrices

$$\begin{bmatrix} I & A \\ 0 & I \end{bmatrix}\begin{bmatrix} A & 0 \\ -I & B \end{bmatrix} = \begin{bmatrix} 0 & AB \\ -I & B \end{bmatrix}.$$

The first matrix has units down the diagonal and only nonzero elements in the upper triangle. It represents the operation of replacing each row of the second matrix by the same row plus a linear combination of later rows and therefore does not change the value of the determinant of the second matrix. Thus

$$\begin{vmatrix} A & 0 \\ -I & B \end{vmatrix} = \begin{vmatrix} 0 & AB \\ -I & B \end{vmatrix},$$

and evaluating each determinant, we have

$$|A| \cdot |B| = |AB|.$$

Exercise A.9.1. If the $m \times m$ determinant is taken from rows $i_1, \ldots, i_m$ and columns $j_1, \ldots, j_m$ and its cofactor from the remaining rows and columns, show that the appropriate sign for the product is $+$ or $-$ according as $i_1 + j_1 + i_2 + \ldots + j_m$ is even or odd.

Exercise A.9.2. Show that all the terms in the definition of the determinant are enumerated with their proper signs and hence establish the Laplacian expansion.

Exercise A.9.3. Let L be an $n \times n$ orthogonal matrix and M and N two of its submatrices of orders $m \times n$ and $(n - m) \times n$, that is,

$$L = \begin{bmatrix} M \\ \cdots \\ N \end{bmatrix}.$$

Show that

$$(M'M + \lambda I)^{-1} = \frac{(N'N + \lambda I)}{\lambda(1 + \lambda)}.$$

A.10. Latent roots and vectors of a symmetric matrix

If A is any matrix and x a column vector, then $y = Ax$ is another column vector. If we think of x as the coordinates of a point P and y of Q, then the matrix A transforms the point P into the point Q. It is a homogeneous transformation for it takes the whole line OP into the line OQ. It is interesting to inquire whether there are lines which are not changed by the transformation. If Q lies on the line OP, $y = \lambda x$, where λ is some scalar. Thus

$$Ax = \lambda x = \lambda Ix$$

or

$$(A - \lambda I)x = 0. \tag{A.10.1}$$

By the theorem stated above this will only have nontrivial solutions if

$$|A - \lambda I| = 0. \tag{A.10.2}$$

For an $n \times n$ determinant this gives an algebraic equation of the nth degree

$$|A| - \lambda(tr_{n-1}A) + \lambda^2(tr_{n-2}A) \ldots (-)^{n-1}\lambda^{n-1}(trA) + (-)^n\lambda^n = 0 \tag{A.10.3}$$

where

$$trA = tr_1A = a_{11} + a_{22} + \ldots + a_{nn}$$

and tr_pA is the sum of all the $p \times p$ determinants that can be formed with diagonal elements that are diagonal elements of A. For a 3×3 matrix this is a cubic giving three values of λ and three corresponding directions that are invariant under the transformation. These three values are called latent roots or characteristic values of the matrix and the corresponding x are latent or characteristic vectors. The convenient, but rather ugly, words eigenvalue and eigenvector are also used.

If A is a symmetric matrix with real elements, its latent roots are all real. For, if they are not, they must occur in pairs of complex conjugates. Let λ be such a root and $\bar{\lambda}$ its complex conjugate. Then $Ax = \lambda x$ and $A\bar{x} = \bar{\lambda}\bar{x}$. Multiplying the first equation by $\bar{x}'$ and the second by x' we have

$$\bar{x}'Ax = \lambda\bar{x}'x \quad \text{and} \quad x'A\bar{x} = \bar{\lambda}x'\bar{x}$$

However, since A is symmetric,

$$x'A\bar{x} = \bar{x}'Ax \quad \text{and} \quad \bar{x}'x = x'\bar{x},$$

so that by subtraction

$$(\lambda - \bar{\lambda})\bar{x}'x = 0.$$

Since x is nontrivial $\lambda = \bar{\lambda}$ or λ is real.

If A is symmetric the latent vectors associated with distinct latent roots are orthogonal. Let x and y be the vectors associated with two distinct roots λ and μ. Then

$$Ax = \lambda x \quad \text{and} \quad Ay = \mu y$$

However, multiplying the first by y' and the second by x'

$$y'Ax = \lambda y'x \quad \text{and} \quad x'Ay = \mu x'y.$$

Again, by the symmetry of A, the two left-hand sides are the same and $x'y = y'x$. Hence, subtracting

$$(\lambda - \mu)x'y = 0$$

and since $\lambda \neq \mu$, $x'y = 0$, or the directions associated with the roots are orthogonal.

Exercise A.10.1. Show that if L is an orthogonal matrix $L'AL$ has the same latent roots as A.

A.11. Canonical form of symmetric matrices and quadrics

Consider first the case of distinct latent roots λ, μ, ν and let the associated column vectors be ξ, η, and ζ. Since only the ratio of the components of ξ, η, and ζ can be determined from the homogeneous Eqs. (A.10.1), we may fix their magnitudes by requiring that the sum of their squares be 1, that is, $\xi'\xi = \eta'\eta = \zeta'\zeta = 1$. Since λ, μ, and ν are all distinct, ξ, η, and ζ are mutually orthogonal, that is,

$$\eta'\zeta = \zeta'\xi = \xi'\eta = 0.$$

The partitioned matrix

$$L = [\xi \mathrel{\vdots} \eta \mathrel{\vdots} \zeta] \tag{A.11.1}$$

will then be an orthogonal matrix. The matrices A and L are conformably partitioned for multiplication and

$$AL = [A\xi \mathrel{\vdots} A\eta \mathrel{\vdots} A\zeta] = [\lambda\xi \mathrel{\vdots} \mu\eta \mathrel{\vdots} \nu\zeta] \tag{A.11.2}$$

Further the matrices L' and AL are conformably partitioned and

$$L'AL = \begin{bmatrix} \xi' \\ \cdots \\ \eta' \\ \cdots \\ \zeta' \end{bmatrix} [\lambda\xi \mathrel{\vdots} \mu\eta \mathrel{\vdots} \nu\zeta] = \begin{bmatrix} \lambda\xi'\xi & \mu\xi'\eta & \nu\xi'\zeta \\ \lambda\eta'\xi & \mu\eta'\eta & \nu\eta'\zeta \\ \lambda\zeta'\xi & \mu\zeta'\zeta & \nu\zeta'\zeta \end{bmatrix}$$

$$= \begin{bmatrix} \lambda & \cdot & \cdot \\ \cdot & \mu & \cdot \\ \cdot & \cdot & \nu \end{bmatrix} \tag{A.11.2}$$

on account of the orthogonality of ξ, η, and ζ. This is the diagonal or canonical form of the matrix A.

If λ, μ, and ν are not all distinct, we must proceed more slowly. Let ξ be the latent vector associated with λ and suppose that it has been normalized so that $\xi'\xi = 1$. Then we can construct an orthogonal matrix with ξ as the first column as follows. Let y be any vector not proportional to ξ and let $\xi' y = \alpha$. Then the vector $y - \alpha\xi$ is certainly orthogonal to ξ for

$$\xi'(y - \alpha\xi) = \alpha - \alpha\xi'\xi = 0.$$

Normalize this vector to give

$$\eta_1 = \frac{y - \alpha\xi}{\{(y' - \alpha\xi')(y - \alpha\xi)\}^{1/2}}\,.$$

Again, let z be any vector not linearly dependent on ξ and η_1 and let

$$\xi' z = \beta, \qquad \eta_1' z = \gamma.$$

Then $z - \beta\xi - \gamma\eta_1$ is orthogonal to both ξ and η_1 and may be normalized to give a column vector ζ_1. Let

$$L_1 = [\xi \,\vdots\, \eta_1 \,\vdots\, \zeta_1].$$

Then

$$AL_1 = [\lambda\xi \,\vdots\, A\eta_1 \,\vdots\, A\zeta_1]$$

and

$$L_1'AL = \left[\begin{array}{c:c:c} \lambda & \xi' A\eta_1' & \xi' A\zeta_1 \\ \hdashline 0 & \eta_1' A\eta_1 & \eta_1' A\zeta_1 \\ \hdashline 0 & \zeta_1' A\eta_1 & \zeta_1' A\zeta_1 \end{array}\right]$$

However, since A is symmetric $A' = A$ and so $L_1'AL_1$ is symmetric and $\xi' A\eta_1 = \xi' A\zeta_1 = 0$. The matrix $L_1'AL_1$ thus has the form

$$\left[\begin{array}{c:c} \lambda & 0 \\ \hdashline 0 & B \end{array}\right]$$

where B is a symmetric 2×2 matrix. However, the latent roots of $L_1'AL_1$ are the same as those of A so that the latent roots of B must be μ and ν. We can find an orthogonal 2×2 matrix M that will transform B into diagonal form (see Ex. A.11.1). Let

$$L_2 = \left[\begin{array}{c:c} 1 & 0 \\ \hdashline 0 & M \end{array}\right]$$

Then

$$L_2'(L_1'AL_1)L_2 = \left[\begin{array}{c:c} \lambda & 0 \\ \hdashline 0 & M'BM \end{array}\right] = \begin{bmatrix} \lambda & \cdot & \cdot \\ \cdot & \mu & \cdot \\ \cdot & \cdot & \nu \end{bmatrix},$$

and the orthogonal matrix $L = L_1L_2$ acts as in Eq. (A.11.2).

We have seen that $y = L'x$ represents a rotation of axes. In the new coordinate system the equation of the quadric $x'Ax = 1$ becomes $y'(L'AL)y = 1$. However, if L is chosen as above, $L'AL = \Lambda$, the diagonal matrix (A.11.2), and

$$y'\Lambda y = \lambda y_1^2 + \mu y_2^2 + \nu y_3^2 = 1. \tag{A.11.3}$$

In this coordinate system the shape of the surface can be readily described (see Fig. A.6), for it is immediately evident that the surface is symmetrical about

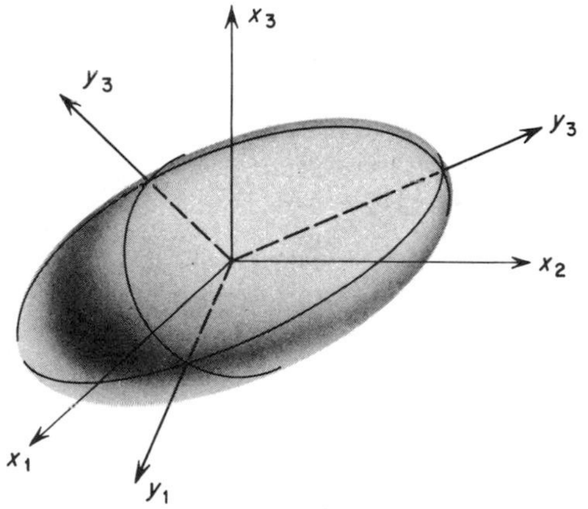

Fig. A.6

the coordinate planes. If λ, μ, and ν are all positive, then the section by any plane parallel to a coordinate plane is an ellipse. In fact, writing

$$\lambda = a_1^{-2}, \qquad \mu = a_2^{-2}, \qquad \nu = a_3^{-2}$$

we have the familiar equation for an ellipsoid

$$\frac{y_1^2}{a_1^2} + \frac{y_2^2}{a_2^2} + \frac{y_3^2}{a_3^2} = 1 \tag{A.11.4}$$

with semi-axes a_1, a_2, a_3. If one of the latent roots is negative, the surface is an hyperboloid of one sheet and, if two are negative, it is an hyperboloid of two sheets. There are various other special cases which it is not our purpose to look into here since the ellipsoid is the case of interest for applications. The important thing to see is that the principal axes of the ellipsoid are given by the latent vectors and that a rotation of axes can be found which makes these the new coordinate axes. The lengths of the principal axes are proportional to the reciprocals of the square roots of the latent roots.

Exercise A.11.1. B is a symmetric matrix

$$\begin{bmatrix} \alpha & \beta \\ \beta & \gamma \end{bmatrix}$$

with latent roots μ and ν. Find the latent roots and normalized vectors in terms of α, β, and γ and show how the orthogonal matrix M reduces the matrix to canonical form.

Exercise A.11.2. Show that $\alpha u^2 + 2\beta uv + \gamma v^2$ is never negative if μ and ν are both positive.

A.12. Stationary properties

The family of surfaces $x'Ax = c$ is a set of similar ellipsoids increasing in size with c if the latent roots of A are all positive. For points on the surface of the unit sphere, the value of the quadratic form $x'Ax$ will vary and we are interested in finding its stationary values. To do this we may use Lagrange's method of undetermined multipliers in which the problem of finding the stationary values of $x'Ax$ subject to $x'x = 1$ is replaced by that of finding the stationary values of $x'Ax - \lambda x'x$ with no restriction and later determining the λ so that $x'x = 1$. Now

$$S = x'Ax - \lambda x'x = x'(A - \lambda I)x \tag{A.12.1}$$

That is,

$$S = (a_{11} - \lambda)x_1^2 + (a_{22} - \lambda)x_2^2 + (a_{33} - \lambda)x_3^2 \\ + 2a_{23}x_2x_3 + 2a_{31}x_3x_1 + 2a_{12}x_1x_2, \tag{A.12.2}$$

remembering that A is symmetric. Thus

$$\frac{1}{2}\frac{\partial S}{\partial x_1} = (a_{11} - \lambda)x_1 + a_{12}x_2 + a_{13}x_3$$

and so

$$\frac{1}{2}\begin{bmatrix} \partial S/\partial x_1 \\ \partial S/\partial x_2 \\ \partial S/\partial x_3 \end{bmatrix} = (A - \lambda I)x. \tag{A.12.3}$$

However, for a stationary value these three derivatives must all be zero and we return to the characteristic equation (A.10.1). It follows that there are three possible values of λ, the latent roots and three corresponding vectors x. We note also that the three values of λ are the stationary values of $x'Ax$, for with the latent vector x

$$x'Ax = x'(\lambda x) = \lambda x'x = \lambda.$$

Geometrically this corresponds to finding the three values of c for which members of the family of ellipsoids $x'Ax = c$ just touch the unit sphere. It is clear from their symmetry that the points of contact will be the end of the axes of the ellipsoid and the values of λ thus correspond to the values of c for which the three axes are of unit length.

An equivalent problem is to look for the stationary values of the distance from the center to a point on the surface of the ellipsoid $x'Ax = 1$. This time we would seek the stationary values of $x'x$ subject to $x'Ax = 1$, or, by Lagrange's method, the unrestricted stationary values of $x'x - \mu x'Ax$. Arguing in the same way as before, we would have to solve the simultaneous equations

$$(I - \mu A)x = 0, \tag{A.12.4}$$

which is only possible if

$$|I - \mu A| = 0. \tag{A.12.5}$$

It is evident that the values of μ given by this equation are the reciprocals of the λ given by $|A - \lambda I| = 0$. However, the equations (A.12.4) are really the same as before, for they are homogeneous and therefore unchanged if we multiply through by $-\lambda$ to give $(A - \lambda I)x = 0$. It follows that the directions of the latent vectors will be the same. This time we notice that the stationary values of the distance $r = (x'x)^{1/2}$ are the values of $\mu^{1/2} = \lambda^{-1/2}$, for

$$x'x = x'(\mu Ax) = \mu x'Ax = \mu.$$

Exercise A.12.1. Find the points on the ellipsoid at which the normal passes through the center.

Exercise A.12.2. Show that if the greatest value of $x'Ax$ on the unit sphere is λ, then the greatest value of $x'AAx$ is λ^2. Generalize this.

Appendix B

Implicit functions and Jacobians

If y is a function of x satisfying the relation

$$F(x, y) = y^2 - 2xy - x^5 = 0, \tag{B.1}$$

we can solve the quadratic for y and obtain

$$y = f(x) = x \pm x(1 + x^3)^{1/2} \tag{B.2}$$

The first formulation is called an *implicit* definition of the function $y(x)$, but in the second y is given as an *explicit* function of x. In many cases it is not possible to give a simple explicit representation from an implicit relation $F(x, y) = 0$ yet we want to treat y as a function of x. What is needed, therefore, is some assurance that y does exist as a function of x and a precise statement of the conditions under which this will be true. Then the function can be handled with confidence and if no explicit formulas are available, numerical methods can be resorted to. A certain care is needed, for in the example given above we really have two explicit functions, corresponding to the positive and negative signs, from the one implicit relation. Accordingly, we have to restrict attention to the vicinity of a particular point. For example, the values $x = 2$, $y = 8$ satisfy the relation $F(x, y) = 0$ but only one of the explicit formulas. We could assert, however, that there exists a function $y = f(x)$, which satisfies

$$F(x, y) = F(x, f(x)) = 0$$

and for which $f(2) = 8$, even though we could not obtain the expression

$$f(x) = x + x(1 + x^3)^{1/2}.$$

This is useful if we want to treat x as a function of y, say $g(y)$, for here no explicit formula exists. At the point $x = -1$, $y = -1$ (or $x = y = 0$) we run into trouble for here both formulas (B.2) are valid. We shall obtain criteria for such points as these.

We shall prove several general theorems that are required in the text and give some specific applications. The proof of the simplest theorem will be given in detail, but the later ones will not be burdened with ϵ and δ's.

Theorem 1. Let $x = x_0$, $y = y_0$ be a pair of values satisfying $F(x, y) = 0$ and let F and its first derivatives be continuous in the neighborhood of this point. Then, if $\partial F/\partial y$ does not vanish at $x = x_0$, $y = y_0$, there exists one and only one continuous function $f(x)$ such that

$$F(x, f(x)) = 0 \quad \text{and} \quad f(x_0) = y_0.$$

Proof: If $\partial F/\partial y - F_y$ is not zero, let us suppose for definiteness that it is positive. Then, since F, F_x, and F_y are continuous we can find a box about the point (x_0, y_0)

$$|x - x_0| \leqq \delta, \qquad |y - y_0| \leqq \delta,$$

within which F, F_x, F_y are all continuous and

$$F_y > \eta > 0.$$

η is some positive number that bounds F_y below and within the region we can take a number ξ that bounds F_x above, that is,

$$F_x < \xi.$$

Now within this region

$$F(x, y) = F(x, y) - F(x_0, y) + F(x_0, y) - F(x_0, y_0),$$

since $F(x_0, y_0) = 0$. Then applying the first mean value theorem to each of these differences,

$$\begin{aligned} F(x, y) &= (x - x_0)F_x(x_0 + \theta(x - x_0), y) \\ &\quad + (y - y_0)F_y(x_0, y_0 + \theta'(y - y_0)) \\ &= A(x, y)(x - x_0) + B(x, y)(y - y_0). \end{aligned}$$

$A(x, y)$ and $B(x, y)$ are continuous functions and throughout the region

$$A(x, y) < \xi, \qquad B(x, y) > \eta.$$

Now let ϵ be a positive number, $0 < \epsilon < \delta$ and ζ be the smaller of the two numbers δ and $\epsilon\eta/\xi$. Then in the interval

$$|x - x_0| < \zeta,$$

we ask for the number of roots of y that lie between $y_0 - \epsilon$ and $y_0 + \epsilon$. The first term is less than $\xi\zeta$ in absolute value and the second term greater than $\eta\epsilon$ in absolute value when $y - y_0 = \pm\epsilon$, so that the second term determines the sign of F. Thus

$$F(x, y_0 - \epsilon) < 0 \quad \text{and} \quad F(x, y_0 + \epsilon) > 0$$

and there must be one root between these limits for any given x. In fact, this root is unique for F_y is positive between these limits. Let this root y be called $y = f(x)$. Then we have a unique solution. As

$$x \to x_0, \qquad y \to y_0 = f(x_0),$$

so that the function is continuous. Hence we have a single continuous function defined within a small region

$$|x - x_0| < \Delta = \text{Min}\left(\delta, \frac{\eta\delta}{\xi}\right)$$

(since ϵ can take values arbitrarily close to δ).

We have only gotten an element and perhaps only a small element of the function, namely, the part within $|x - x_0| < \Delta$. However, we notice that the conditions of the theorem are satisfied for any point x_1 within this region. We can therefore begin again to construct the function $y = f(x)$ from the point x_1, y_1 and hope to extend the region. In fact, we shall always be able to extend the region until we reach a point where $F_y = 0$ and then there is no unique solution to be found.

In the example first quoted we have $x_0 = 2$, $y_0 = 8$

$$F(x, y) = y^2 - 2xy - x^5$$

$$F_x = -2y + 5x^4, \qquad F_y = 2(y - x)$$

Thus in the region $|x - 2| < 3 - \alpha$, $|y - 8| < 3 - \alpha$ about (x_0, y_0) we have $F_y > 4\alpha > 0$, however small $\alpha > 0$ may be. Let

$$\delta = 3 - \alpha, \qquad \eta = 4\alpha \quad \text{and} \quad \xi = 5^5$$

(which is certainly greater than F_x). Then if $\epsilon = 3 - 2\alpha < 8$, ζ is the smaller of $\delta = 3 - \alpha$ and

$$\frac{\epsilon\eta}{\xi} = \frac{(3 - 2\alpha)4\alpha}{5^5}.$$

The second of these is extremely small if α is small so that by going too near the limit at which $F_y = 0$ we are evidently running into trouble. It were better to start with a more modest region

$$|x - 2| < 1, \qquad |y - 8| < 1$$

then $F_y > 8$ and $F_x < 391$. Even so ζ must be less than about $\frac{1}{50}$, since the upper bound of F_x is still large.

This theorem may be extended in a number of ways. For example, the proof of the case where one function $F(x, y, z) = 0$ defines a unique function $z = f(x, y)$ through a given point x_0, y_0, z_0 provided that F_z does not vanish goes through in precisely the same kind of way. A more general theorem follows.

Theorem 2. Between the $m + n$ variables $x_1, \ldots, x_m, y_1, \ldots, y_n$ there is a system of n relations

$$F_j(x_1, \ldots, x_m;\ y_1, \ldots, y_n) = 0, \qquad j = 1, \ldots, n.$$

If these equations are satisfied for $x_i = x_i^0$, $y_j = y_j^0$ and the F_j are continuous with continuous first partial derivatives in the neighborhood of these values and

$$\frac{\partial(F_1, \ldots, F_n)}{\partial(y_1, \ldots, y_n)} = \begin{vmatrix} \dfrac{\partial F_1}{\partial y_1} & \cdots & \dfrac{\partial F_1}{\partial y_n} \\ & \cdots & \\ \dfrac{\partial F_n}{\partial y_1} & \cdots & \dfrac{\partial F_n}{\partial y_n} \end{vmatrix}$$

does not vanish there, then there is one and only one set of solutions

$$y_j = f_j(x_1, \ldots, x_m)$$

which are continuous, satisfy the relations $F_j = 0$ and for which $y_j^0 = f_j(x_1^0, \ldots, x_m^0)$.

Proof: If we assume the theorem true for $(n - 1)$ equations and prove its truth for n equations, then by induction we can rise to the general case from $n = 1$. Since the Jacobian does not vanish at least one of the cofactors of its last row must not vanish and for convenience we may take this to be

$$\frac{\partial(F_1, \ldots, F_{n-1})}{\partial(y_1, \ldots, y_{n-1})} = 0.$$

Then since the theorem is assumed for the case $(n - 1)$, we can solve the first $(n - 1)$ equations in the form

$$y_k = \varphi_k(x_1, \ldots, x_m;\ y_n), \qquad k = 1, \ldots, (n - 1),$$

and the φ_k are continuous. Then substituting in the last equation we have

$$\Phi(x_1, \ldots, x_m;\ y_n) = F_n(x_1, \ldots, x_m;\ \varphi_1, \ldots, \varphi_{n-1}, y_n) = 0.$$

If the derivative $\partial\Phi/\partial y_n$ does not vanish, then this can be solved to give $y_n = \varphi_n(x_1, \ldots, x_m)$ and then substitution gives $y_k = f(x_1, \ldots, x_m) = \varphi_k(x_1, \ldots, x_m;\ \varphi_n)$, $k = 1, \ldots, (n - 1)$ and $f_n = \varphi_n$. However,

$$\frac{\partial\Phi}{\partial y_n} = \frac{\partial F_n}{\partial y_1}\frac{\partial \varphi_1}{\partial y_n} + \cdots + \frac{\partial F_n}{\partial y_{n-1}}\frac{\partial \varphi_{n-1}}{\partial y_{n-1}} + \frac{\partial F_n}{\partial y_n}$$

The $\partial\varphi_k/\partial y_n$ are calculated from

$$\frac{\partial F_1}{\partial y_1}\frac{\partial \varphi_1}{\partial y_n}+\dots+\frac{\partial F_1}{\partial y_{n-1}}\frac{\partial \varphi_{n-1}}{\partial y_n}+\frac{\partial F_1}{\partial y_n}=0.$$

$$\frac{\partial F_{n-1}}{\partial y_1}\frac{\partial \varphi_1}{\partial y_n}+\dots+\frac{\partial F_{n-1}}{\partial y_{n-1}}\frac{\partial \varphi_{n-1}}{\partial y_n}+\frac{\partial F_{n-1}}{\partial y_n}=0.$$

However, solving these equations and substituting in $\partial\Phi/\partial y_n$ we have

$$\frac{\partial \Phi}{\partial y_n}=\frac{\dfrac{\partial(F_1,\dots,F_n)}{\partial(y_1,\dots,y_n)}}{\dfrac{\partial(F_1,\dots,F_{n-1})}{\partial(y_1,\dots,y_{n-1})}},$$

and since neither term on the right-hand side is zero, $\partial\Phi/\partial y_n \neq 0$. This proves the theorem by induction.

The particular form in which this theorem is most needed is the inversion of a functional transformation. If $m = n$ and F_j has the form

$$g_j(y_1,\dots,y_n) - x_j = 0,$$

where the g_j are continuous with continuous derivatives then the theorem takes the following form.

Theorem 3. If $x_j = g_j(y_1,\dots,y_n)$, $j = 1,\dots,n$, are n continuous functions of the variables $y_1,\dots,y_n$ with continuous first partial derivatives, and if the Jacobian

$$J=\frac{\partial(x_1,\dots,x_n)}{\partial(y_1,\dots,y_n)}$$

does not vanish, then the transformation from y to x can be uniquely inverted to give

$$y_i = f_i(x_1,\dots,x_n).$$

Exercise B.1. Carry out the proof of Theorem 2 in detail to show that

$$F(x, y, z;\ u, v) = 0,$$
$$G(x, y, z;\ u, v) = 0$$

gives $u = f(x, y, z)$ and $v = g(x, y, z)$ if

$$\frac{\partial(F, G)}{\partial(u, v)} \neq 0.$$

Index